SpringerWienNewYork

Alexius J. Hebra

The Physics of Metrology

All about Instruments: From Trundle Wheels to Atomic Clocks

SpringerWienNewYork

Professor Alexius J. Hebra
Charleston, SC
USA

SpringerWienNewYork is a part of
Springer Science + Business Media
springer.at

Typesetting: Thomson Press (India) Ltd., Chennai, India
Printing and binding: Strauss GmbH, 69509 Mörlenbach, Germany

Printed on acid-free and chlorine-free bleached paper
SPIN: 12206773

With 226 Figures

Library of Congress Control Number: 2009935065

ISBN 978-3-211-78380-1 SpringerWienNewYork

Acknowledgements

Had Pheidippides, ancient epitome of the self sacrificing herald, died of exhaustion before he could utter his famous words "We have won!" in front of Athens elected archon, the message would have faded before its echoes had a chance to change history. Three little words deciding the outcome of a game of "all or nothing" much like a book's publication sounds the author's message to the world rather than let it fade away unheard.

Which brings to mind those whose criterion and diligence made that the message on the pages of this book doesn't rest there forever, but got off to reach the minds of those concerned. And herewith go my heartfelt thanks to those at SpringerWienNewYork who dedicated their time and efforts to help my writing to overcome that threshold, in particular Mr. Stephen Soehnlen, Editor Natural Sciences, and Mr. Edwin W. Schwarz, Project Management.

Within my personal sphere my thanks go to my son in law, David Lench, for helping me with his professional knowledge and experience in hammering out my publishing contract with Springer; and likewise to all those who put up with my work related eccentricities and offered their support and input whenever asked for: My wife Gerda von Hebra and our daughter Renata von Hebra Lench, who both helped in reading proof pages and offered valuable comments on engaging and readability of my text. My sons Alexandre and Andre, and my grandchildren Daniel, Julie, and Ashley Lench; Melissa von Hebra; and Alexander and Andre von Hebra, whose patience used to be unduly strained by the priorities I extended to my work.

Thanks also to Mr. Scott Orlosky, Marketing Manager at BEI Industrial Encoder Division, for his valuable explanations on incremental and absolute encoders as well as the related binary and Gray code conversions.

And last but not least, I thank the country that became my new home for opening its vast markets and opportunities to my work and activities.

Preface

When Albert Einstein's assistant alerted him that present year's examination papers had the same questions as last year's, Einstein responded: "True, but this year all answers are different."

Likewise, this year's measuring instruments might be different (and hopefully better) than last year's, but the principles underlying their operation remain the same. You may set up an instrument by pressing buttons on its face plate, or click on virtual buttons popping up on your monitor; you may read the measured values on old fashioned dials, or from digital readouts. But the methods for obtaining those values are dictated by the immutable laws of nature. What changes are our ways of interpreting what those laws have to tell.

The concept of measurement brings us to the eternal question of the "appropriate" units of measure, which has split the world in two camps: Those based on the British *Imperial System of Weights and Measures* (what a complicated way of saying "ours"!), and the other side, the *metric units*, which were updated and unified a few decades ago into the *International System of Weights and Measures*.

It may surprise that both, the US customary system and the metric system, refer to the meter as the universal unit of length, but the truth of the matter is that the United States gave up on their proprietary standards of length and mass in 1893, and based their customary units on the meter standard bar and the kilogram standard mass. Following some minor corrections, our inch became exactly 25.4 millimeter (0.0254 meter), which makes the foot exactly 0.3048 meter. This yields the whole number relation of 5 inches = 127 millimeter.

The meter, originally intended as the ten millionth part of the meridian quadrant[1], lost this distinction when updated surveys revealed that distance as equal to 10,001,954.5 meter. But by then, the classical meter standard was already far too entrenched in commerce and industry worldwide as to allow for updates. And the latest definition of the meter, 1/299792458 of what a ray of light travels in vacuum in one second, has been derived so as to match the de facto length of the meter standard bar – not the other way round. Anyway, it would be hard to imagine that somebody out to check the accuracy of a caliper, micrometer, or gauge block, would do so by comparing his tools with the length of the meridian quadrant or the velocity of light, while NIST keeps all standard references at your fingertips, so to say.

1 the distance from the equator to the pole.

Metric units are taught in our schools because formulas and equations are more transparent with them then in our foot/pound system. And conversion is simple. At the gas station, you get 3.785 liter for the price of a gallon. And on the scale in the grocery store the pound stands for 0.454 kilogram. Better still, the kilowatt-hours of electricity in your monthly utility bills are the same in all systems of measure.

No big deal. But wait, and forgive me for warning that things get worse – read somewhat more complex – before they get better. The headache of native foot pound users is the metric unit of force. Accustomed as we are to the pound stepping in for mass and force as well, we find it strange that mass should come in kilogram, force however in *newton* (N). Blame the space age, that made the weight of one kilogram of mass only 1/6 kg on the moon and 0.40 kg on planet Mars. So the "honest pound" (as some of the anti-metrics call it), isn't so honest anymore if you take it on a ride into space. What's invariable is the *mass*, be it of the pound or the kilogram or any other quantity. The weight, actually the gravitational force on a piece of mass, is location dependent.

So, the International System of Weights and Measures changed from gravitation defined units to a unit of force built on inertia rather than the weight. The unit of newton is thus the inertial force of 1 kg of mass being accelerated by 1 meter per second squared. Under earthbound conditions, that means that 1 N = 1/9.80665 kilogram of force.

Whenever you look at the math in this book, remember that an input of mass in kilogram gets you force in newton. Also that pressure results in newton per square meter (called pascal or P), and not in psi, atmospheres, or mm Hg. So far, you needn't worry about the rest. Much like you learn to drive a car by driving one rather than by brooding over instructions, familiarity with metric units will follow the lecture of this book pretty much by itself.

Metrology keeps our civilized world alive and kicking. Precise measurements make that the replacement parts for our cars always fit, that doorframes can be bought off the shelf, and that a five pound bag of sugar contains five pounds. Like a friendly genie, metrology has been around all the time to help – and yes, sometimes control – our lives. Let's let the genie out of the bottle and contemplate her features!

Table of contents

Illustrations

1 The measurement of length

2 Angles and arcs

3 Clocks and the measurement of time

4 Velocity and acceleration

5 Force, mass, weight, and torque

6 Vibrations

7 Thermodynamics

8 Pressure

9 Density of solids, liquids, and gases

10 Light and radiation

11 Acoustics

12 Electrical and electronic instruments

13 Automation – instruments that think

1 The measurement of length

The great Pyramid of Gizah towers over the Egyptian desert. In modern units, it is over 725 feet, or 230 meters tall. The Ancient Egyptians, though, used neither feet nor meters, but cubits – the length from the tip of a middle finger to the elbow. In terms of cubits, the Great Pyramid stands on a base of sides 439.80, 440.18, 440.06, and 440.00 cubits. How did they measure these distances? Suggestively, $439.82 = 140\pi$. A wheel of one cubit diameter has a circumference of π cubits. Roll this 140 times, and you will have marked out the side of the pyramid. If the Egyptians used an early version of a perambulator or trundle wheel, their devices were accurate to 0.1%!

No wooden artifacts have been preserved from antiquity into our times, which may explain that we lack direct proof of an Egyptian wooden device for measuring distance. Hieroglyphs do show people using ropes with regularly spaced knots to measure. These ropes were conserved by rubbing them with a mixture of resin and beeswax. But attempts to measure a building the size of the pyramid with hemp ropes would have been plagued by the effects of the applied pull, changing air temperature and humidity. Copper bars would have shrunk in the cold of the desert nights and expanded in the heat of day.

Trundle wheel

The followers of Pharaoh's use of a rolling disk was likely due to wood's low thermal expansion, of 3 to $4 \times 10^{-6}/°C$, about 1/3 that of steel. Trundle wheels similar to Fig. 1.1 have been in constant use ever since. Early in the eighteenth century, John Bennet[1] used fine brass to craft trundle wheels with multiple scales in units of *poles* (equal to the rod, of 16.5 feet or 198 inches in length), *furlongs* (220 yards), and *miles* (5280 feet). The wheels had 30.51 inches diameter, which made their perimeter a straight 99 inches or half a pole, so that the device advanced by one pole for every two revolutions of the wheel.

In the clockmaker's tradition, the dials of Bennet's waywisers resemble clock faces, complete with large and small hand, and a third, still smaller hand on a separate scale, like the seconds dial on traditional pocket watches. This allowed for readouts in the British length units of that period.

Trundle wheels of up to six feet diameter were used in surveying because bigger wheels are insensitive to bumps and ditches. While small wheels measure the

1 British clock and instrument maker.

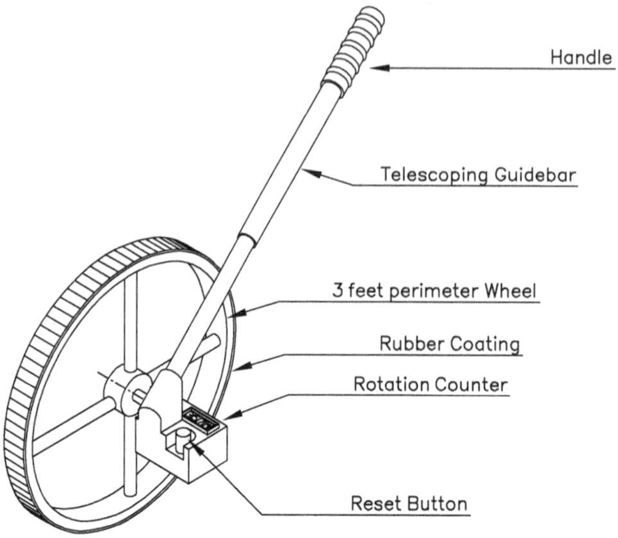

Fig. 1.1. Trundle wheel for length and distance measurements

inner perimeter of a ditch they roll through, wheels which radius exceeds the width of the ditch bridge it from edge to edge and size its width. Eighteenth-century trundle wheels, called Waywisers, whose structure resembled a lightweight wheelbarrow, came from England and remained in use in the United States for another one hundred years. In more recent times, simple trundle wheels are being sold in hardware stores for a host of uses, such as pricing of fences and preliminary layouts of lots and buildings.

The familiar automotive odometer uses the principles of the trundle wheel, for it picks up the revolutions of the vehicle's wheels and, through reducing gears of appropriate ratio, converts their count into miles traveled. For instance, a car with properly inflated P185/75-14 tires, of 24.925 inches external diameter, drives $24.925\pi = 78.304$ inches or $78.304/12 = 6.525$ feet per turn of its wheels. With 5280 feet to the mile, that gets us $5280/6.525 \approx 809$ revolutions per mile. Therefore, a typical odometer, which readouts advance by one mile for every ten revolutions of the instrument's driveshaft, has to be actuated over a $10/809 = 1{:}80.9$ gear reducer to read miles. Part of this reduction is provided by the typically 1:3 stepdown of the differential.

If you drive with low tire pressure, the odometer renders exaggerated mileage readings, because the effective radius of the wheel, from hub to pavement, is below standard and the wheel has to turn more often to cover a given distance.

The updated version of a trundle wheel (Fig. 1.1) typically employs a circular disk of 5.730 inches diameter, which rolls over $5.730\pi = 18$ inches per turn. Adding an 18 teeth ratchet and counting the "clicks" would give rolling distances in inches per click, but mechanical counters have since taken the place of Bennet's precision dials. Thus the trundle wheel has changed from Bennet's *analog* devices into a simple brand of today's *digital* instruments.

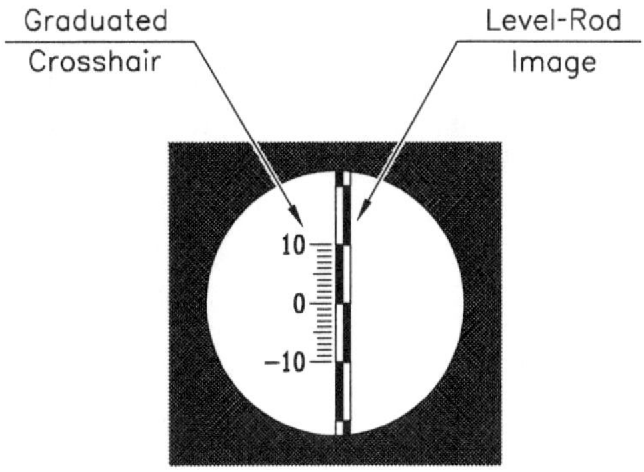

Fig. 1.2. Reticule-based distance measurement

Chains and tape measures

The temperature sensitivity of metallic measuring tools might be the reason that their use spread no sooner than the invention of the thermometer and the availability of data on thermal expansion of metals. As the first thermometers, then called "thermoscopes," turned up in the early sixteen hundreds, Aaron Rathborne's personal "Decimal Chayn" was reported on in 1616. The Rathborne chain of ten links, which added up to the length of one pole or 16.5 feet, was followed by Edmund Gunter's 100-link chain of four poles (66 feet) overall length. The nineteenth-century "Engineer's Chain" was made of one-foot links and measured fifty or one hundred feet overall. A modern Keuffel and Esser chain of this type is made from No. 12 gauge steel wire links, brazed shut, and carries a center spring hook for dividing the chain – if needed – into two identical halves.

In surveillance work, thermal expansion must be precalculated and taken into account. Where money is not an issue, Invar, an alloy of 36% nickel, 0.2% carbon and 63.8% iron, with its coefficient of expansion (1.26×10^{-6}) almost ten times lower than that of steel, makes for tape measures that can be used – in most cases – without worrying about temperature.

Steel measuring bands were first fabricated in 1853 by James Chesterman (1792–1867) of Sheffield, England, and became the forerunners of present days pretensioned and springloaded steel tapes.

Distance measurement

Where tape measures become too short and trundle wheels too laborious, several types of instruments, from simple to sophisticated, are available for the job.

If high accuracy is not essential, a telescope with a crosshair scale in the eyepiece, along with a surveyor's staff (also called "stadia") will do. In Fig. 1.2, ten divisions on the telescope's eyepiece scale coincide with one 10 cm long field on the staff, if posted 10 meters away. Move the staff to 20 meters, and the image of one shaded area will coincide with only 5 crosshair scale divisions. At 100 meters, scaling of staff and crosshair coincide. In short, the distance of the staff and the apparent size of its scales in the telescope are inversely proportional to each other.

Although this method lacks precision, it excels in simplicity. In a similar approach to measuring the distance to neighboring stars, Astronomers use the diameter of the orbit of Earth in ways of a staff. With 149.6 million kilometers from the Earth to the Sun, our planet's position in the galaxy shifts by

$$2 \times 149.6 = 299.2 \text{ million kilometers in six months,}$$

causing an equivalent drift in the apparent positions of the stars. In 1838, Friedrich Wilhelm Bessel (1784–1846) was first in reliably measuring this motion, called parallax, on the star 61 Cygni in the constellation of Cygnus, the Swan. Though only few fixed stars locate close enough to show parallactic displacements, this method created a specific astronomic length-unit, the *parsec*, equal to the distance from where the radius of the orbit of Earth is seen one arcsecond wide. With $360 \times 60 \times 60 = 1{,}296{,}000$ arcseconds to the full circle, the circumference of a circle with one parsec radius is $1{,}296{,}000 \times 149.6 \times 10^6 = 193.88 \times 10^{12}$ kilometers, which makes its radius equal to $(193.88 \times 10^{12})/2\pi = 30.857 \times 10^{12}$ km. And with 9.460550×10^{12} km to the light-year (ly), this leads to the conversion

$$1 \text{ parsec} = \frac{30.857 \times 10^{12}}{9.460550 \times 10^{12}} = 3.26 \text{ ly} .$$

The unit we got here seems enormous, and yet, no fixed star locates within the one parsec range. The Sun's closest neighbor, Alpha Centauri on the southern celestial hemisphere, is 1.33 parsecs or 4.34 light-years away.

Triangulation

The efforts of ancient Egyptians in re-appropriating the fertile land along the Nile every year after the waters of the river swept their landmarks away mark the first steps in the development of geometry and mathematics, which much later in history inspired generations of scientists in the Hellenistic age, such as Euclid, Archimedes, Eratosthenes, and Heron of Alexandria. The latter was first to come up with a formula for the area of a triangle based on the lengths of its sides.

Since the triangle is the building block of all other polygons, Heron's formula can be used for multifaceted areas as well. For instance, a quadrilateral can be split into two triangles, the pentagon into three, the hexagon into four, and an n-sided polygon into $n - 2$ triangles. If you own a piece of land bordered by the sides a, b, d, and e, you can find the square feet of your property by measuring the length of one of the diagonals, c, and apply Heron's formula to the resulting triangles $a - b - c$ and $d - e - c$. For a start, call the circumference of the first triangle $2s$, so that

$$s = \frac{1}{2}(a + b + c) ,$$

and use Heron's formula for the area A of said triangle as follows:

$$A = \sqrt{s(s-a)(s-b)(s-c)} .$$

Repeat that process for the second triangle, add the two areas, and – voilà – you got the square feet of your four-sided land without buying a theodolite.

Although we remember Heron of Alexandria as a prolific writer on geometry and mechanics, including pneumatics, we aren't even sure about his name. Some called him Heron, others Hero, and his date of birth has not been documented.

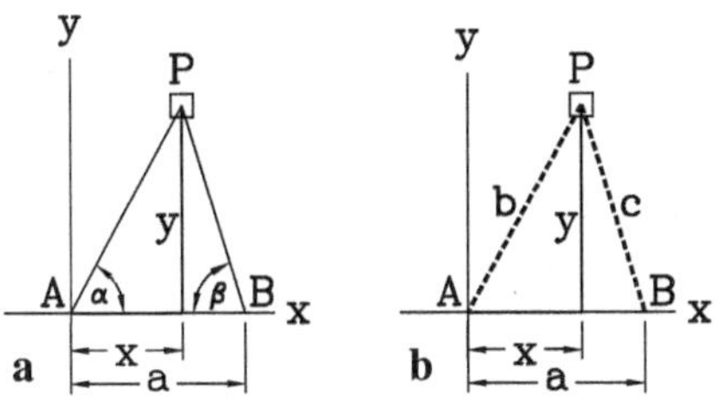

Fig. 1.3. a Triangulation. **b** Trilateration

Historians place his lifetime between 250 BC and 150 BC. He might have been a Greek or an Egyptian with Greek education, but his mark on history remains, whatever his lifetime and nationality!

Modern times surveyors use an arbitrary baseline between two landmarks, A and B in Fig. 1.3, to describe the position of an object P by its plane coordinates, x and y.

We have two distinct methods to determine an object's position: *triangulation* and *trilateration*. Triangulation is defined as (take a deep breath) "measurement of the position of an isolated point in space accomplished by constructing a scalene triangle with a known base and known angles of the two adjacent legs." In Fig. 1.3 a, the coordinates x and y of point P are derived from the length of the baseline, a, and the angles α and β, measured from points A and B to P. The geometry of the figure allows for direct deduction of the relations

$$x \tan\alpha = (a - x)\tan\beta\,,$$

which yields

$$x = a\frac{\tan\beta}{\tan\alpha + \tan\beta} \tag{1.1}$$

and likewise

$$y = a\frac{\tan\alpha}{\tan\alpha + \tan\beta}\,. \tag{1.2}$$

Make A and B the milestones on a straight section of highway, and P the location for, say, a planned gas station, and you get x and y as the coordinates of P relative to the origin of the stretch of highway you selected as baseline.

If the baseline AB is drawn in the direction east-west, the geographical coordinates of the object P can be found by converting x and y into degrees of longitude and latitude and summing their values to longitude and latitude of point A. Keep in mind that for a place in North America, x and *western longitude* are counted in opposite directions, which makes the values of x negative.

The degree of *latitude* equals 60 nautical miles, which makes $1'$ (minute of arc) of latitude equal to one nautical mile. In metric units, the degree is 111 km. These values are not exact; due to the oblate shape of the Earth, they vary slightly with the observer's location.

By contrast, the degree of *longitude* equals that of latitude along the equator only, but shrinks toward the poles proportional to the cosine of the latitude of the observer's location. Here again, this relation gains complexity if we consider the ellipsoidal shape of the globe, and reads for this case

$$L = \cos\varphi(111.320 + 0.373\sin^2\varphi)\,,$$

where L stands for the length of one degree of longitude in kilometers, and φ for latitude.

Surveys for housing developments and other short-distance projects are usually conducted under the supposition of a flat Earth, and the Eqs. 1.1 and 1.2 apply. Farther reaching projects must account for the spherical shape of the globe and use the math of spherical trigonometry for the deduction of the equivalents of said equations. Continent wide measurements need to count in the oblate shape of the Earth, and for highest accuracy, even the local irregularities in the shape of the Earth' surface. Such complexities turned up in the case of the French engineer and surveyor P. F. André Méchain, entrusted with measuring the length of the meridian quadrant. Méchain's work was to create the natural basis of the metric length unit, the mètre, which later became the fundamental unit of the metric system, the meter. Although Méchain did his math correctly, his network of triangulations wouldn't close at a single point because he apparently underestimated the magnitude of some deviations of the shape of the globe from an ideal ellipsoid.

The concept of triangulation reaches back into the late fifteen hundreds and apparently stems from the Danish astronomer Tycho Brahe. Since the telescope hadn't yet been invented at Brahe's times, all his observations were done with visors sliding along enormous quarter-circular scales. Nonetheless, his measurements of the positions of the planets on the sky were of such precision that they allowed for Johannes Kepler's deduction of his laws on the motion of celestial objects.

Theodolites

Present-day instruments for measuring the angles α and β, called theodolites (or dioptra in older texts) (Fig. 1.4), rely on a sighting telescope with crosshairs in the

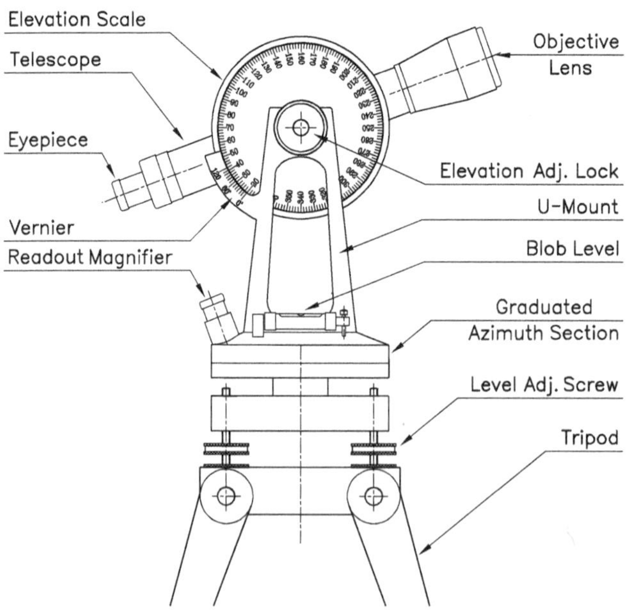

Fig. 1.4. Theodolite

eyepiece, mounted in a U-shaped support frame. It can be rotated horizontally and tilted vertically. The angles of rotation and tilt are read through extra eyepieces, sometimes called microscopes, on graduated measuring circles with usually one minute of arc precision.

Before use, the instrument is set up and precisely leveled on a sturdy tripod. Some types of theodolites have a compass encased in their base, but the "magnetic declination" of the needle – an angular deviation from true North stemming from the different locations of magnetic and geographic poles – makes a compass a poor match for the accuracy of the Sun's meridian transit. Presently, magnetic North locates in the area of Prince of Wales Island in the Arctic Archipelago north of Canada. However, its position changes over the years.

Updated models of the theodolite with optoelectronic encoders replacing the graduated measuring circles offer digital readouts with typically one arcsecond of accuracy. But all that gadgetry is hidden inside the telescope's U-mount, which is why a classical instrument, with all its parts in the open, has been chosen for Fig. 1.4.

The invention of the theodolite is credited to Leonard Diggs of Kent for his 1571 description of such an instrument in his book *Pantometria*. But only the circular dividing machine, constructed in 1775 by Jesse Ramsden (1735–1800) made the crafting of scales for true precision-instruments possible.

In 1782, when British cartographers intended to catenate the locations of the Royal Observatory in Greenwich, England, and the Observatory in Paris, France, by triangulation, Jesse Ramsden was appointed to build a theodolite with the necessary accuracy. His instrument, nicknamed "The Great Theodolite," now housed in the Greenwich Museum in England, weighs some 200 pounds and sports a three-foot diameter horizontal circle. But the two mightiest theodolites ever, also fitted with 36 inches wide scales and five scale reading microscopes, where built in the early eighteen hundreds for the "Great Trigonometric Survey" on British India. So heavy were those behemoths that it took twelve men to carry one around.

In 1843, with Andrew Scott Waugh in charge of the project, surveyors focused on the distant Himalayan peaks in the north of India. Hidden by clouds and haze most of the time, those peaks could rarely be viewed from the expedition's point of observation down in the lowlands, some 100 miles distant. Nevertheless, Radanath Sikhdar, in charge of the trigonometric evaluation of the survey's data, reported in 1852 the discovery of "the highest mountain in the world," then designated Peak XV. Waugh himself renamed it four years later to Mount Everest[2] after Sir George Everest, Waugh's predecessor as chief surveyor.

While triangulation measures the angles of the lines of sight from two observers to the target, *trilateration* uses the distances b and c between observers and target (Fig. 1.3 b). For this case, we derive

$$b^2 = x^2 + y^2 \quad \text{and} \quad c^2 = (a - x)^2 + y^2$$

2 29,028 feet high.

which yields

$$x = \frac{1}{2a}(a^2 + b^2 - c^2)$$

and

$$y = \sqrt{b^2 - x^2} \qquad \text{or} \qquad y = \frac{1}{2a}\sqrt{(2ab)^2 - (a^2 + b^2 - c^2)^2}\,.$$

Optical distance meters

The principles of triangulation and trilateration pop up in everyday items, such as the optical rangefinders we find in photographic cameras, where a miniature telescope (Fig. 1.5) shows the object's reflections in a pair of face coated mirrors; one set at 45°, and the other pivotable around its 135° homing position. The formers coating is semi-transparent, the other's solely reflective. Through the eyepiece the viewer sees two distinct images, which merge into one when the adjustable mirror is rotated by $\varphi = \alpha/2$ off its 135° base position. α is the angle between the lines of sight from the object to each mirror and thus depends on how far the object is away. With a for the spacing of the mirrors, and D for the distance of the object, we get from Fig. 1.5

$$\tan\alpha = \frac{a}{D} \quad \text{and} \quad \alpha = \arctan\frac{a}{D}$$

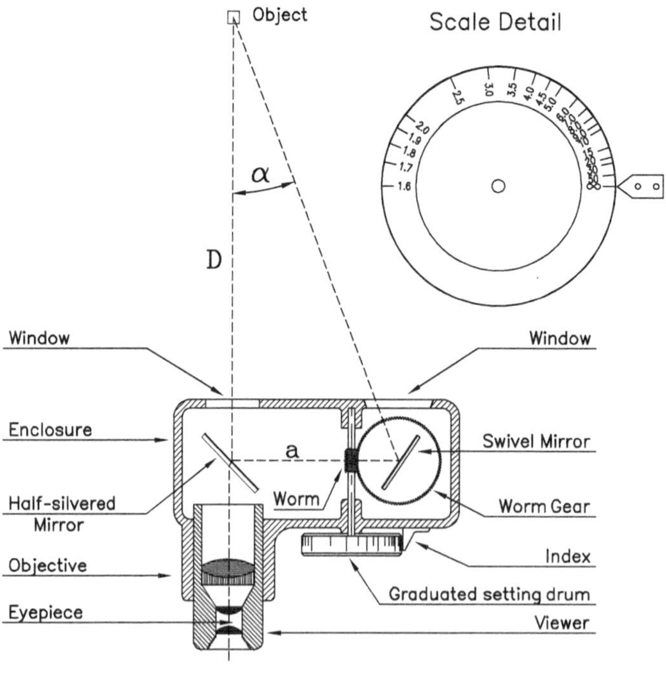

Fig. 1.5. Rangefinder

and figure the correct rotation angle of the mirror in degrees as

$$\varphi = 135 + \frac{\alpha}{2} = 135 + \frac{90}{\pi} \arctan \frac{a}{D}.$$

The circular unevenly spaced scale shown in Fig. 1.5, upper right, is derived from this equation.

With range finder cameras, you don't have to physically read such a scale, as a cam governed drive mechanism transfers the angular displacement of the pivoting mirror to the spiral groove mount of the objective lens and moves it forth and backward into focus. Sharpness of the shot depends on how well the profile of the cam has been matched with the optical characteristics of the objective lens.

Correct focusing – manual or auto – is the basic and most important condition for shooting good pictures. The most expensive objective lens, if improperly focused, might be outperformed by its counterpart in a disposable camera. Few photographs from the era of plates and films match the sharpness of Anselm Adams' classical landscapes, shot with a view camera from nearly a hundred years ago.

Reflective distance meters

Much like rangefinders, reflective distance meters too measure distance by triangulation. The reflection of a focused laser beam by the target is projected by a stationary objective lens onto a PSD[3] image sensor. The position of the resulting point of light on the PSD changes with the instrument's distance from the target, as can be seen in Fig. 1.6, where the light spot strikes the sensor farther down the further away the target is.

A gradually darkening light filter in front of the sensor is used to convert the light spot's position into an electrical signal. Again in Fig. 1.6, the beam from target B goes through the thickest region of the filter, of highest opacity, while the beam from target C crosses the filter's thinnest and therefore lowest opacity zone. As the PSD converts light intensity to voltage, the readout from the beam reflected by B will be much less than that from C.

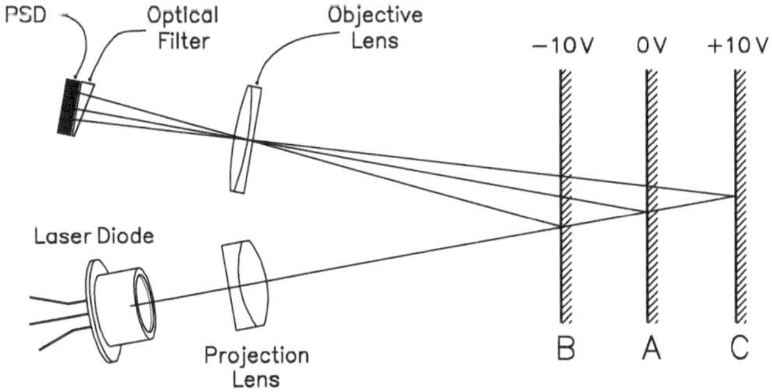

Fig. 1.6. Reflective distance meter

3 position sensitive device.

If the location of target A has been selected for the light spot to centrally strike the image sensor, the instrument's readout is set to zero and the distance of A becomes the "Instrument Constant." Other positions, such as C and B, are counted from there up and down, as exemplified by the voltage outputs marked on each of the three differently set targets in Fig. 1.6.

If all that sounds too good to be true, it is. Making this principle into a usable instrument involves a series of highly complex design features, due to a number of factors, other than the target's distance, are paramount for the voltage output of the PSD. They include the intensity of ambient light, the target's reflectivity, and the losses of brilliance along the light path.

The effect of ambient light is eliminated by flashing the light source sequentially on and off. This makes its output an alternating voltage (AC), while the voltage stemming from ambient light is continuous (DC). Those two components are separated by a capacitor, which impedance $R_C = 1/2\pi fC$ is frequency dependent. R_C is low for high frequency AC, but tends toward infinity for DC with $f = 0$.

Differences in the target's reflectivity, which reach from 0.05 for black paper to 0.92 for fresh snow, with porcelain (0.75) and Stellite (0.62) in between, are dealt with by using a light-sensing semiconductor to convert a sample of the reflected light into a control voltage that sets the intensity of the light source inversely to the target's reflectivity. This also takes care of distance contingent variations in the intensity of the reflected beam. A signal processor (let's think of it as a chip) stores information on all those sources of error, and uses them to correct and linearize the output voltage accordingly.

Reflective distance sensors with ±0.001 inch precision are available in various ranges from 0.375 inches to 6 feet and more. Beyond simple distance checks, their applications include the alignment of CNC[4] machine tools and MRI medical imaging devices; further position recognition, depth of hole and thickness measurements, checks on sag of beams and plates, and filling levels in tanks.

Beam-modulation telemetry

Beam-modulation telemetry, also known as *tone ranging*, measures an object's distance by sending an intensity-modulated light-beam toward the target and derive the time difference between emitted and reflected beam from their mutual phase shift. It is used in constructional surveys and for high-accuracy distance metering in general, but also in object recognition devices in warehousing and in collision avoidance systems for pilotless vehicles.

Some commercial instruments offer millimeter precision over the range of several kilometers, while short range high accuracy instruments are made to operate within 5 meters on diffusely reflecting (noncooperative) targets, and up to 20 meters if a mirror-like reflector (cooperative target) is being used.

The technique of deriving an object's distance from the time difference be-

4 computer numerical control.

tween sending a signal and receiving its reflection is self evident. We estimate the distance of a thundercloud by dividing the time between seeing a bolt and hearing the thunder by the velocity of sound (341 m/s). Radar takes the same approach with radio waves, and the GPS satellite ranging system figures the distance between satellite and ground station from the time it takes radio signals to travel that distance. A technique hard to emulate "back home," because electromagnetic waves, including light, propagate at such frightening speed[5]. For instance, the time between triggering a flash and seeing its reflection from a 30 meter (ca. 100 feet) distant mirror would amount to 0.2 millionth of a second (0.2 μs) – obviously impossible to clock. But such minute time intervals can be detected by comparing them with the cycle time (T) of a rapidly flashing light source.

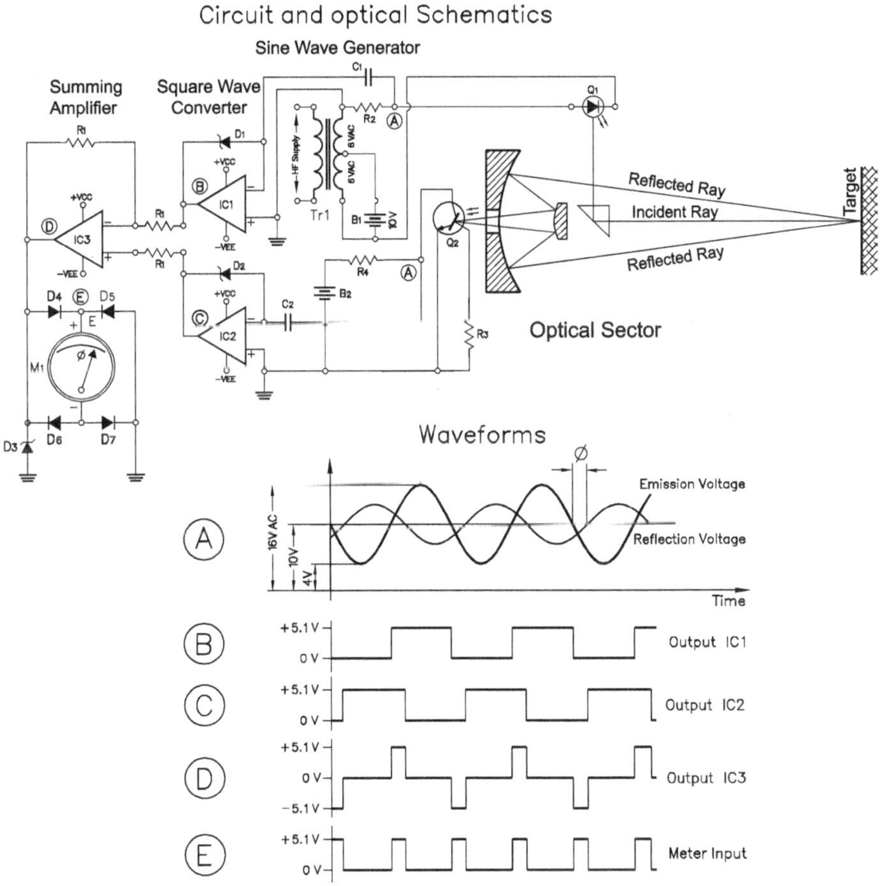

Fig. 1.7. Tone ranging distance meter

5 $c = 300{,}000\ \mathrm{km/s}$.

Intensity modulated (flashing) light is nothing new. Even domestic light bulbs would flicker at 120 cycles per second if it weren't for the thermal inertia of their filaments. By contrast, LEDs and laser diodes can be made to change their light output in step with the voltage of their power supply over a wide range of frequencies. If AC powered, they lighten up for the positive half cycle and darken throughout the negative half cycle. Yet full cycle operation can be obtained with a bias voltage from an auxiliary power source, B1 in Fig. 1.7, applied to the center tap in the secondary windings of the 1 : 1 transformer Tr1, which primary is powered by a 12 Vp/p AC supply of high frequency. In that case, a +10 V bias makes the output oscillate from +4 V to +16 V in sinusoidal mode (graph A in Fig. 1.7).

Propagating with $c = 3 \times 10^8 \mathrm{m/s}$, a light pulse arrives at an a meter distant reflecting target in $t = a/c$ seconds, and takes an equal interval of time to bounce back – a total of $2t$. From f for the beam's modulation frequency, you get the period of oscillation as $T = 1/f$, and the ratio of phase difference to cycle time as $2t/T$ or $2af/c$. For sinusoidal oscillations, this term converts to the phase angle φ (in radians) between emitted and reflected pulses by multiplication with 2π, the length per cycle of the basic sineline, which leads to

$$\varphi = \frac{4\pi af}{c} \,.$$

Inversely, we get

$$a = \frac{c\varphi}{4\pi f}$$

for the distance of the object.

For a pulse frequency of, say, 200 kHz, and the phase angle of $\varphi = 0.75$ radians (as in Fig. 1.7), the distance of the reflecting object follows as

$$a = \frac{(3 \times 10^8)0.75}{4\pi(200 \times 10^3)} = 89.52 \text{ meter} \,.$$

Practical application of such theories affords to place both the light emitter and receptor along the same line of sight without shading one from the other. The going solution is to have the receiver gather a bundle of light wide enough to allow for the in line introduction of the light source without seriously darken the image of the reflection. Astronomy buffs may recall such a trick from Newtonian reflectors, where a 45° mirror reflects the light from the primary parabolic mirror into the eye-lens without popping up in the image. Likewise, the prism in Fig. 1.7, upper right, does not disturb the beam that reaches the photodetector. Rather, the bundle of light that circumvents the prism reaches the parabolic mirror virtually intact and is focused by a small convex mirror on the receptor – a phototransistor.

For deduction of the phase angle φ from the voltages generated by the phototransistor and the one powering the light emitting laser diode Q1, they are first converted into square waves of mutually identical amplitude (5.1 V for the case in point) by the operational amplifiers IC1 and IC2, operating in saturation mode with their output clipped by the feedback Zener diodes D1 and D2. As shown in graphs B and C, the outputs of the op-amps are in counterphase because reflections are in counterphase with the beam they stem from.

Op-amp IC3, wired as a summing amplifier, adds up the voltage outputs from IC1 and IC2 into a staggered square wave which, biased by Zener diode D3, converts into pulsed AC as in graph D that the full wave rectifier, consisting of silicon diodes D3 through D7, makes into a series of pulses (graph E). Their width – proportional to the phase difference – is shown on galvanometer M1.

Global Positioning System

The Global Positioning System, sketched schematically in Fig. 1.8, is based on *trilateration*. Twenty-four communication satellites cruise at 20,200 kilometers above the Earth in six orbital planes of 55 degrees inclination. This particular size of orbit makes those satellites' time of revolution equal to 12 hours sidereal time, so that a set of six satellites is always high enough over the horizon for reception of their radio signals on Earth.

Each satellite carries its own atomic clock, which emits timing signals with close to absolute precision. Ground stations – such as the GPS receivers in cars, boats, and elsewhere – generate their own timing signals in ways similar to crystal controlled wristwatch mechanisms. If all the clocks within the entire GPS were in synchronism, a receiver on the ground would find each satellites' clock late by the time interval the signal takes to reach it. Since radio waves propagate with the speed of light ($c = 300{,}000$ km/s), this time delay Δt is indicative for the transmitting satellites' distance, s, according to

$$s = c \times \Delta t = 300{,}000 \times \Delta t\,.$$

This formula is used by the receiver's built-in computer to calculate the distances s_1, s_2 and s_3 of each of three satellites by gauging the delay of their clock signals relative to the receiver's clock.

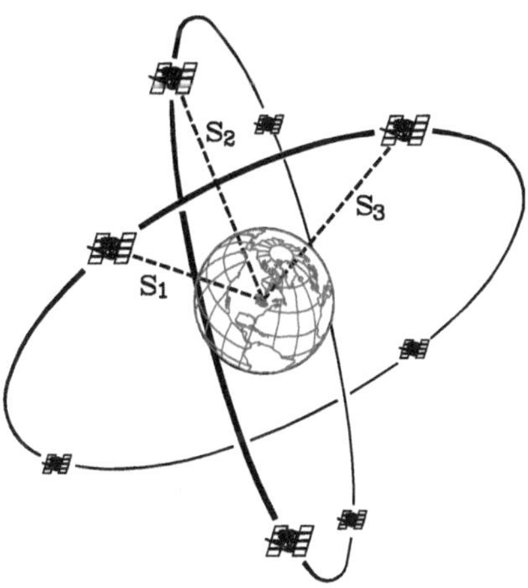

Fig. 1.8. Global Positioning System

At the same time, the receiver's computer derives each satellite's momentary position in space from orbital parameters stored in memory, and taking it from there, uses s_1, s_2 and s_3, to check its own position on the ground.

That would do the trick if the receivers' clocks were as precise as those in the satellites. But as it is, small differences will invariably build up and distort the results. The more so, since the signals' travel times from satellite to receiver are in the order of magnitude of about 25 milliseconds. Thus, the accuracy of the satellites' atomic clocks, of three thousandth of a microsecond, or three nanoseconds, must somehow be matched in our humble GPS ground receivers.

Regardless the problem's complexity, designers came up with an amazingly simple solution: Instead of using signals from three satellites, they use all four, which we may number A, B, C, D. As we have seen, three satellites, such as A, B, and C, suffice for locating a ground station (Fig. 1.8). With an extra satellite, D, we get a total of four ground station pilings, stemming from the groups of satellites, including

$$A, B, C; \quad A, B, D; \quad A, C, D; \quad B, C, D.$$

In the best of both worlds, all those readings would point at identical locations, but in reality, minor inaccuracies of the ground stations' clocks will lead to four slightly different results.

But rather than bemoan the inaccuracies, each receiver's computer sets its clock tentatively forward or backward until all the four sets of positional coordinates match. That synchronizes the receivers' clocks with the satellites' atomic clocks, and warrants the system's amazing precision.

Ruler and gage blocks

Back here on solid ground, we often express distance in units once derived from the dimensions of human body parts. The *inch* is roughly the width of a thumb; the *foot* unit is based on the size of a real foot. The *yard* was allegedly the distance between the tip of a man's nose and the tip of his outstretched middle finger. And the Ancient Egyptians used the *cubit*, from the elbow to the fingertip, subdivided into *palms* (the width of a typical human hand), and further broken down into *fingers*.

The trouble with all of these is that – at least until human cloning becomes widespread – we humans come in a variety of shapes and sizes. My thumb is different from yours, and so my version of the inch, or the cubit, will differ from yours. Buying and selling, though, afford unified measures. And with the introduction of standard length units came the ruler, the tape measure, the surveyor's chain, and a whole host of simple measuring devices.

The ruler, something we've used since early schooldays, is a magnificently simple device. It is a standard length, a foot, divided up into standard divisions, the inch, the half inch, and so forth. If a class of children have to measure the distance between two points, their accuracy is limited by the spacing of the smallest division on their ruler, the precision of the scales, and last but not least, the accuracy of their observation. A ruler marked with millimeters lets you find the length of an object with millimeter precision, but the tenth of millimeters remain guesswork. How can scientists and engineers be so sure about the correctness of their measurements on such and even still tinier distances?

We are familiar with the standard meter bar, cast of platinum-iridium alloy and treasured in the basements of the Bureau International de Poids et Measures in Sèvres, near Paris. The choice of an underground location for what was to become the worldwide standard of length was far from accidental, but for reasons

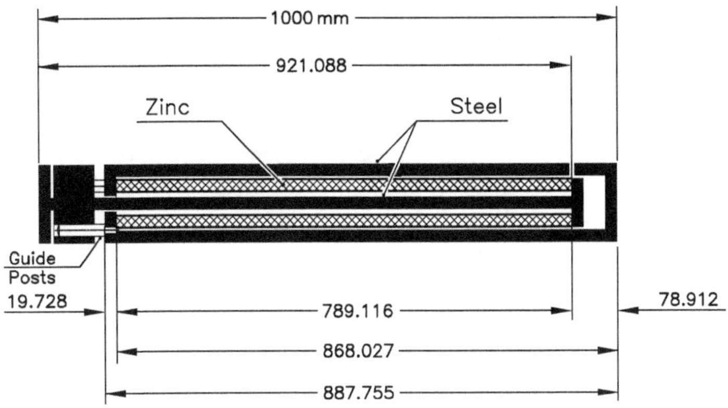

Fig. 1.9. Temperature-stable meter bar

of – no, its not what you think: Air raids were still a century in the future, and the meter bars sheltered existence was merely supposed to protect it from temperature fluctuations that would make it stretch in the heat of summer and shrink in winter.

A bar of invariable length could be made to the principles of the compensated pendulum[6] from a combination of rods of metals with different coefficients of thermal expansion, such as steel with $11.6 \times 10^{-6}/°C$ and zinc, with $26.3 \times 10^{-6}/°C$. In Fig. 1.9, raising temperature would make the center bar and the outer housing, both made of steel, expand toward the right, and the zinc pipe toward left. Herein, the effective length of the steel parts must be 26.3/11.6 times the length of the zinc tube. The dimensions shown in the illustration comply with this condition for an overall length of 1000 mm.

That we stick to the original meter bar rather than replace it by an invariable length standard comes from the fact that every additional component of a measuring device carries with it an equal number of error sources. For the case in point, the danger of recrystallization of the zinc as well as the steel components come to mind, let alone the danger of condensed moisture and the related electrolytic corrosion between different metals.

But those in need of a precise length standard will hardly travel all the way to Paris to no better end than to compare their bars with the original. Highly precise length standards are commercially available under the name of *gage blocks, Johansson gauges*[7]. The standard set includes 81 gage blocks of different length, and allows for over 100,000 combinations. Blocks of highest precision, labeled AAA, have their length lapped to ±0.00005 mm tolerances.

Crafted of steel, ceramics or tungsten carbide, they are made to the master

6 we recall from grandfather's clock.

7 slip gauges.

gage blocks of the National Institute of Standards and Technology, NIST, which holds a collection of laser calibrated standards. Some six thousand customer owned gage blocks per year are going through recalibration to nanometer (a millionth of a millimeter) precision.

The mating surfaces of gage blocks are lapped to such perfection in flatness and surface finish that once properly stacked (wrung), they stick together until manually separated. If left unattended for a few months period, a process called "cold welding" – actually the migration of atoms across the mating surfaces – makes the blocks inseparable.

Caliper

Gage blocks are used to calibrate other, less basic measuring instruments, including the one that can be found in virtually every engineer's toolbox: the caliper (Fig. 1.10). The surprising precision of early calipers stems from the invention of a seventeenth-century French scientist, Pierre Vernier, which allowed for readouts within 1/256 and on later models 1/1000 of an inch.

Calipers with millimeter scales show the exact number of tenths of a millimeter on a dedicated scale on the slide, called vernier, which typically divides the space of 9 millimeters into 10 parts, each of them 0.9 mm long. If you open the caliper by 0.1 mm, the first mark on the vernier scale coincides with the first mark on the main scale. Slide to 0.2 mm, and the second mark on the vernier coincides with a mark on the main scale, and so on. In short, we read full millimeters on the instrument's main scale at the zero mark on the slide, and then the tenth of millimeters on the vernier scale by finding the mark that aligns with a mark on the main scale.

A great variety of vernier scales is possible. For instance, the vernier of the caliper in Fig. 1.10 divides the space of 19 mm into 20 parts, which allows for readings of 1/20 or 0.05 mm. The upper scale has 1/16 inch spaced marks, and its vernier divides the space of 15/16 inches into 16 parts; thus, the first after the zero mark on the vernier falls short by $1/(16 \times 16) = 1/256$ inches, which is the instrument's readout precision on the inch scale. All said, the resolution of a vernier scale is found by dividing the smallest unit of the main scale by the number of divisions on the vernier.

The caliper in Fig. 1.10 a is set to accurately 30 mm. Conversely, in Fig. 1.10 b, the zero of the vernier scale falls somewhat to the right of the 16 mm mark on the main scale, and the mark that most closely coincides with one on the main scale locates between the marks 2 and 3. Taken as 2.5, this indicates 16.25 mm for the diameter of the shaft being checked.

The inch scale on that same instrument shows 1/16 inch, which the Vernier divides once again into 16 parts, each of $1/(16 \times 16) = 1/256$ inches. Thus we read at the zero mark 10/16 or 5/8 inches. With the vernier's fourth mark coinciding with one on the main scale, this adds up to $10/16 + 4/256 = 40/64 + 1/64 = 41/64$ inches, the equivalent of 16.27 mm. The difference between this and the 16.25 readout on the millimeter scale typifies the limits of precision inherent in the vernier system. Otherwise, you could draft a vernier scale that divides 99 mm in 100 parts, hoping

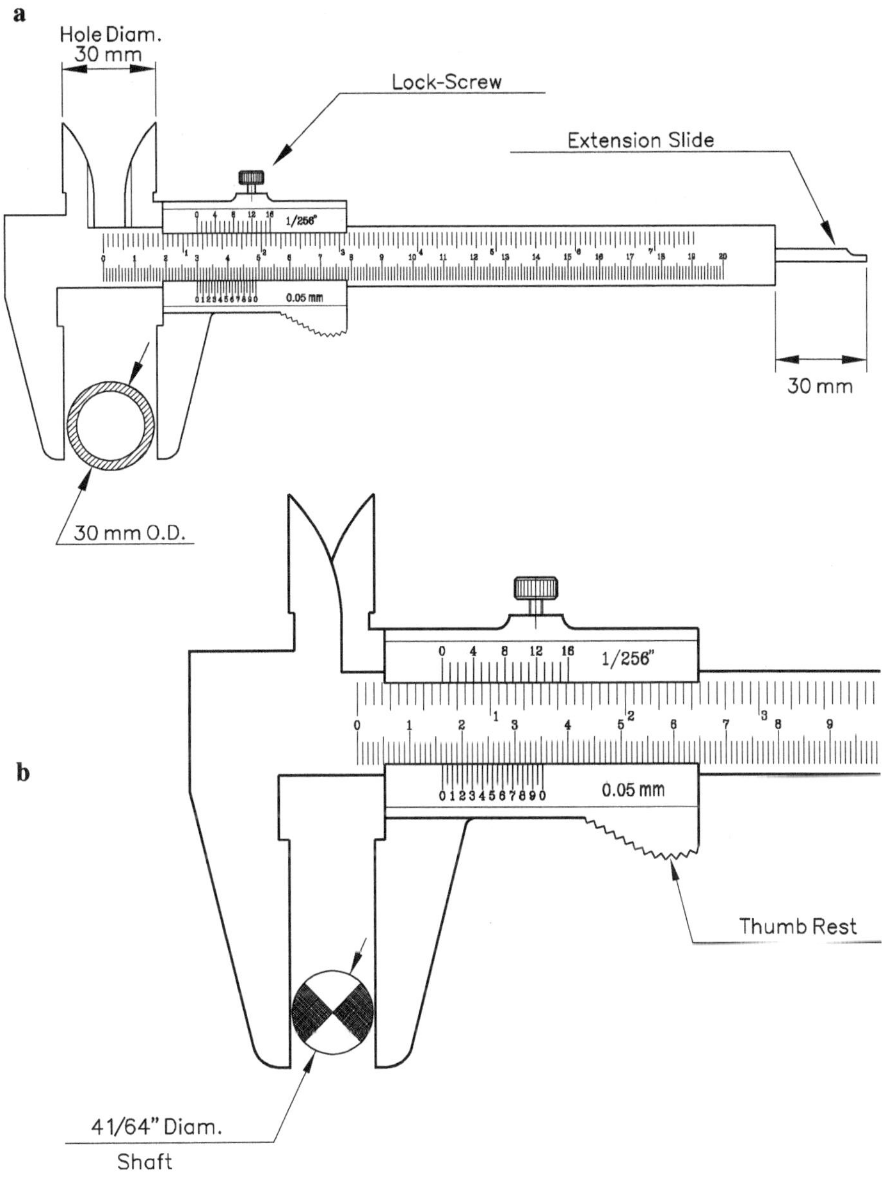

Fig. 1.10. Caliper with vernier scales for mechanical work

to read 1/100 or 0.01 mm if it weren't that on such a scale, you could no longer be sure which marks coincide.

That led to the development of calipers outfitted with dials (refer to Fig. 1.12) and later, in the age of the "chip", to numerical readouts. The latter have the inch/millimeter conversion built in and accessible with the touch of a bottom. A second bottom nulls the reading on the position you wish – such as both legs of the instrument in close contact.

Though that takes care of the limitations of readout precision, it still leaves us with distortions caused by the unpredictable manual pressure on the caliper's jaws.

Micrometers

Therefore, a more rugged device, the micrometer (Fig. 1.11), is used for measurements of 0.01 mm of accuracy. Sir Joseph Whitworth (1803–1887), founder of the first standardized thread system in history, the Whitworth thread, showed a set-screw based measuring device in a London exposition in 1851. Modern micrometers consist of a spindle with a precision-ground thread of 0.5 mm pitch (in metric models), which rotates in a fixed nut. Each turn of the spindle makes it advance by 0.5 mm, so that 1/50 of a rotation equals a 0.01 mm advance. Mounted on the spindle is a sleeve (thimble) with a scale engraved around its perimeter for reading the distance between the anvil and the front end of the spindle with 1/100 mm precision, as needed in critical machining operations, such as sizing seats for ball bearing and other parts with press or gliding fits.

Users must keep in mind that the half millimeter advance of the spindle with each rotation of the thimble makes that a reading of, say, 10 mm and 33 thimble divisions could stand for openings of 10.33 or 10.83 mm. Which value is right depends on whether the edge of the thimble stands on the lower or – as in the illustration – the upper half of the millimeter graduation.

Micrometers are available in ranges from zero to 25 mm, from 25 to 50 mm, from 50 to 75 mm, etc. Precision ground calibration pins of 25, 50, 75 mm, etc., are used for setting the correct zero point of all but the smallest models, which start at

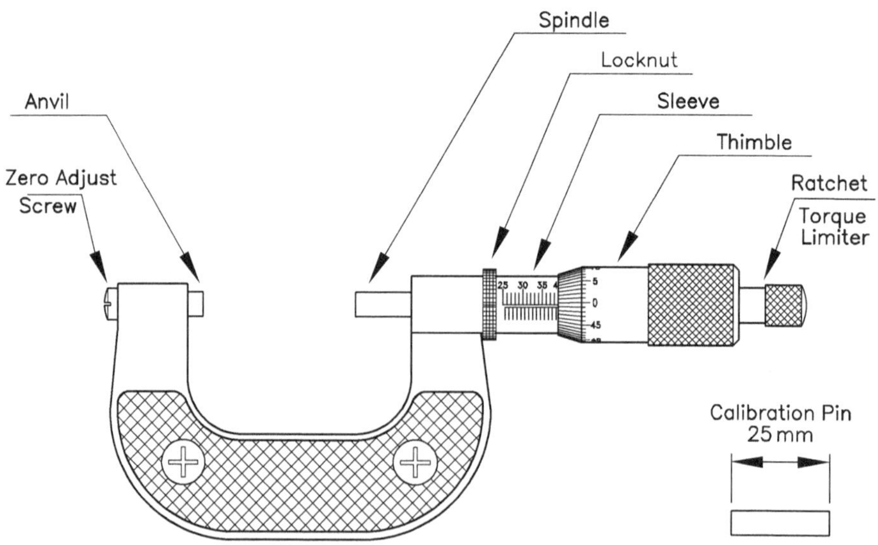

Fig. 1.11. Micrometer of 25 to 50 mm range

zero anyway. The micrometer per se is the same for all sizes; what changes is the size of the C-frame. Inch based micrometers use a pitch of 40 threads to the inch, while the scale around the thimble has 25 divisions, which makes their resolution $1/(40 \times 25) = 1/1000$ inch per division This complies with American drawing standards, concerning the sizing of dimension lines in 1/1000 rather than fractions of the inch.

Dial indicators

While micrometers give the overall dimensions of the measured object, dial indicators (Fig. 1.12) show the deviation of the object's actual size from it's nominal[8] dimension. They are available with resolutions of either 1/100 or 1/1000 mm, or still 0.001 inch and sometimes 0.0005 inches in our system. The scale of the instrument in Fig. 1.12a is numerated clockwise and counterclockwise. In the first case, it reads the instrument's central stem's upward displacement, and in the second case, downward. A knurled ring around the front of the housing is what holds the dial in place, herewith allowing for manual adjustment of the angular position of the zero mark. The knob on the lower right locks the ring in the selected position. The allowable tolerances can be adjusted with a pair of sliding indexes on the ring's perimeter.

The mechanism of a dial indicator (Fig. 1.12b) is based on the rack and pinion principle. The central stem's left side is teethed and geared to an interacting pair of compound gears that ultimately drive the needle.

Despite the mechanism's simple design, there is a caveat. Gears machined to a perfect mutual fit would frct and eventually coldweld together like joined gauge

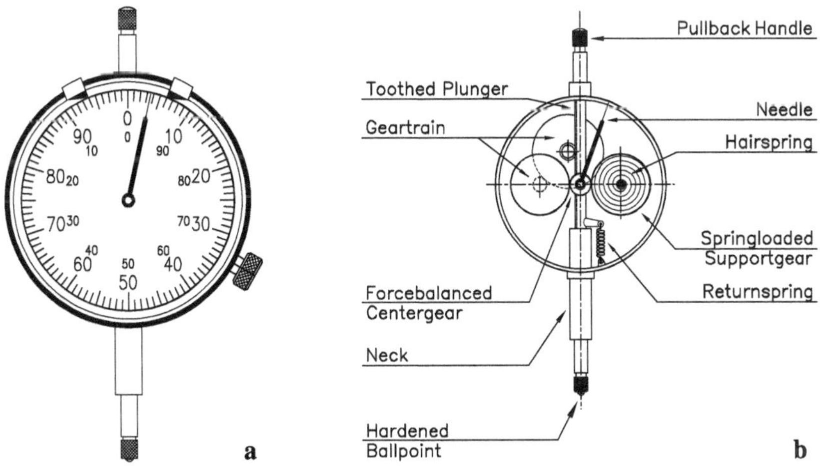

Fig. 1.12. Dial indicator. **a** Front view. **b** Needle drive mechanism

8 (desired).

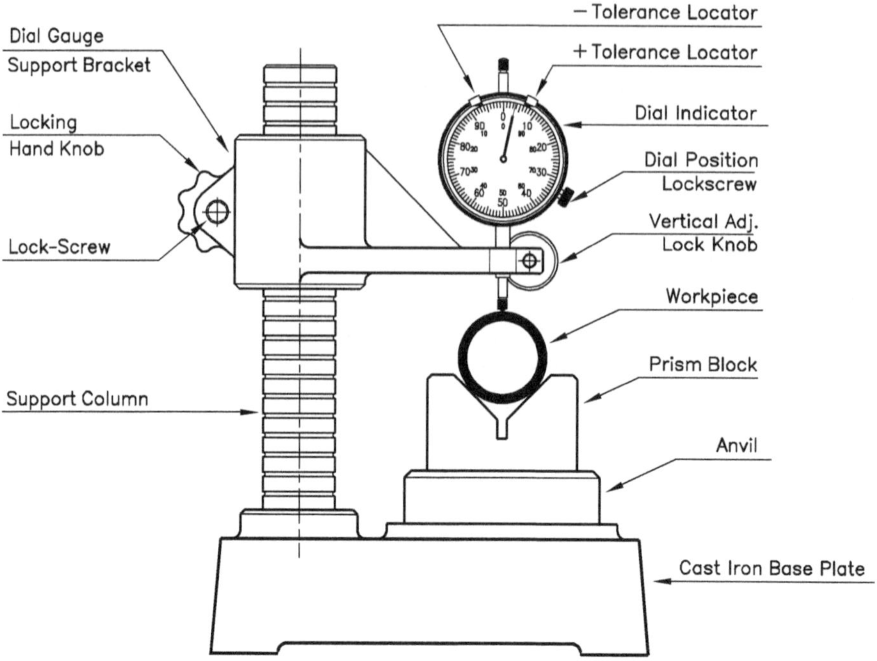

Fig. 1.13. Jig with dial indicator set up for measuring round objects

blocks. That's why it was so hard to make noiseless gearboxes for our cars. Like it or not, gears shatter because a certain amount of clearance between tooth spaces and teeth width is indispensable for them to work. In automotive applications, it takes highly sophisticated gear cutting techniques to bring that noise down to comfortable (or should I say "bearable?") levels.

In a dial indicator's gear train, that inescapable clearance, multiplied by the overall ratio of the gear train, would make readouts vary wildly. One solution to the problem of gear shatter is in the use of a pair of identical, face to face mounted gears, spring loaded against each other. That maintains the teeth of the driving gear in permanent close contact with those of the driven gear, readily adapting to any imperfections in gear cutting. That "clearance-free" match prevails for as long as the spring's preloading exceeds the load the gears transmit, a condition that makes the system unfit for power applications, as it would lead to high frictional losses.

Increased gear friction would cause sluggish response of the indicator needle to deviations of the stem. An alternative is to positively engage only one shoulder of the gears tooth profile. This approach is shown in Fig. 1.12b, where a hairspring preloads the spur gears into consistently opposing the upward motion of the stem. Besides, the added spur gear keeps the needle and pinion shaft free of lateral loads and the corresponding bending stress.

One of the most common applications of dial indicators is in fixtures like Fig. 1.13, which anvil can be outfitted with all types of work-holding jigs, such as the V-block for round workpieces in diameter checks. The illustration shows the dial indicator nulled in for a cylindrical object of precisely 30 mm diameter that rests in the V-groove of the base block (prism table). The reading shown as 3.6 divisions on the scale of 0.01 mm per division suggests that the diameter of the part under scrutiny is $3.6 \times 0.01 = 0.036$ mm above the desired value of 30 mm, namely, 30.036 mm, yet still within the tolerance limits of ± 0.06 mm marked by the tolerance locators.

Dial indicators are the instruments of choice in serial production. Checking the diameter of a thousand bushings with a micrometer would take hours, but can be done in seconds per piece with a dial indicator mounted in the fixture shown above. Approval or rejection depends on whether or not the dial indicator's needle comes to rest between the pair of adjustable indexes on the rim of the dial. Their spacing is set to include the tolerance of the part. Setting of the tolerance locators at ± 0.06 mm makes parts from 29.94 through 30.06 mm diameter acceptable, while parts smaller than 29.93 or bigger than 30.07 mm are rejected. The operator needn't even read the dial but merely checks whether or not the needle is within the space between the indexes.

2 Angles and arcs

Predominant amongst leaders with the courage to fight prejudice and superstition ranks Dom Henrique el Navegador, Henry the Seafarer (1394–1460), fourth son of King Johannes I of Portugal. Under his reign, atavistic fears of the open oceans, rooted in antiquity, were overcome and forgotten.

The vision of a disklike Earth kept the Phoenician, Grecian, and Roman navies confined to the Mediterranean Sea and the Atlantic coastline because he, who dared to navigate farther out, would be swept over the waters' edge at the borderline of the earth-disk. Such believes prevailed into the second century CE until the Greek astronomer Ptolemy rejected the entire flat Earth concept in favor of a sphere. His work, the "Almagest", became the "astronomer's bible" from antiquity through the Middle Ages. But fears of sailing into the unknown resurfaced under new propositions, and at Dom Henrique's times focused on the zone between the tropics of Cancer and Capricorn, where the Sun's passage through the zenith twice a year was expected to scorch everything below. The region's waters were imagined as boiling hot, salty slurries, which no ship could plow through. To make things worse, maelstroms capable of swallowing the mightiest of vessels, and sea monsters never beheld before, were thought to populate those inhospitable seas. Not even the landing of Portuguese ships on the then uninhabited island of Madeira, and subsequently (from 1427 through 1452) in the Azores, changed such visions.

The latitude of Cape Boujdour (then Cabo Bojador) – less than three degrees north of the tropic of Cancer – made this speck of land on the coast of the Western Sahara to be known as "the edge of the sea of darkness". Worse, not all of the sailors' tales were superstitious. Some referred to hard facts, such as the region's ever-present fog, desolate coastlines, and uncharted reefs and sandbanks.

It took Henry the Seafarer over a decade to convince his naval crews to try their luck. Offering generous compensations to the first seaman to round Cape Bojador, the king succeeded in 1422 to send off a fleet, but to no better end than seeing his ships prematurely retreat and sail home. Undisturbed, Henry made fifteen more such attempts, until in 1434 his skipper Gil Eanes, rather than fighting the many uncertainties of the cape, sailed west into the open ocean, then south, and finally back east, and found – surprise, surprise! – a close resemblance to the world north of the cape. Thus, Gil Eanes, Henry's former squire, lifted the region's evil spell and with it, the fear of the open ocean, which for so long had locked the European population into the confines of their continent.

Nevertheless, 54 more years had to pass until Bartholomeu Diaz sailed around the southernmost expansion of the African continent, which he baptized "Cape of Storms", more optimistically renamed "Cape of Good Hope". The cape's circumnavigation opened the first sea-route to the Far East, which led

Vasco da Gama into the Indies in 1498, six years after Columbus' voyage with the same objective had brought the discovery of the Americas instead.

Only when you set foot on deck of a reconstruction of those times' vessels can you imagine what it must have been to sail those fragile and primitive wooden shells for weeks, and sometimes months, through stormy weather and tropical heat. Often enough, marines had to survive on moldy biscuits, tepid water, and sooner or later suffered from scurvy, the malady caused by lack of vitamin C. Still more mind-boggling is the thought that early explorers were guided by nothing but the most rudimentary navigational instruments.

Finding a ship's position boils down to the measurement of angles, because the familiar terms "latitude and longitude" refer to the shortest angular distances of your geographic location from the equator and, respectively, from the Greenwich meridian. Latitude can be derived from the positions of celestial objects, such as the Sun and a number of guide stars, and the classical navigational instrumentation consists of devices designed to do just that. One of the oldest is the Latitude Hook of the Polynesians, a piece of split bamboo tied to a loop of string. Along this line, the Arabs used the *Kamal*[1], a wooden tablet with a string attached at its centerpoint. To measure the height of a celestial object, the user secured the string loosely between his front teeth and pulled the tablet outward until its edges covered the angle from the horizon to the object. The free length of the string became a measure for the altitude of the target star, and was marked with a knot.

Navigators used to assure themselves of a "happy return" by aiming the Kamal at their homeport at some celestial reference point, such as one of the brighter circumpolar stars at the time of its lowest position, and tying a "Home Knot" into the string of the Kamal. In principle, the way back from the port of call could be found by sailing north or south to some location where the Kamal zeroed in on the Home Knot, and then follow a straight west or east course that – safe for contingencies – would lead into the ship's homeport.

In the third century BC, Eratosthenes of Kyrene derived the difference in latitude between the cities of Alexandria and Siene (today's Aswan) on the day of summer solstice from the angle between the Sun and the zenith in those two places. Knowing the distance between the cities as fifty "camel days" (the time a camel caravan took for the trip), Eratosthenes used his results for the first ever evaluation of the circumference of the Earth, which matched with modern figures within less than one percentage point.

By the same token, we can find the latitude of our geographical position from the height of the celestial North Pole over the northern horizon. Polaris, the North Star, lingers about one degree of arc off the pole, and for good measure, we have to shoot its height twice, at eleven hours fifty-eight minutes of time difference, and average the results.

Intuitively, one would be tempted to set that time interval at half a day, but the Earth completes 366 full turns per year while the Sun rises only 365 times

1 also used by navigators on ships trading between the Middle East and Southwest India.

because her yearly transit around the Zodiac leads to the loss of one day per year. Hence the definition of star-day for the time of rotation of the Earth around its axis, along with the concept of star-time, which astronomers use to set the right ascension of their instruments.

Having a bright star like Polaris so close to the celestial north pole is a blessing that lasts for a few centuries only. Precession – a gyroscopic swing of the Earth's axis – makes the celestial poles trace a circle of 23.5 degrees of arc radius in 25,800 years. At present, that favors us 21st century's earthlings by making the celestial pole nudge toward the North Star to within half a degree of arc. Christopher Columbus would have loved such a nature-given navigational tool, but at his times, Polaris stood almost eight degrees off the pole. This might have been one of the admiral's reasons for using the Sun as point of reference, regardless of the complications caused by the Sun's wandering over the sky, while the celestial pole retains its position.

The identity of the height of the celestial pole (α) and the latitude of the observer's location makes that the highest point of the celestial equator locates at an angle of $90 - \alpha$ above the horizon. That's where the Sun stands at noon on the days of the equinoxes. At summer solstice (around 22 June) the Sun locates 23.5 degrees above the celestial equator; and at winter solstice (around 22 December) by the same measure below. For the dates in between, tables of the Sun's declination (its position above and under the celestial equator) have been available since antiquity. Such tables, named "Calendar Scales," are often found engraved on the back of planispheric *astrolabes* and other historical instruments for maritime navigation.

Cross staff angle measurement

The cross-staff, a simple but fairly accurate device from Columbus' times, also known as "Baculus Jacob" or "Jacob's Staff" (Fig. 2.1), is credited to the mathematician, astronomer, and bible commentator Levi ben Gershon (1288–1344). Unlike other navigational tools which measure the height of a reference point on the sky above the horizon, the cross-staff is designed to check the angular distance between two celestial objects. It consists of a square bar, called *staff*, and a sliding crossbar, the *transom* or *limb*. Much like in the case of the Kamal, the navigational objects whose angular distance was to be measured got viewed from the rear end of the staff over the knife-edges of the cross slide (Fig. 2.1). Results could be read on a scale engraved along the length of the staff. This scale is not linear like that of a ruler, but has successively narrower spacing (Fig. 2.1). With $2y$ for the length of the crossbar, and x for the distance from the left end of the staff to the centerline of the cross bar, the angle 2α, subtended by the viewing end of the staff and the knife edges of the crossbar, can be figured from $\alpha = \arctan(y/x)$, an inverse tangent function. The geometrically constructed scale in Fig. 2.1, marks where your eye should be for sizing the desired angle. But in reality, not the eye but the crossbar slides along the staff. Therefore, the scale as shown must be engraved inverted along the length of the staff, making that the angles subtended by the crossbar's edges become smaller the farther out it is moved.

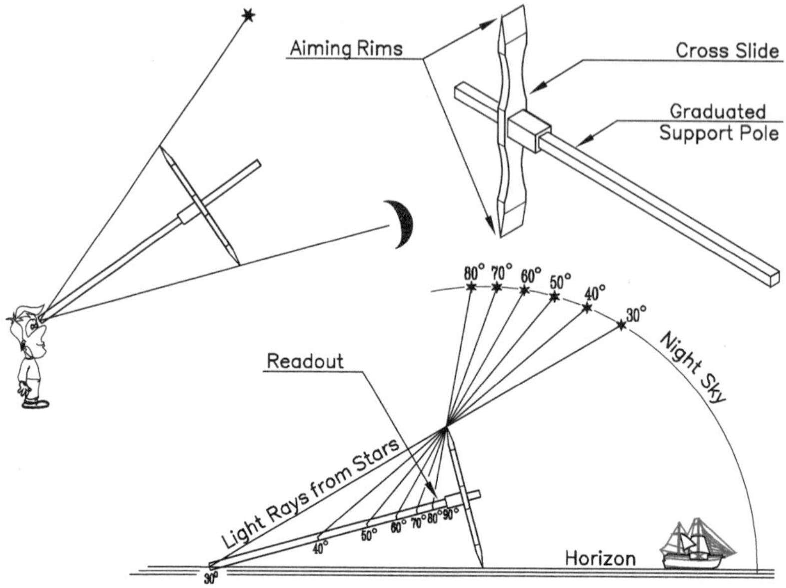

Fig. 2.1. Baculus Jacob's (cross) staff, an early instrument for angular measurements

Later versions of the instrument came with four sets of transoms of different size, each related to by an appropriate scale engraved on one of the staff's four sides. Notably, the roots of the saying "shooting the stars" may reach back to the similarity of aiming the cross-staff and aiming a crossbow or rifle.

Astrolabe and quadrant

The astrolabe (Fig. 2.2) is a rather elaborated navigational instrument from earlier times. It consists of a cast brass ring engraved with circles in degrees, equipped with a rotating sighting mechanism, the alidade, for viewing the sun or any other point of reference on the sky. Grim stories abound of marines losing their eyesight from exposure to direct sunlight, though the coincidence of the shadows of the alidade's front and rear sight may have been used alternatively for correct setting.

Letting the instrument hang from its hinged (Cardan) carrying loop, it self-aligns by its own weight with the vertical.

The planispheric astrolabe is credited to Hipparchus of Bithynia (around 130 BC), though its first description stems from John Philoponus of Alexandria (AD 490–ca. 570). The Muslim conquest of Spain led to the introduction of the astrolabe in Europe, where it first appears in a tenth-century manuscript by Llobet of Barcelona. All the great explorers, such as Columbus, Magellan, and Drake, used astrolabes for navigation. Chaucer's Treatise on the Astrolabe (1391) is remembered as the first scientific text written in English rather than Latin.

The quadrant (Fig. 2.3), one of the earliest and simplest instruments for finding the altitude of a celestial reference point above the horizon, consisted of a graduated quarter circle with two sighting vanes along the 90° line, and further a plummet

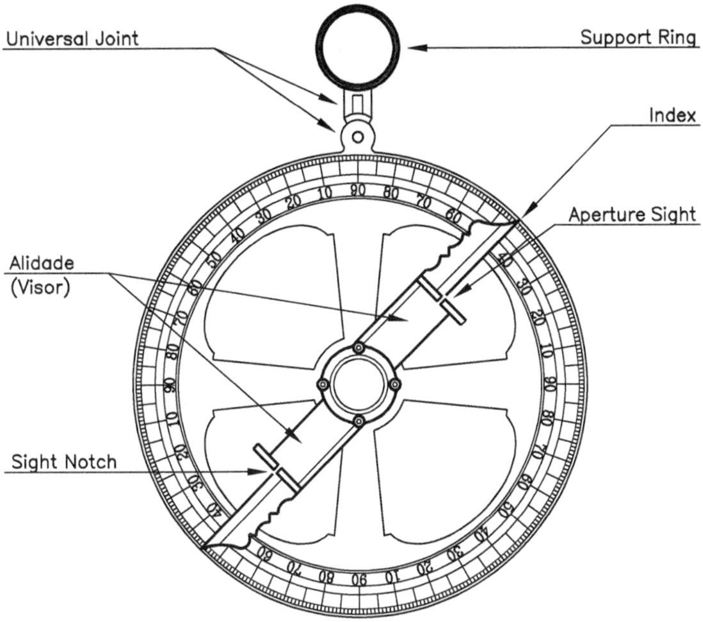

Fig. 2.2. Astrolabe

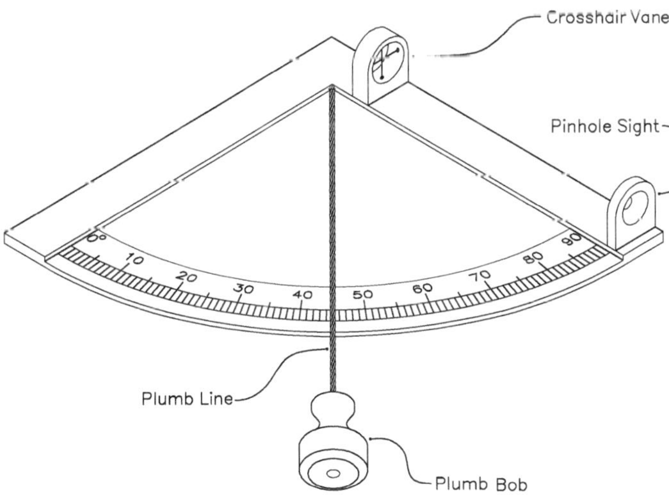

Fig. 2.3. Mariner's quadrant

suspended from the circle's centerpoint. Aiming the quadrant at the target through the vanes makes that the plummet line marks the altitude of the target above the horizon. The moment the user had the target in his sight, he held the string with two fingers pressed against the dial and kept it there while turning the quadrant into the most convenient position for reading the scale.

But sighting celestial objects without the help of a spotting telescope has its own drawbacks. The crosshair-equipped vane in Fig. 2.3 would work by day, but needs to be somehow illuminated for observations on the night sky. And as the observer's eye focuses either on the crosshair or on the target, one of the two will always seem blurred. Tycho Brahe avoided this kind of problems in the course of his highly precise measurements of the planets' positions by making his graduated circles with radii of several meters, enough for relaxed viewing of crosshair and object alike. That same idea may have inspired Basal Ringrose, navigator of Bartholomew Sharp's ship on a voyage around South America in the 1680s, to construct quadrants two feet and a half wide. By contrast, those times standard models were 10 to 12 inch wide, forcing their users to change their focus repeatedly from infinite to about 1 foot or less, and back.

The back-staff (Fig. 2.4), invented in 1594 by the English captain John Davis, also known as the Davis Quadrant, allows the simultaneous observation of the horizon and the Sun.

The back-staff's design splits the full 90° circle of the quadrant into two sectors: the rather large 30° circle with a pinhole vane for aiming at the horizon, and a smaller 60° circle with a shadow mask which the operator adjusts so that the edge of its shadow from the Sun centers on a screen in the circle's center. A pinhole in

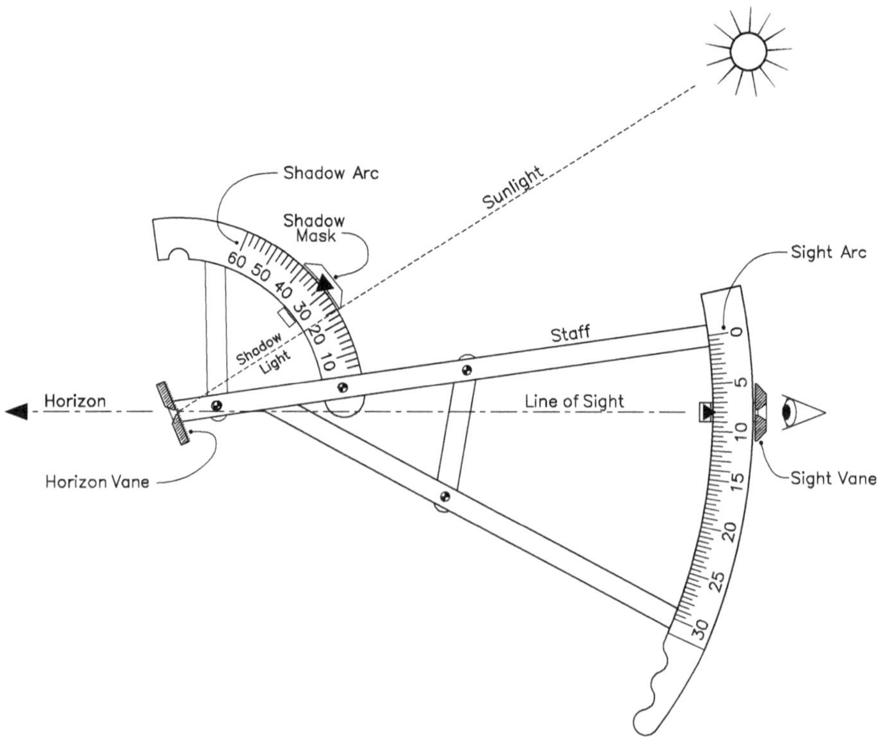

Fig. 2.4. Davis quadrant

the screen's center allows it to double as the front vane in aiming toward the horizon. Hence the term "horizon-vane."

The smaller size of the shadow circle reduces the overall footprint of the instrument and makes for a sharper contour of the shadow. That's important because the apparent diameter of the sun, of about half a degree of arc, makes the shadow's boundary into what's known as *penumbra*. We remember the term from eclipses of the moon, where totality is long preceded by the moon's gradual darkening on its transit through the Earth shadow's penumbra.

Down here, the penumbra manifests itself as a gradual – vulgo "blurred" – transition from bright to dark. Again, the width of that "twilight zone" is given by the sun's apparent diameter of 30 minutes of arc. The real neat shadows we sometimes see on the wall of our garage stem from the reflection of the sun on a polished convex surface, such as domed glass objects or curved, chrome-plated car accessories. But short of such optical tricks, the only way to sharpen the contour of a shadow is by reducing the distance between shadow mask and screen until the equivalent of half a degree becomes small enough to be neglected. For the case of a 6 inch radius circle, the width of a 0.5° wide strip becomes 0.052 inch or 1.3 mm – not too accurate but as good as it gets while keeping the spacing of the degree marks on the circle wide enough for easy readout.

With the instrument aimed at the horizon and the shadow mask properly placed on its circle, the position of the Sun is given as the sum of the readouts from each of the two circles. This makes the positioning of the horizon sight on the big circle arbitrary, unless the Sun stands near the zenith, in which case a position at or near the 30° mark is called for.

As the Davis Quadrant released the sailors from the "eyeing the Sun" dilemma and enhanced the precision of measurements, it got readily accepted throughout the seafaring community and remained a standard onboard accessory into the end of the 17th century.

Sextant

In later times and into the recent past, the sextant (Fig. 2.5), invented by Isaac Newton in 1742, became the seafarers instrument of choice. With the use of optical elements, such as front surfaced mirrors and a small telescope with sun filter, the sextant allows for viewing both reference points in one session. "Shooting the Sun" with a handheld sextant remained the way of checking the latitude of a ship's position, until satellite oriented navigation took over.

As the name implies, the sextant uses one sixth of the full circle (60°), but allows to measure angles of up to 120°, because images from a tilting mirror swing by twice the mirror's rotation angle. Similar instruments of a range of 90°, such as the octant, were described in principle by Sir Isaac Newton and foreshadowed by Robert Hooke in 1644.

Navigation aside, astronomers use the sextant along with a pan of mercury to measure the height of celestial objects over the horizon. Since the surface of a bath of liquid mercury models the surface of Earth better than anything else, the angle between a star and its image in such a "mercury mirror" equals twice the

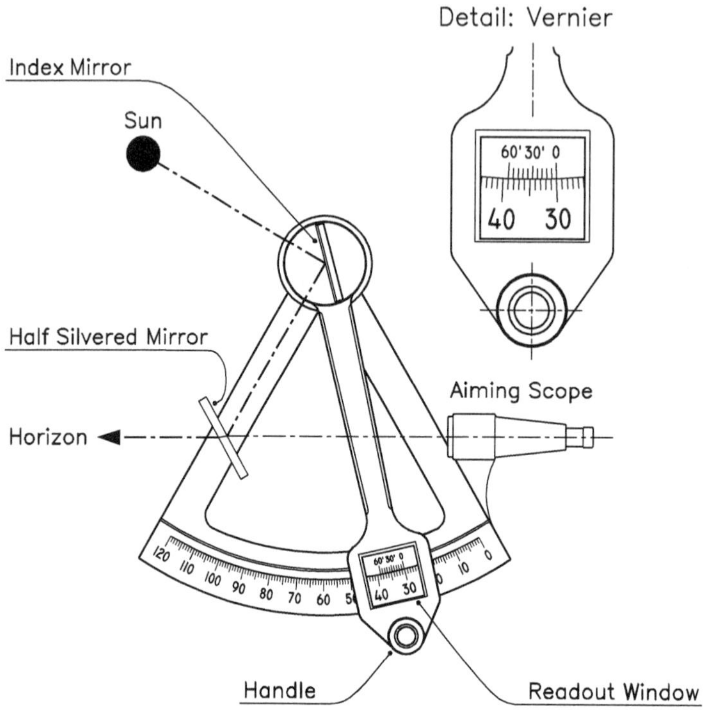

Fig. 2.5. Sextant with vernier readout

star's elevation over the horizon. Artificial horizons are essential where the landscape blocks the view at the geographical horizon. But even in coastal areas, where the sea renders a well-defined line of horizon, the mercury bowl gives better precision, partly because it circumvents aberration of light rays by atmospheric refraction, strongest at low elevations. However, the bowl must be large and deep enough to keep surface tensions from distorting the evenness of the liquid metal surface.

Levels

On the shop floor, however, a bath of mercury is not the only, and neither the most convenient way of defining the horizontal. A carpenter's level does that by means of the force of buoyancy, discovered by the Grecian philosopher and mathematician Archimedes while dwelling in a steaming bathtub. In open containers, the surface is level and a floating object may come up anywhere, but an air bubble in an arced glass vial always seeks the highest point of the arc. This is the principle of the spirit level, a sealed, slightly curved tube, mounted in a wooden or plastic body. The instrument's precision depends on the bending radius of the vial. The greater that radius, the less the vial's curvature, and the more sensitive an instrument you get.

To check a level's accuracy, place it on a flat horizontal surface, such as a machine shop's measuring plate. Set the plate so the bubble zeros in, and then invert

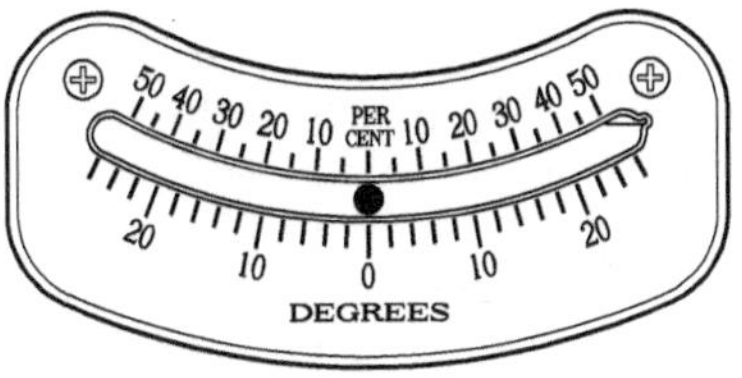

Fig. 2.6. Spirit inclinometer

the level's position. Of course, the bubble should return to the center mark of the vial in both cases. If not, slightly tilt the plate until the deviations of the bubble in both orientations are opposed but of equal magnitude. At that point, the table is set horizontally and the bubble's aberrations from center position indicate the readout error of your leveling tool. If it has setscrews, adjust for position-independent zero readings.

Box type levels, such as carpenters and mason levels, are for flat surfaces, and their sensitivity is typically 0.5 millimeter per meter, or about 0.03 degrees of arc. The highly sensitive stride type levels have a V-groove in their base to rest on round objects, such as machine shafts and surveyors telescopes.

Inclinometers and laser levelers

While bubble levels function on the principle that an air bubble's lift carries it toward the highest point in its casing, a popular type of inclinometers (Fig. 2.6) is based on gravitational forces. To this end, the vial is arced downwards rather than upwards, so that a little solid ball inside tends toward the lowest instead of the highest point. There is still fluid in the vial, but its function is to damp the motion of the ball and avoid erratic displacements.

Considering that the motion of a rolling ball suffers higher frictional losses than an air bubble, precision of this type inclinometer is not meant to compete with that of traditional levels, but their range is usually far greater. Applications thus include fast changing tilts, as they occur in vehicles up and down an incline, or on board a reeling ship or boat.

In Fig. 2.6, the lower scale reads degrees, and therefore is regular; the upper, drawn to a tangent function, shows percent of incline or gradient, that is the ratio of the elevation difference between two points and their horizontal distance. For instance, a 10% incline means a one-foot rise per ten feet distance, or an angle of $\arctan 0.1 = 5.71°$. Note that a 100% slope is only 45° steep, and that the gradient of a vertical precipice is infinite (∞).

Laser levelers are sensitive bubble levels mounted in the handle of a laser pointer – a combination that makes them handy tools for mechanics, engineers, home inspectors, and interior designers. Mounted on a tripod, they can be used with surveyors' staffs for finding the height differences between two or several locations.

Camera levels house the spirit under a slightly domed glass surface, and thus level equally in all directions, a job that otherwise would take two vial levels at right angles with each other.

Protractor and the carpenter's bevel

Last but not least, let's not forget the venerable protractor (Fig. 2.7) and the carpenter's and machinist's bevel or miter gauges (Fig. 2.8). Protractors made

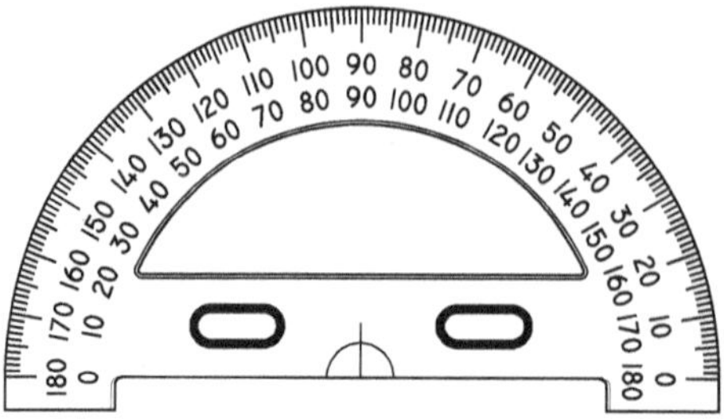

Fig. 2.7. Protractor

part of the "must have" outfit of machine and architectonical designers and draftspersons in times bygone, when drawings were made on a drawing board with ruler, compass, triangles and – not to forget – protractors. The principal circle of a protractor counts counterclockwise, as it is standard in trigonometry. But most protractors scales show clockwise markings too.

The Carpenter's Bevel (Fig. 2.8) is a combination of a precision protractor with an adjustable T-square. Good quality bevels have the scales on ruler and protractor engraved rather than imprinted. The pivoted crossbar is set at the desired angle and held there by a lockscrew. Likewise, the ruler is locked in place with the screw lock assembly that shows in Fig. 2.8 inside the semi-circular cutout of the protractor. Special applications include tracing other than 90° cuts, and depth measurements.

Sine-bar

The sine-bar (Fig. 2.9) is used for angular measurements with the high degree of precision afforded in machine shop work and principally in tool and die making. It excels by its simplicity. All there is to a sine-bar is a hardened, ground and lapped steel bar with precision ground pins of identical diameters mounted on each end. Used with gage blocks (accurately ground and lapped steel blocks of

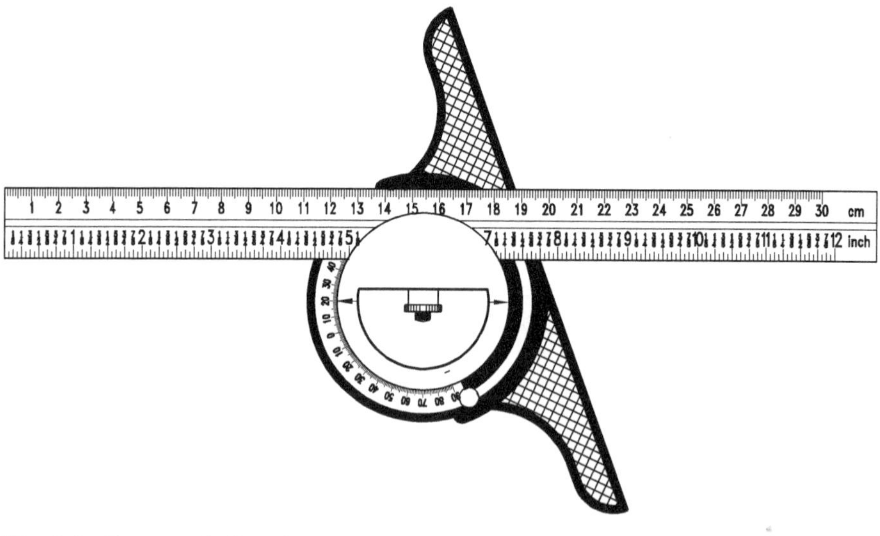

Fig. 2.8. Carpenter's Bevel

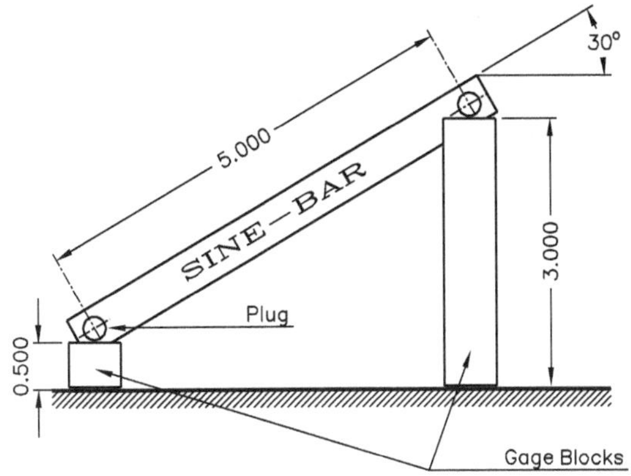

Fig. 2.9. Sine-bar

standard length), the sine-bar can be set at any desired angle. For instance, a typical sine-bar of five inches center distance between pins, supported left and right on gage blocks of one inch and three and a half inches respectively, will settle at a thirty-degree angle with the tabletop, because: $\sin \alpha = (3.50 - 1.00)/5 = 0.5$, for which your calculator (or trig tables) show $\alpha = 30°$.

Sine bar angle measurements are precise to five minutes of arc or better if made on a machinist's *flat plate* as base plate. Flat plates are very thick and stable steel or hard stone plates with flatness and surface finish within the thousandth of a millimeter range. And yet, the amazing precision of the sine-bar pales in the face of digital techniques.

Digital theodolites and encoders

Digital theodolites (Fig. 2.10) measure angles with arc-seconds precision, and motorized models, controlled with a joystick, sometimes reach 0.1 arc-seconds of accuracy. That's the equivalent of dividing the degree of arc into 36,000 parts, which is why counting pulses gives so much better resolution than the reading of scales.

The amazing machines which do just that are linear and rotary encoders. If you think you never saw one, think again, and while you are on it, search your desk for an old type mouse. Not the furry type, but the one in the ergonomic housing and, for a tail, a cable connecting to your computer. Remember the day when you took heart and unscrewed the retaining ring on the bottom of that gizmo to take out the hard rubber sphere for a cleansing? Behind was a pair of rotary encoders: Two tiny disks with typically 32 radial perforations near the perimeter. While spinning, they block and clear alternately the path of light beams between built in light sources and sensors, which convert those flashes into electrical pulses at the rate of 32 pulses per rotation. A chip converts them into binary signals, which go to the computer for processing and controlling the cursor's position on the screen.

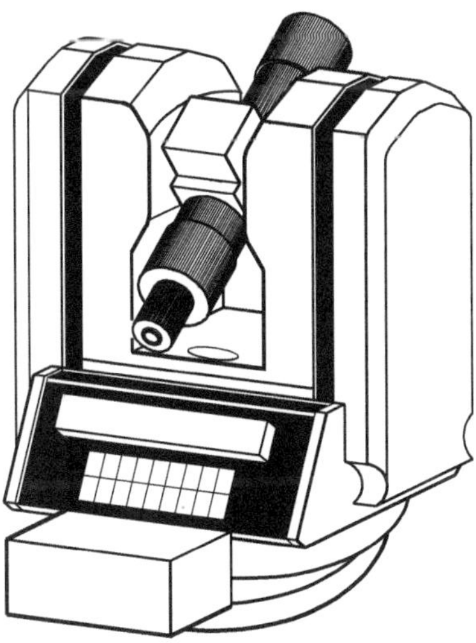

Fig. 2.10. Digital theodolite

Those gradated disks are oriented at right angles to each other and have their shafts linked to a pair of friction wheels (rollers), which touch and are driven by the rubber ball. This way, the rollers sense the space and direction of the ball's rotation, and herewith the displacement of the mouse in the x and y directions. It all sums up to a couple of rotary encoders, driven by the displacement of the mouse on the mouse-pad; the mouse-pad acts as a rack and the ball as the matching pinion.

Rotary and linear encoders have become core elements in machine design and motion control as the links between automated systems and the computers that control them. In desktop computers, input pulses come from the keyboard and the mouse, output pulses go to monitor and printer. Control computers have several input and output terminals but are functionally similar to our desktops. Inputs are picked up from sensors, such as limit switches, light sensors, proximity switches, and principally encoders, strategically located in machinery and material handling systems. Output terminals connect to actuators, which convert the control computers' signals into metered motion.

The typical optical encoder operates with a beam of light passing through a gradated glass disk into a photoelectric sensor. This latter generates one electric pulse for each of the disk's divisions (Fig. 2.11). Unlike the traced scales on protractors, circles on rotary encoder disks are divided into alternately dark and bright fields. Glass disks with radial markings are common, but metal drums with divisions etched around their perimeter are used for very large diameter requirements. Since their scales are read by reflection, they alternate highly polished and matte finished fields. The first reflect like mirrors, while the latter disperse light.

Photoelectric sensors, such as photoresistors, photodiodes, and phototransistors detect the dark and light fields by changes in their internal resistance; from tenths of kilo-ohms (kΩ) in the dark to a few ohms in bright light.

Infrared light emitting diodes (LED) are the device of choice in encoder design because of their short reaction times, minimal generation of heat, and extended operating life. Unlike incandescent lamps, LEDs can be turned on and off in microseconds.

As an encoder disk rotates, opaque and transparent fields alternately block the light from the source, which makes the photoelectric receptors, such as the photodiodes in Fig. 2.12c, change their resistance. Powered by a direct current (DC) source, these changes generate pulses from zero volt to $+E$ and back to zero.

Though the traces left on the face of an oscilloscope by pulsating DC and

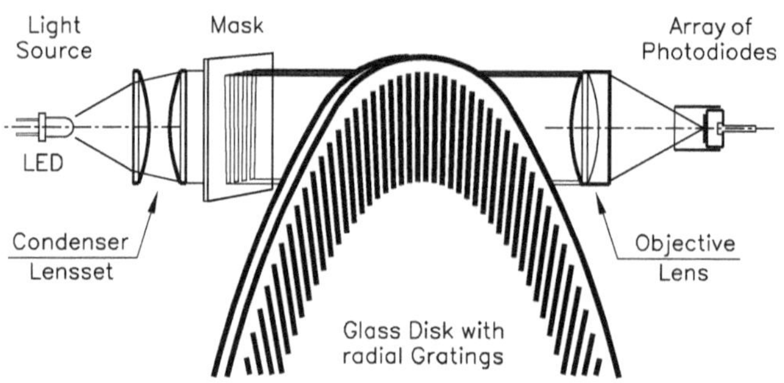

Fig. 2.11. Encoder disk and optical readout system

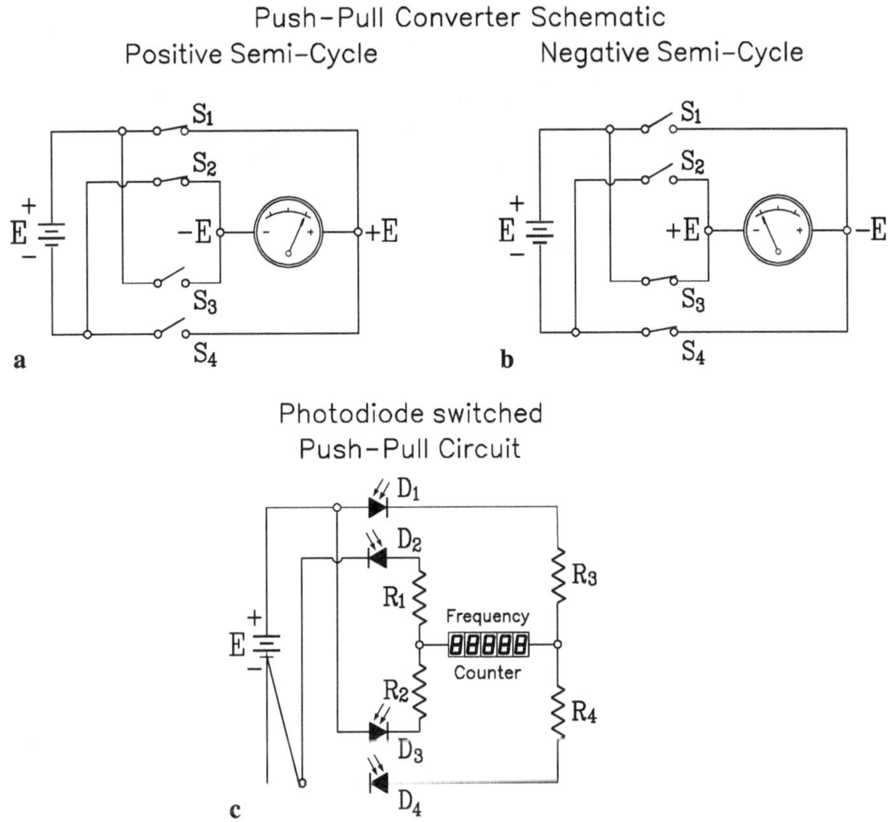

Fig. 2.12. Encoder output circuits

square wave AC look similar, alternating current differs by varying from zero volt to $+E$, back to zero, down to $-E$, and back to zero again. But pulsed DC can be converted to AC with the switch controlled circuit shown in Fig. 2.12 in its two possible states: outputting $+E$ volt (Fig. 2.12 a), and outputting $-E$ volt (Fig. 2.12 b). Thus, the circuit interchanges the connections of the load in every cycle from $+E$ to $-E$ and back, generating square wave AC. For higher frequencies, semiconductor switching devices replace the mechanical on/off switches, as in Fig. 2.12 c, where the four photodiodes "read" the position of the gradated disk. Each passing division, including a dark and a light field, generates one cycle of alternating current at the output terminals, capable of driving a counter or frequency meter, or still command a process control computer.

Each photodiode is located behind a mask with windows the size of one field on the disk. As the disk rotates, waxing and waning bright areas are illuminated. Figure 2.13 a shows the size of the illuminated area for each of the instrument's four photodiodes, D_1 through D_4, sequentially for twenty-four consecutive disk positions, much like a movie strip would depict the process.

If the response of the diodes were linear, i.e., proportional to the incident light, we would get a triangular output, like the thin line on the plot in Fig. 2.13 b. In

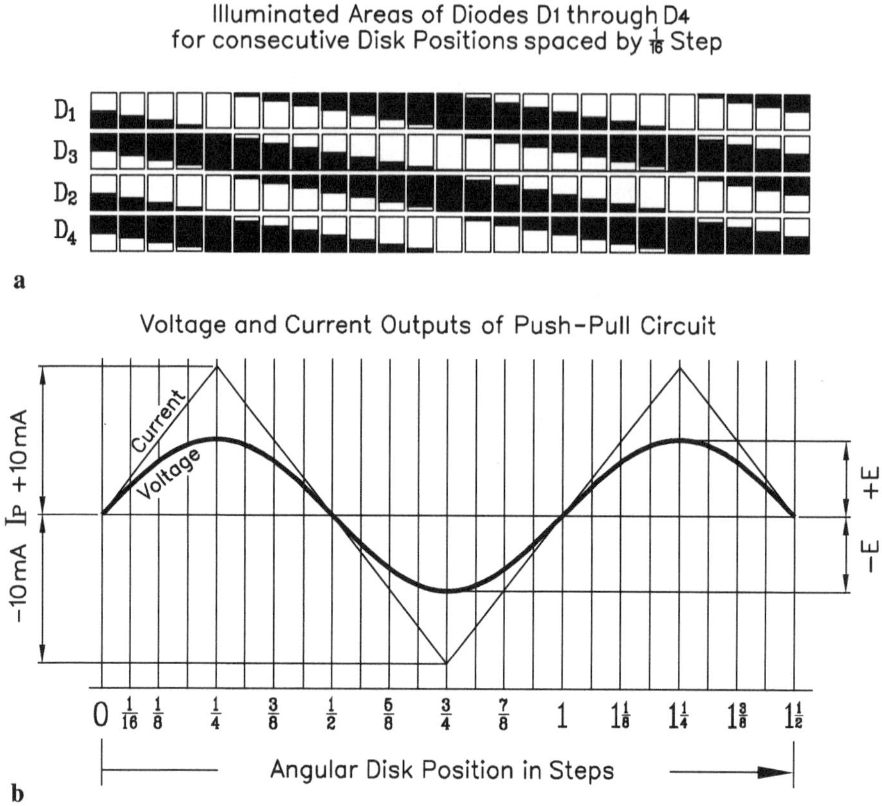

Fig. 2.13. Output current and voltage in function of the degree of diode illumination

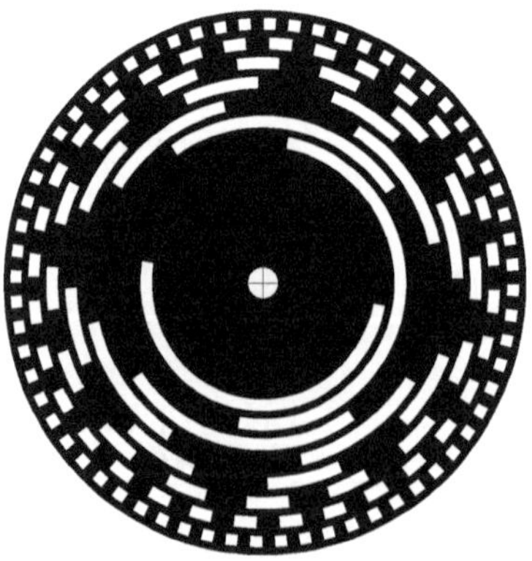

Fig. 2.14. Absolute encoder disk. Courtesy of Mr. Scott Orlosky, BEI Ind. Encoder Div.

reality, the devices' sensitivity to changes of illumination is highest from dark to some light, and far less from bright light to somewhat brighter light. This cuts off the peaks of the triangular curve and leaves the approximately sinusoidal alternating output, shown as a dark line.

The type of encoder discussed so far outputs one pulse for each increment of rotation, and thus has been termed "Incremental Encoder," opposed to the "Absolute Encoder," which shows the encoder shaft's angular position with relation to an absolute zero point. Absolute encoders provide their data in binary format, read directly from the encoder disk along a radius outward from the center. To this end, the disk of an absolute encoder (Fig. 2.14) holds a number –

call it n – of tracks, the outermost divided into 2^{n-2} fields of a transparent and an opaque sector (64, or 2^6 in Fig. 2.14). Outward from the center, each subsequent track holds double the number of fields of the one before, that is 1, 2, 4, 8, 16, 32, 64. Each track is read by its correspondent photodetector.

The term "binary" evokes in computer-literates (isn't everybody these days?) an on/off based code, developed in search for a mathematical counterpart to the operation of an electrical switch. *1* stands for "contacts closed", and *0* for "open". The natural binary code maintains the tenets of our decimal number system. The same set of rules that allows us to write the number 358 in the form of $3 \times 10^2 + 5 \times 10^1 + 8 \times 10^0$ makes the binary number 01111 into

$$0 \times 2^4 + 1 \times 2^3 + 1 \times 2^2 + 1 \times 2^1 + 1 \times 2^0 = 8 + 4 + 2 + 1 = 15\,.$$

In data transmission, natural binary has its shortfalls. For instance, the next higher number in the sequence above, 16, converts to 10000, so that the one step transition from 15 to 16 requires the shift from four *ones* into four *zeros*. Since data transmission proceeds at a constant rate, the time span needed for such massive shifts would have to be allotted to every step, including single digit changes such as from 110 to 111 (from 6 to 7). On the other hand, numbers as distant from each other as 7 and 15 (0111 and 1111) differ by only one bit, so that an early or late transition could lead to errors with the potential of causing a piece of artillery to swing by 180° and fire into the wrong camp.

The solution came with the development of a specific binary code, called the Gray code[2], introduced by Frank Gray. NIST defines the Gray code as "an ordering of 2^n binary numbers such that only one bit changes from one entry to the next." This suggests a reshuffling of the natural binary numbers through codification.

In general terms, codification may be as simple as: A = 1, B = 2, C = 3, etc., a system that even laypersons wouldn't have much trouble cracking. For improved safety, we could modify it into A = 17, B = 21, C – 2, and so on, in an arbitrary pattern for the 26 letters of the alphabet, yet without repetition of numbers. Only those in possession of the key can transpose the hidden message in such a row of numbers. Thus, we may understand the Gray code as coded binary, which Table 2.1 shows for the numbers from 0 to 7.

Note that the least significant digits in natural binary numbers alternate between 0 and 1 in each step (column 4), while the corresponding column for the Gray code alternates in arrays of two identical digits each. This doubles the width of the windows on the encoder disk, so that the 64 fields on the outermost track on the Gray code disk houses the equivalent of 128 fields in any other code.

Compared to incremental encoders, the added complexity of absolute encoders has its rewards: If the machine is shut down with the encoder disk at, say 90°, the incremental encoders would reset it, while an absolute encoder picks up at 90° when the machine comes back on.

2 proposed by Bell Telephone's G. R. Stibitz.

Table 2.1. Gray code conversion

Number in decimal	Natural binary code			Gray code		
	First digit	Second digit	Third digit	First digit	Second digit	Third digit
0	0	0	0	0	0	0
1	0	0	1	0	0	1
2	0	1	0	0	1	1
3	0	1	1	0	1	0
4	1	0	0	1	1	0
5	1	0	1	1	1	1
6	1	1	0	1	0	1
7	1	1	1	1	0	0

A further advantage of absolute encoders is their immunity against stray signals. If electrical noise corrupts encoder data, an incremental encoder does not make up for it in subsequent counts, while an absolute encoder resumes the correct positioning as soon as the disturbance has been overcome.

Encoders read angles with far higher precision than other devices and do it faster. And yes, the machine converts the data into commands and implements them on other machines reliably at speeds we humble human beings could never hope to match.

But let's not forget the unidentified Babylonian who, four thousand years ago, thought of the sexagesimal number system, which led to the division of the circle into 360 degrees.

What inspired him? His brainpower.

What motivated him? Necessity.

Creating an idea out of necessity is the human power that machines will never emulate, and that's why we will always be their masters.

3 Clocks and the measurement of time

Would the impact of a falling tree still produce sound if nobody were around to hear it? Would the sky still be blue if it weren't for people looking up?

The problem here is hidden in the definition of concepts. If sound is defined as periodic pressure variations in air, then the falling tree makes sound, whether or not we are there to listen. But if you define sound by its stimulation of the human auditory system, the unwatched tree tumbles in silence. And if blue is defined as electromagnetic oscillations at a certain wavelength, the sky will always be blue. But if defined by its effect on the human retina, the beauty of the skies vanishes with our demise.

But how about time? Would time still pass if we weren't there to count the seconds? We measure an object's length or angle of position by comparing it with other objects, such as a ruler or a protractor; but we cannot directly compare two time intervals, because the first is history before the second begins to tick. Being irreversible, time can only be measured by comparison with a reversible process, such as motion – usually that of a clockwork. We talk time, but measure displacement.

As if time had a life of its own, we neither can grab nor control it. To complicate matters, it lives in one direction only: from the future to the past, so that we age into the future. Squeezed in between past and future is the present, which is where we live and love, although psychologists blame some people for living in the past and others for living too far into the future. But when they tell us to live the present to the fullest, how much time do they allot for such activities?

In other words: How long does the present last?

Taking it as the boundary between future and past would make the present infinitely short, but then, could we survive squeezed into such infinitely short time intervals? Actually, it seems futile to attribute any measurable length of time to the present. Assume it to last for two seconds, and you will instinctively assign one second of that period to the past and one second to the future, and hereby – once again – compress the present into the unsubstantial. In a movie, the present lasts as long as one single frame, typically that twenty-forth of a second the human brain needs to assemble an arbitrary light pattern into a pictorial image. But such a definition of "the present" would link time with human physiology and still leave us with questions about the beginning and the end of time, which tend into the fields of religious believes and philosophical perceptions.

To measure time, we treat it as the parameter in the equations of motion. Humans have been doing that since the day they apprehended the passing of time.

The device we call clock is nothing more than a pair of pointers and a mechanism that keeps them turning at a fixed rate. When we talk about the time it takes to get from our home to the office, we actually compare our move from location A to location B with the motion of a clockwork – a black box made to rotate a pair of hands at the mutual ratio of 1:12.

Readers of Herbert G. Wells' *Time Machine* (1895), or those who assisted the movie, may have wondered in which way the many clocks on the walls of the inventor's workshop could have inspired his ideas on designing a machine that takes him more than 800,000 years into the future. The bitter truth is that this collection of Victorian grandfather clocks wouldn't have helped him the least bit. Even an atomic clock of our space age wouldn't have made a difference. Aspiring time machine builders better bury their hopes that a pair of moving hands could lead them to manipulate the flow of time.

Motion as a measure of time implies that, in the absence of motion – all motion for that matter – time would grind to a standstill. Within the confines of a petrified universe, it would be impossible to tell how long this particular state of immobility has lasted and will last. One second, or a thousand years? It wouldn't matter, because nobody would bite his or her fingernails waiting for the universe to start all over again. If there was a world before motion, it had no definable age until – all of a sudden – something in that torpid universe changed its position and all our troubles began. The sleeping beauty in the fairy tale could as well have slept for a thousand years, because the "closed motionless system" her castle had been enchanted into, lacked a timeframe. Time came back into her reign in the person of the prince whose kisses restored – along with the flow of time – the princess' mobility.

We measure space by comparing the length of one object with the length of another, such as a yardstick; but we use motion to measure time. We set our clocks to the daily motion of the Sun, and then compare the motions of a host of physical processes with the motion of those clocks. The term "clock" goes back to the French vocable "cloche" for bell, "glocio" in Latin, and "Glocke" in German.

In prehistoric times, the length of day became the unit of time, intuitively at first and later standardized as the time interval between two successive meridian crossings of the sun. That held until atomic clocks showed the rotation of Earth as less then perfect. Thus the country's master timekeepers, the International Earth Rotation Service, compensate the earth-ball's fads by unobtrusively smuggling one second in or out of our timescale, whenever the bell tolls, so to say. For instance, $23^h\ 59^m\ 59^s$ on 30 June, 1992 was followed by $23^h\ 59^m\ 60^s$, which preceded – rather than coincided – with 1 July, $00^h\ 00^m\ 00^s$. Watch out, because from that date on, your birth certificate showed you younger than your true age. Well, well, if one single second may not seem worth worrying about, it still symbolizes a dream come true, doesn't it?

The world's first time related document is the Jewish calendar of 3761 BC, but only in the 26th century BC appear the implements for subdivision of the length of day, such as shadow staffs, pillars, and later obelisks and Chinese sundials.

Shadows can be interpreted by their direction and length. If you can believe what we read in the books and see in the movies, Indian chiefs used to set their ultimata with a mark on the ground for the length of a pole's shadow when the allotted time would be over. But unless they had a computer-like memory for the Sun's positions in the sky, a circle around the pole would have served them better. Shadows get longer toward evening, but they also swing by an average of fifteen degrees of arc per hour from west toward east, and likely would miss a lone mark on the ground. In any case, the direction of a pole's shadow – diametrically opposed to the position of the Sun – is a better timekeeper than the shadow's length, which along with the time of day, changes with the seasons. That made sundials the timepieces of choice in antiquity.

While ancient Egyptians employed sundials some 4000 years ago, the most "bang for the buck" in this field came much later with the French Revolution. A convex lens was set to focus the rays of the midday sun on the fuse of one of those times muzzleloader cannons, making it announce high noon with an unmistakable thunder. Undoubtedly the ideal timepiece for those who count only the sunny days of their lives.

Water clocks

The first human-made instrument for time measurement appears in 1530 BC as an artfully decorated water clock (Fig. 3.1 b) built by an Egyptian, Amenemhet, for King Amenophis. If it weren't for the engraved hieroglyphs (not shown in the illustration), those containers could be described – with due respect – by their similarity with today's flowerpots, except that the opening in the bottom was calibrated for the desired emptying period. The ancient Greeks called them Klepshydra (water thief) and along with the Romans used them to set, among other things, time limits to the loquacity of congressional speakers.

If you make such a water thief of thin metal and let it float in a tub, it gradually fills up through the bottom hole, until a certain time later, it sinks. This "subspecies" of the Klepshydra – we may call "Scuttling-clock" – was still being used occasionally in Africa during the 20th century.

A formula for predicting how fast a container empties can be derived from Galileo Galilei's free fall experiments with weights dropped from the Tower of Pisa. With h for the height of the tower, and g for gravitational acceleration

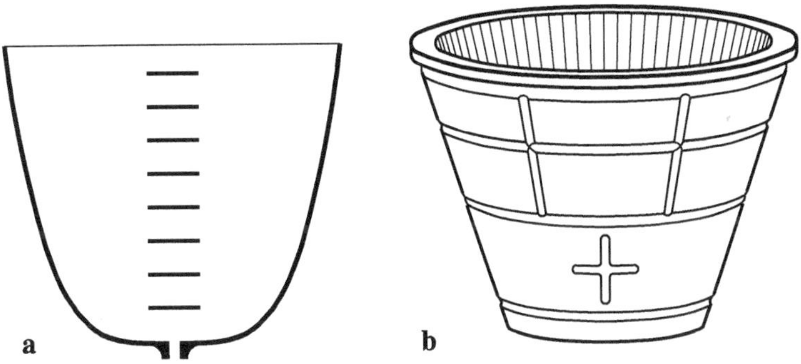

Fig. 3.1. Klepshydra water jar

(9.807 m/s^2), the velocity v at which such objects hit the ground was found to be $v = \sqrt{2gh}$. The same formula applies to a drop of water descending from the liquid surface down to the drip-hole. With the liquid's exit velocity proportional to the square root of the liquid level, the time scale on a Klepshydra had to sport unequal increments for identical time intervals. Heron of Alexandria, whose name we remember from Chap. 1, reportedly criticized the Klepshydra (then renamed clepsidra) for its inconsistent outflow which precludes the reading of fractional periods of time: At half a clepsidra's total emptying time, far more than half of its initial contents have drained.

For an evenly spaced time scale, the vessel shown in Fig. 3.1 a is shaped to a fourth-order parabola. If we set the volume of liquid escaping through the bottom hole equal to the loss of liquid within the container, and R for the radius of the pot at the height h from the bottom up, this volume is $V = R^2\pi \times \Delta h$, where Δh is the drop of the liquid surface in a short time interval Δt.

On the other hand, the exit velocity of liquid passing through the bottom hole is $v = \sqrt{2gh}$ and, with the opening's cross-sectional area A, the corresponding volume becomes $V = vA \times \Delta t$, that is, $V = A \times \sqrt{2gh} \times \Delta t$.

Dividing those two expressions for V by each other gives:

$$\frac{\Delta h}{\Delta t}\frac{R^2\pi}{\sqrt{2gh} \times A} = 1\,.$$

For a regularly spaced timescale, the quotient $\Delta h/\Delta t$, representing the lowering of the liquid level per unit of time (such as one inch per hour), must be kept constant (k). This makes $\Delta h/\Delta t = k$, and our equation becomes

$$\frac{\sqrt{2gh}A}{R^2\pi} = k\,.$$

Squaring the equation and reshuffling its terms leads to

$$h = \frac{(k\pi)^2}{2gA^2}R^4\,.$$

Herein, the value of the fraction depends exclusively on the dimensions of the vessel and may be understood as the clock's "instrumental constant," K. Thus we end up with $h = KR^4$, the clepsidra formula old Heron had in mind.

But expecting a potter to mold flowerpots to a fourth-order equation might have been asking too much. Easier to make is a cylindrical container marked with the nonlinear scale in Fig. 3.2.

Here again, the proportionality of the liquid's exit-speed and the square root of the liquid head allows for designing the scale. Take a container that, filled to 144 mm liquid level, empties in 12 hours. With the zero mark at 144 mm, which incidentally equals $12^2 = 144$, the eleven-hour mark would be at $11^2 = 121$ mm, the ten-hour mark at 100 mm, the nine-hour mark at 81 mm, and so on. Note that the size of the increments follows the progression of odd numbers, such as: $144 - 121 = 23$, $121 - 100 = 21$, $100 - 81 = 19$, etc. Taking it from here, any size

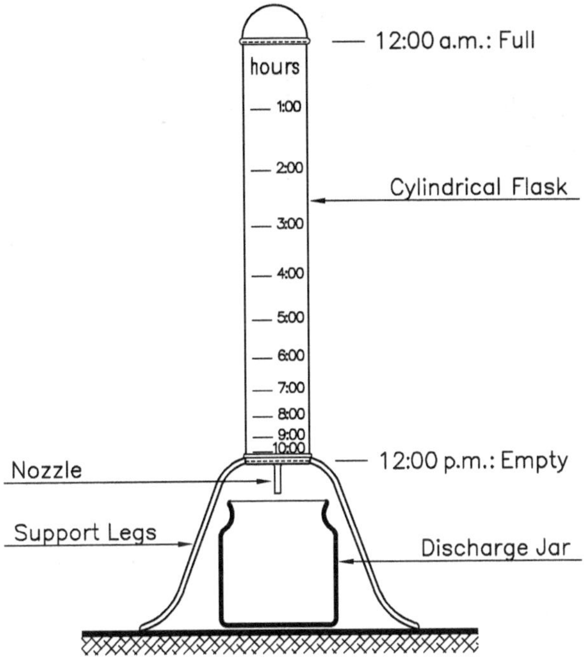

Fig. 3.2. Water clock with cylindrical container and quadratically divided scale

scale can be designed. For instance, for a clock meant to empty within 12 hours if filled to 216 mm, the scale is made multiplying the above values by $216/144 = 1.5$. For a 250 mm container, that factor would be $250/144 = 1.736$.

But since vessels for this kind of clocks had to be blown of glass for the meniscus to be visible from the outside, they appeared much later in history, actually at an age when mechanical clocks were about to take over. But back in the third century BC, machinist Ktesibios of Alexandria, contemporary of Euclid and Archimedes, came up with a design that circumvents the scaling problems by building his clock around a water supply with constant level and therefore unchanging rate of outflow. In Fig. 3.3 this is the funnel-shaped container on the left in which the liquid level always remains at the height of the runoff (drain), while a faucet-controllable influx replenishes the tank's contents. The rate of the clock is controlled manually by lowering or lifting the cone-shaped piston which controls the flow from the funnel into the cylindrical reservoir (Fig. 3.3). Liquid level is picked up by a float, which via an attached rack and pinion drive rotates the clock's single hand over a 24-hour clock face.

Variations of this basic design aim to accommodate to Egyptian and Grecian calendars, which divide the time of daylight into six hours, regardless of the true lengths of day and night. Designing a clock that sets itself with the seasons sounds like a formidable challenge, but ancient engineers took it in stride. Their solution was to keep the clock itself running at an unvarying pace, and use different readout scales over the year. A drum, set to rotate on its vertical axis by $1/365$ turn per day, replaced the clockface (Fig. 3.4). Specific timescales could have been drawn for every month of the year – wide spaced scales for the long summer days

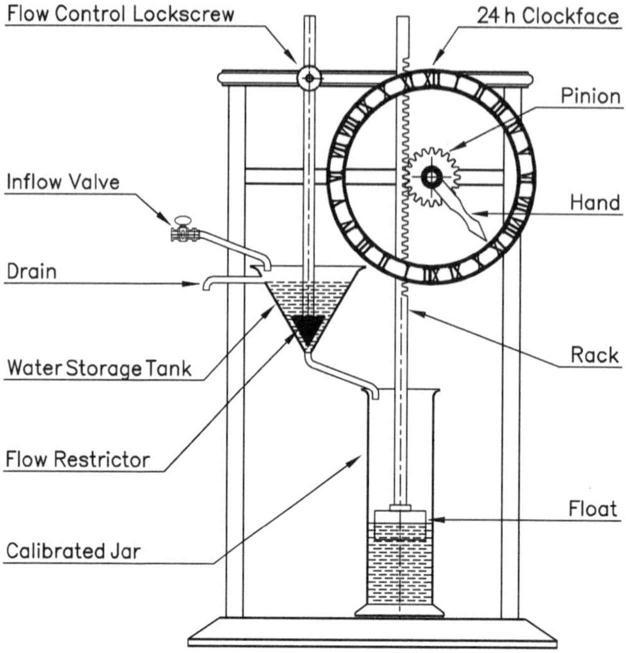

Fig. 3.3. Clepsydra, the constant flow waterclock

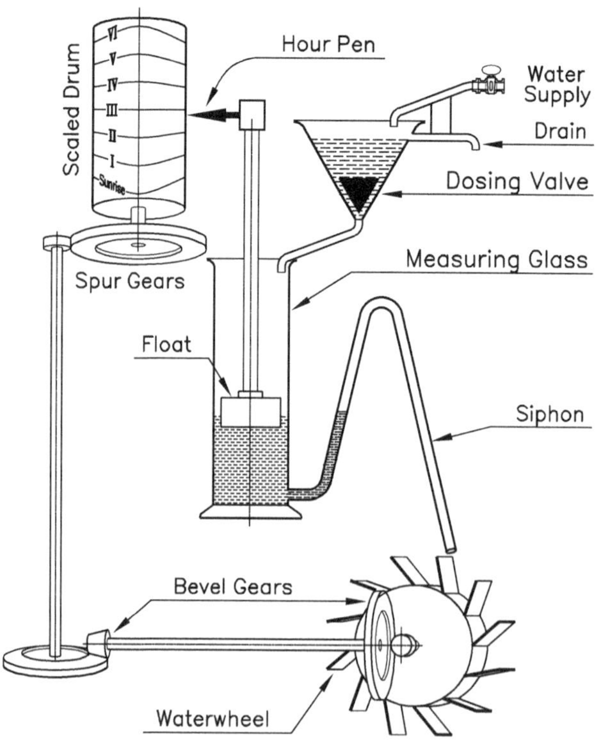

Fig. 3.4. The too complete water timing system: fact or fantasy?

and narrow scales for the short daylight hours of winter. But rather than overcrowding the cylinder with individual scales, antique watchmakers drew curves interconnecting the hour marks from scale to scale without showing the scales themselves. The resulting curves were essentially sine-lines, except for the straight three o'clock noon mark, which however could be interpreted as a sine-line with zero amplitude.

In an age of cheap labor, one would expect a butler or slave taking care of the daily settings of such readout drums, perhaps by inching up a 365 teeth ratchet-gear by one tooth per day. But some ancient drawings and sketches offer far more ambitious ideas, such as the auto-reset feature in Fig. 3.4, supposed to reset the clock automatically for a new operating cycle the moment the holding tank is full.

To this end, a siphon was supposed to empty the reservoir in one big gulp as soon as the liquid level reaches the siphon's upper bend; pretty much like the sustained discharge from a wall mounted overhead tank of old water closets is triggered when a pull on the chain momentarily flushes water into a downward leading inverted U-tube (compare Fig. 8.4). As kind of fringe benefit, the siphon's discharge is still being used to power a water wheel just long enough to rotate – over an intricate gear-train – the readout drum by $1/365$ of the full circle. The Geneva Drive of classical movie projectors, used to make the filmstrip jump from frame to frame, could have done the job error free, but was still waiting to be invented.

The saying "if it sounds too good to be true, it probably is" may well apply to such ancient sophistication. Not that the inventors hadn't acted in good faith. They just missed to mark their drawings with comments such as "workable" or "failed" after testing their prototypes – if there ever was such a thing. In the case in point, the clock's siphon would not function like that of a water closet. The latter soaks the holding tank empty because the initial surge (at the pull on the chain) suffices to completely fill the cross section of the discharge pipe, making the descending water column act like the piston of a suction pump. Unfortunately, that wouldn't be the case of a siphon fed by the metered amounts of liquid that drive the water clock. Rather, the overflow would, drop by drop, rinse down along the walls of the siphon tube.

All said, we shouldn't blame ancient engineers and scientists for unrealistic ideas. First time designs fail quite often to provide the desired results, even in our times, but we usually scrap them before they get a chance to make it into history.

On the other hand, a majority of ancient ideas make good sense. Figure 3.5 a, left, shows an ingenious idea for a consistent flow rate water supply from Heron of Alexandria. The proper drainpipe, shaped like an inverted U, is mounted on a float. With the descending leg of the U-tube longer than the one leading up, the water-head between intake and discharge of the U tube remains unchanged, regardless of how full the tank actually is. Once the outflow of water has been triggered by suctioning the first mouthful from the discharge tube, it continues at constant rate, independent of how full the tank actually is. If the outflow from the siphon is gathered in a cylindrical measuring cup, the liquid level therein is proportional to the time of day and could be read from an appropriately spaced linear scale, engraved in the wall of the tank, or on a water gage.

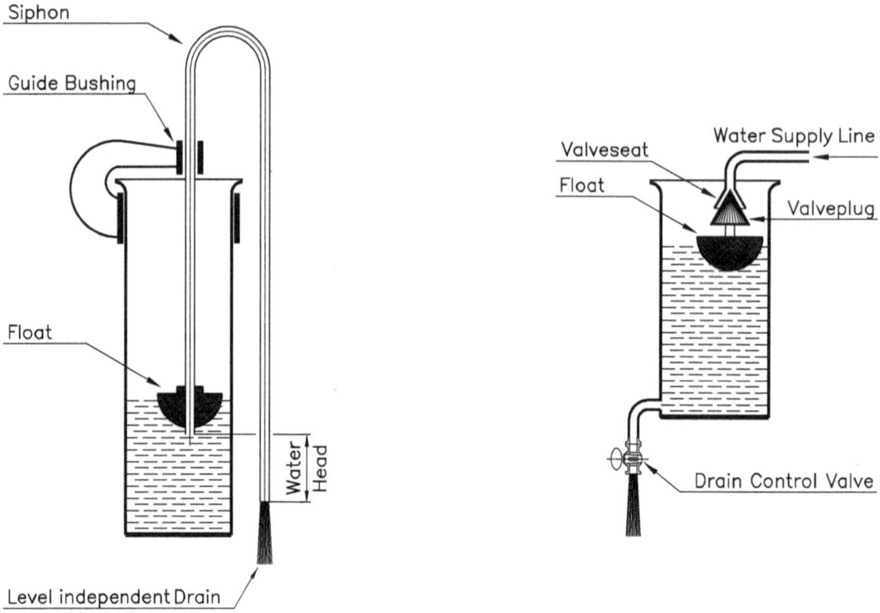

a **b**

Fig. 3.5. a Heron's and **b** Csesibius' designs of a constant flow water supply

Csesibius' solution to the same problem (Fig. 3.5 b) is straight forward: Much like the level control in an automotive carburetor, he uses a float controlled needle valve for a consistent liquid level.

Gravity-driven timepieces

The variety of sometimes quite intricate mechanisms, which had become the order of day in the Hellenistic era, makes us wonder why mechanical clocks hadn't entered the picture right there and then. But the concept of measuring time by stepped rather than continuous motion, as generated by the pendulum and the balance wheel, was still missing.

The two types of gravity-driven clocks from the third century BC shown in Fig. 3.6 are built around the same principle, but defer in ways of showing the hours of day. The one in Fig. 3.6 a has the time scales marked on the staffs, while the other employs a circular dial, shown here in rear view with counter-clockwise hour marks. Either type is powered by the weight of a hollow metal cylinder slipped on the main shaft, which descends under its weight, either by unwinding the suspension cords coiled around its axis (Fig. 3.6 a), or by rolling the length of the shaft (Fig. 3.6 b) down the staffs which must have been installed with a certain backwards slant.

What kept the cylinders from spinning down in seconds is hidden inside: a number of vans with small pinholes at their bases, holding a measured volume of water. Figure 3.7 a shows the cross-sectional view of the cylinder at rest, with

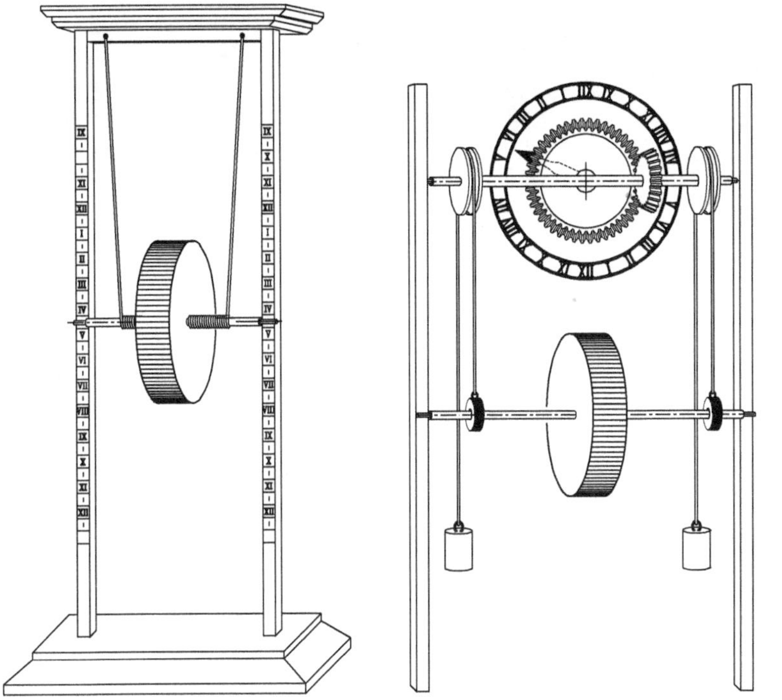

a b

Fig. 3.6. **a** Heron's and **b** Csesibius' clockwork driven by its own weight

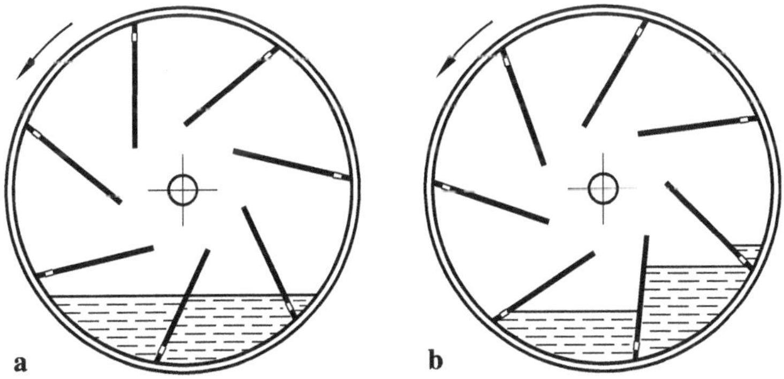

a b

Fig. 3.7. Hydro-speed controller

equal water level in all flooded chambers, while Fig. 3.7b depicts what happens when the cylinder turns and hereby lifts the water-filled chambers close to halfway up, the point where the torque from the weight of the water counterbalances the gravitational torque from the cylinder's unrolling. From here on, the cylinder can only rotate at the rate the higher chambers empty into the ones below.

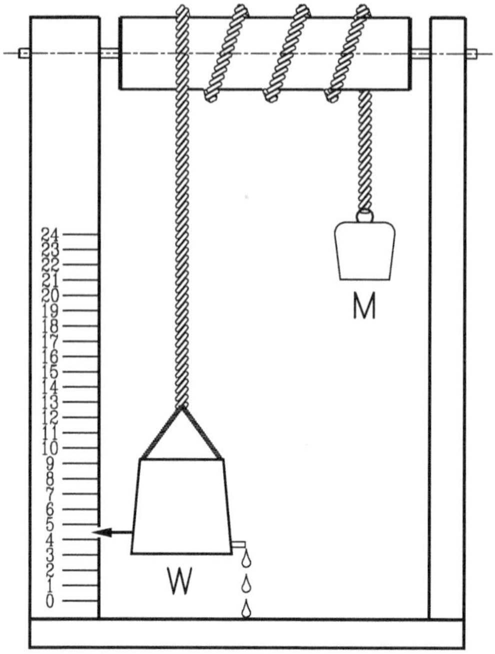

Fig. 3.8. A stillborn hybrid clock design

A much later design (from 1615), called "Walgeuhr," is shown in Fig. 3.8: A water-filled container (*W*) and a weight (*M*) initially counterbalance each other over a rope wound around a reel. As the gradual outflow from the spud of container (*W*) makes it lose weight, it gets lifted by the counterweight the higher the lighter it gets, which makes its position indicative for the time passed.

If *W* were constant, that machine would operate like an "Atwood" free fall demonstration apparatus and make the weights descend and, respectively, rise, in accelerated motion, according to

$$\frac{a}{g} = \frac{M - W}{M + W},$$

there g stands for gravitational acceleration ($9.807\ \mathrm{m/s^2}$), and a for upward acceleration of W and likewise the downward acceleration of M.

Make for instance $W = \frac{9}{11}M$ to get

$$\frac{a}{g} = \frac{1 - \frac{9}{11}}{1 + \frac{9}{11}} = \frac{\frac{2}{11}}{\frac{20}{11}} = 0.1\ .$$

With $W = \frac{9}{11}M$, the machine would thus demonstrate free fall in a gravitational field of $1/10$th the strength of that on Earth.

In the case of the Walgeuhr, the pail would start its accelerated upward motion as soon as it gets lighter than the counterweight M, resulting in a parabolic time scale, as we had it in Fig. 3.2. Spacing between time marks would however widen even more than in Fig. 3.2, because the bucket's acceleration increases as it empties. Actually, the clock would start at a snail's pace but shoot up as it gains height.

Apparently unaware of such subtleties, historical sources show the clock with a linear scale. Here again, a nice idea, but one that would not work as expected had it ever gotten off the paper.

Altogether, water clocks of whatever design have an inherent problem in common: the accentuated change of the water's viscosity with temperature, from $1.794\ \mathrm{mm^2/s}$ near freezing to 1.011 at room temperature 68 °F (20 °C), and 0.659 in sweltering heat 104 °F (40 °C). In short, water clocks tend to lag back in wintertime and go fast in the hot season, the more so in times when life in a heated room was a luxury and air conditioners had yet to be invented.

Candle clocks

This might have motivated Father Beda in England in AD 700 to count his hours with candles engraved with hour-marks (Fig. 3.9). Alfred the Great, the Anglo-Saxon king (871–899), picked up that idea in AD 875 and timed his daily schedule with the help of six candles of four hours burning time each. One eight-hour period was assigned to administrative duties, another to studies, meals and sleep, and the third to prayers.

As candle clocks caught on, the addition of a delicately crafted weight, chained to a pin, allowed their use as alarm clocks (Fig. 3.9, lower left on the candle): the sleeper drove the pin into the candle at the notch for wake-up time. When the approaching flame softened the candle's wax around the pin, the weight dropped down, hit the saucer, and a (hopefully) melodic ring woke up the clock's keeper.

So the story goes. But on the other hand, Germany's most famous poet, writer,

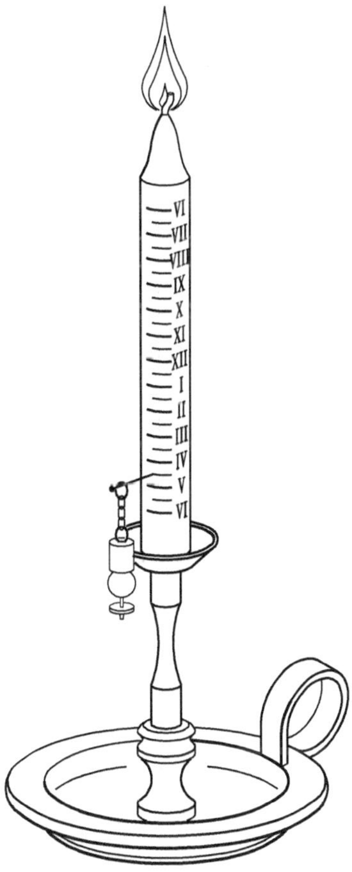

Fig. 3.9. Candle clock

Fig. 3.10. Oil lamp clock

and scientist, Johann Wolfgang von Goethe, composed as late as in the early eighteen hundreds the two-liner:

Wüßte nicht, was sie besseres erfinden könnten,
als dass die Lichter ohne Putzen brennten.
(Best of inventions they could pick:
No need to trim the candles' wick).

Wicks, then spun from fibers of flax, moos, and rush, did not shed their charred sections, which had to be cut off manually every half-hour or so. The user of a candle alarm clock would have to get up twice per hour to prevent a drooping extra piece of burned wick from conducting the flame downward and release the clapper ahead of time.

Sixteenth century's "Fire Clocks," the high-tech version of candle clocks, avoided such shortfalls by using oil lamps as time keepers (Fig. 3.10). A scaled cylindrical glass jar holds the oil for a twelve-hour burn. The wick reaches into the jar through a hole in the bottom, narrow enough to keep oil from leaking out. Capillary action makes that oil creep through the length of the wick into the burner.

Devoid of the help of glass forming molds, jars for fire clocks where made by vertically spinning and inflating a hollow slab of semi-liquid glass, so that gravity took care of stretching it into a near cylindrical shape, resembling a prolate spheroid. The unavoidable waist bulge must have prolonged the hours at half-full and shortened those at the beginning and the end, though no consumers' complaints have been filed so far.

Sandglasses

While the design of water and fire clocks aimed mostly at time spans of twelve or twenty-four hours, sandglasses and hourglasses targeted shorter periods. A Grecian third-century BC bas-relief, mounted on the wall of the Roman Mattai Palace, shows Morpheus, deity of dreams, leaning on a sandglass. But since such details might have been added in the course of a 17th-century restoration, the Greek's knowledge of sand-clocks remains an open question.

A 1379 inventory of Emperor Charles V lists three clocks and one sandglass, and yet, the Italian author Martinelli still calls the sandglass an innovation in 1665. Whatever its origins, the sandglass has made it into the computer age as the icon for "busy" and into our kitchens as egg timer for those who want theirs soft boiled. In her historical book, *I sailed with Columbus* (Harper Trophy 1992), based on Columbus' logbook, Miriam Schlein relates as one of the ship boy's principal duties the flipping of the ship's "ampoletas" (hourglasses) precisely when the last grain of sand had passed through the neck. The event was announced by singing a religious verse to remind the officer on duty to mark the half-hour on the traverse board.

Like a specter from alchemists' kitchens, sand-clocks resurfaced in the 20th century for use in corrosive atmospheres, such as laboratories producing and handling acids. And they are still found in saunas.

Sandglasses are simple devices, but you better think twice before crafting your own. A receipt from 1339 prescribes the sand to be made of "ground black marble, boiled nine times in wine" (presumably to etch away nicks and burrs and leave nicely rounded grains). Lead-containing sands are praised for their auto-lubricating features. Instructions on shaping and finishing the waist sections for smooth flow and least wear are more demanding still. Aspiring watchmakers had to prove their ability to correctly size and shape those sections for full hour, half-hour and quarter-hour models.

Like in Egyptian water clocks, "half the sand down" does not mean that half of the clock's period has passed. Each sand-clock measures its particular time span, but no fractions of it. That led to brainteasers such as how to measure 16 minutes with a pair of hourglasses of, respectively, 7 and 12 minutes runtime? (Solution: Let both complete one run and turn them over. After 14 minutes, the 7 minutes glass has completed two turns, and the 12 minutes glass is 2 minutes into its second turn. Invert it right away and it will run for the two minutes left to complete the desired sixteen).

The world's tallest sandglass is reportedly found in the Japanese Sand Museum in the city of Nima. It stands six meters tall and runs for an entire year. At least as interesting as the clock itself should be the physique of the man in charge of flipping it at year's end.

The variety of time-keeping devices which preceded the event of a mechanical clock doesn't necessarily mean that the idea didn't come up earlier. Rather, as shown for the "Walgeuhr" of 1615, it was that weight-driven mechanisms tend to run at accelerated pace, because the energy, fed into the clockwork by a descending weight, accumulates and makes the weight gain speed. For a rate of gain of, say, one foot per second for every inch-pound of energy input, the final velocity at time t becomes $v = 1t$ and the distance covered $s = \frac{1}{2}t^2$, a parabolic progression. To make a weight descend at an unchanging speed, the inflow of energy must be throttled.

We could get rid of excess energy by consistent braking, such as by air friction acting on a fast rotating paddle wheel. But since their consumption of energy varies with air temperature and barometric pressure, such devices are found for controlling the striking mechanism in classical pendulum clocks rather than the clockwork itself.

The proverbial "quantum leap" in horology came with the control of clockworks by oscillating motion, commanded either by gravitational or inertial forces.

Mechanical clockworks

No other than Gerbert d'Aurillac, the later Pope Silvestre II, is credited with the introduction, in AD 996, of mechanical clockworks. But it took until 1344 for the Italian astronomer Giovanni Dondi to come up with a weight-driven, swing weight regulated clock; surprisingly, Dondi controlled the pace of his clock with the inertial forces of a horizontally swinging bar rather than the gravitational forces that keep a pendulum in motion.

Inertial forces surge in any piece of mass being accelerated or slowed down: in speeding up or braking a car; a test driver's maneuvering his car through a slalom; the kickoff of a dragster. Although inertial forces can have any direction in space, their vertical components are often overshadowed by gravitation and we perceive principally their effects in the horizontal plane.

In 1973, measurements of astronauts' body mass through inertial forces in the weightless environment inside Skylab, the first American space station, confirmed a host of so far earthbound discoveries, such as the proportionality of the period T of a torsion pendulum and the square root of its mass.

The foliot escapement

Giovanni Dondi's inertial force escapement (Fig. 3.11) employs the *foliot*, an end-weighted swing bar, to stabilize the pace of the clockwork. The center shaft of the foliot, called the *verge*, hangs suspended from the clock's frame by a rope, preferably spun from silk, to minimize friction. Torque from the weight driven pinwheel moves the foliot forward and backward by alternately pushing and releasing one or the other in a pair of dogs on the shaft. Thus, the pin that engages with the upper pallet induces the verge to move clockwise, and the one engaging the bottom pallet acts in the opposite direction. Set at a wide enough angle relative to each other (about 100°), only one pallet at a time is being driven, resulting in the alternate motion of the verge.

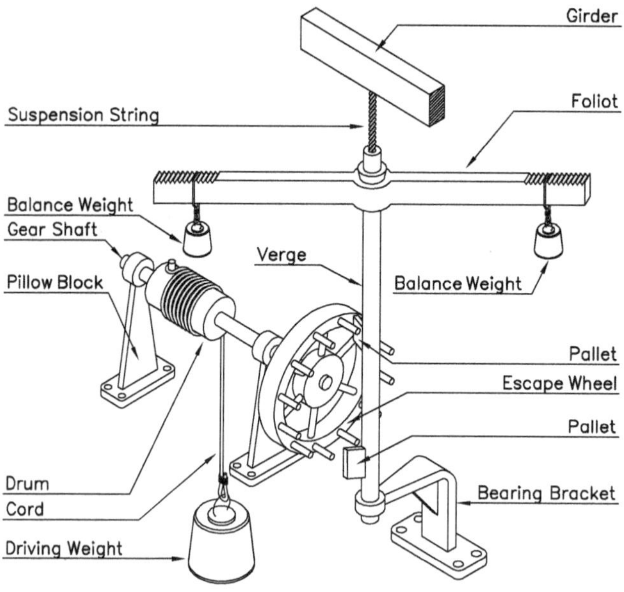

Fig. 3.11. Dondi's inertial escapement

The shortcoming of the foliot escapement is its swinging in forced mode, as the frequency of the balance bar is commanded by the torque of the pinwheel. By contrast, the swing-mass in modern clocks is spring loaded and allowed to oscillate at the system's natural frequency with the period of $T = 2\pi\sqrt{I/c}$, where I stands for the balance wheel's moment of inertia, and c for the torsional constant of the return spring.

Nevertheless, foliot clocks remained standard for about three hundred years. In 1577, Jost Bürgi's introduction of the minute hand in a clock for the Danish astronomer Tycho Brahe improved the display accuracy 12-fold. Even John Harrison (1693–1776), widely considered the most brilliant horologist of all times, still used a verge escapement with a balance wheel in his precision timepiece "H 4". Applied to the measurement of longitude, it won him part of a 20,000 pound sterling Longitude Prize from the British government.

Pendulum escapement

Real improvement came with Galileo Galilei's discovery of the laws of motion for the mathematical pendulum in 1582, and with them the "isochronism of the pendulum," i.e., the independence of a pendulum's period from the amplitude of its swing. Yet it took him until 1637 to model a pendulum-controlled clockwork.

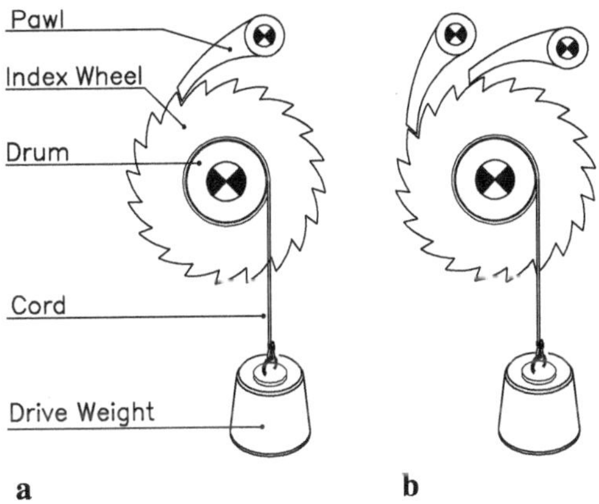

Fig. 3.12. Ratchet drive

Ratchets, as used in hoisting mechanisms to keep the load from back-driving the drum (Fig. 3.12 a), may be seen as the harbingers of the pendulum escapement. A saw tooth gear with a pivoted pawl can turn in one direction only, but spins freely if the pawl is lifted. In the two pawls mechanism (Fig. 3.12 b), momentary lifting of the pawl on the left allows the gear to advance by half of a pitch, and then engage with the pawl on the right. A short disengagement of the latter causes another advance by half a pitch before the gear gets stopped by the left pawl. This basic mechanism resurfaced quite recently in tuning fork clockworks.

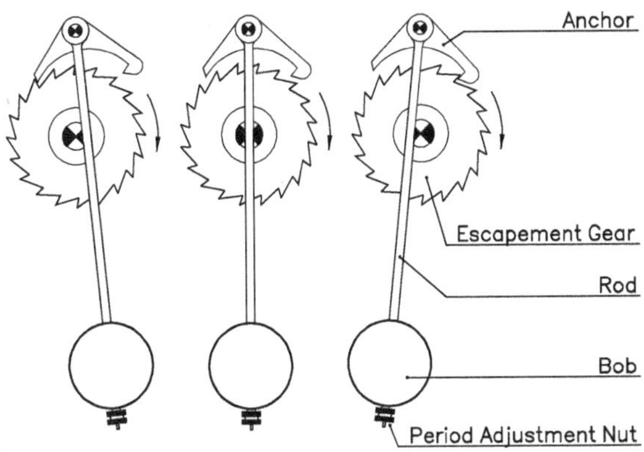

Fig. 3.13. Pendulum escapement

In a pendulum escapement (Fig. 3.13), the left and right pawls are combined in a single unit, the pallet-arm or anchor, which releases one tooth while it blocks the advance of the other, half a pitch further on. Each swing of the pendulum to and fro allows for one tooth advance of the escapement gear.

But that tells only half of the story. While the weight's gravitational force sustains the gear's stepwise advance, the pendulum itself needs an impulse every time it changes direction, much like a swing needs a push every time it comes back if you want the fun to go on. But instead of an external driving force, such as the swing pusher, the pendulum clock derives such impulses from the gear's stop-and-go anchor that combines the two pawls in Fig. 3.12b into one single unit, shaped and located so that inertial forces from the process of stopping the advance of the weight and the drum mount up to a jolt on the anchor and with it, the pendulum.

For higher accuracy, modern clockworks relieve the anchor bearings from carrying the pendulum's weight by hanging it separately from a thin strip of flat spring steel. The blade takes care of the load of anchor and pendulum, and virtually eliminates the inherent frictional losses of conventional bearings, since the energy for flexing the strip to one side is recovered by elastic forces on the way back.

The superior performance of pendulum clocks can be traced to the equations describing the "mathematical pendulum" – a bob on a weightless string. With L for the distance between swivel point and the center of mass of the bob, a mathematical pendulum's period, from left to right and back, is given by

$$T = 2\pi\sqrt{L/g}$$

for small amplitude oscillation, where the approximation $\sin\theta \approx \theta$ is applicable.

Of the three principal parameters of a pendulum, length, mass of the bob, and the amplitude θ of the pendulum's oscillations, this formula contains only one, namely, the length $L = (T^2 g)/(4\pi^2)$ as determinant for the pendulum's period. This independence from two out of three parameters makes the pendulum the ideal timekeeper, as long as it swings within narrowly set limits. Faithful to this condition, Grandfather's clocks' very slow swinging pendulums (usually $T = 2$ seconds) with length of $L = 2^2 \times 9.80665/4\pi^2 = 0.9936$ meter preclude wide swing angles anyway. 0.9936 m is known as the length of the "One second pendulum," which takes one second for the stroke from left to right or vice versa.

Nevertheless, other clock designs, such as the classical Black Forest Clock, use deflection angles far beyond the $\sin\theta \approx \theta$ limits. Under those conditions, any change in the pendulum's stroke causes a change of period, according to

$$T = 2\pi\sqrt{\frac{L}{g}}\left(1 + \frac{1}{4}\sin^2\frac{\theta}{2} + \frac{9}{64}\sin^4\frac{\theta}{2} + \dots\right)$$

The formula contains the basic amplitude formula, $T = 2\pi\sqrt{L/g}$, multiplied by the bracketed infinite progression, which accounts for the effects of greater amplitudes.

But there is always a better way. In 1673, the Dutch astronomer Christian

Huygens developed an amplitude-independent pendulum by sustaining the string between a mirrored pair of cycloid-shaped metal jaws. The wider the swing, the more of the string's length snuggles along the jaws' perimeter, and the effective length of the pendulum gets reduced.

The cycloid is easily imagined as the curve that a chalk mark on the perimeter of a tire describes with the vehicle rolling. But figuring the specific cycloid that reflects the infinite progression in the exact formula for L is something our mathematical forefathers deserve respect for.

Notwithstanding his advanced theories on free swinging pendulums, Huygens' first pendulum clock used a foliot type escapement to keep up the beat. Only the invention of the anchor escapement in 1660 by William Clement did away with the need for the sustained swings of the old foliots and opened the doors for a regulating element operating at its particular frequency of resonance.

We instinctively keep a play-swing in the state of resonance by applying a push at the very moment it changes direction. How far out the swing goes depends on the power we use, but we have little control on the time it takes for coming back for another push. By contrast, the pistons of an internal combustion engine, such as the one in your car, swing in forced mode, dictated by the rpm of the motor. Here is the difference: An engine can be run at any speed, while a swing operates at an unchangeable period.

Still, the use of the escapement to rekindle the pendulum's swing makes for an interdependence detriment to the accuracy of the mechanism. That led in 1721 to George Graham's "dead beat escapement", which moves the point of attack of the impulse maintaining the pendulum's motion from the reversal points to the point of passage through the vertical.

But even the dead beat escapement drains momentum by the amount needed to keep up the pendulum's oscillations. This introduces a certain degree of irregularity into the rhythm of the mechanism, which at the time was well known to clockmakers and made the search for "free swinging pendulum" mechanisms continue throughout the nineteenth century. Still in 1921, W. H. Shortt demonstrated a clockwork controlled by a master and a slave pendulum; the former for "keeping the beat," the latter for driving the clock's hands and providing the impulses that keep the master pendulum swinging.

Yet, atmospheric conditions, such as temperature and barometric pressure, still influenced the pace of pendulum clocks. For instance, a pendulum with a shaft of steel loses one second per day for every two degrees rise in temperature. That inspired George Graham in 1721 to use a jar of mercury for a bob. The mercury's upward expansion with rising temperatures compensated for the pendulum's downward increased length. For obvious reasons however, this idea never made it into the bestsellers list. The gridiron pendulum, invented in 1726 by Harrison, employs a shaft made from metals of markedly different coefficients of thermal expansion, such as steel and brass, mounted in such a pattern that the elongation of one leg compensates for that of the other.

Controlling the effects of barometric pressure proved a more daunting task. In principle, clocks could be sealed into pressure-controlled enclosures, but that remained a pipedream until the 19th century brought the availability of small enough electric motors for winding the clockwork within the confines of such an encasing.

Balance wheel escapement

The need for a portable timepiece led to the development of spring-driven clocks with balance wheel escapement, which work in any position and can be wound without stopping the clock. The balance wheel appeared around 1400. But no sooner than in 1674 introduced Christian Huygens the essential element, a spiral escapement return spring, often called hairspring, to step in for the role gravitation plays in pendulum clocks. The proper (resonance) frequency of spring-powered balance wheels occurs at $T = 2\pi\sqrt{I/c}$, compared to $T = 2\pi\sqrt{L/g}$ for the pendulum. By analogy, I, the moment of inertia of the balance wheel, substitutes for L, the length of the pendulum; and c, the torsion constant of the return spring, takes the place of gravitational acceleration g.

An essential step in the development of a free swinging balance wheel escapement was Robert Hooke's 1666 introduction of an intermediate element, the *lever*, as the link between escape wheel and the oscillating assembly. As shown in Fig. 3.14, the lever is kept immobile during most of its swing while its guide pin slides along the circular section of the hub of the balance wheel. Only when the guide pin dips into a recess in the hub can the lever flip and let the escape wheel advance by half a tooth. This intermediate period of stalling the anchor allows the balance wheel to swing through angles as high as 330° to 360° with periods of typically 1/4 to 1 seconds (Fig. 3.14b). As a further innovation came pellets of low-friction metal inserted at the ends of the anchor.

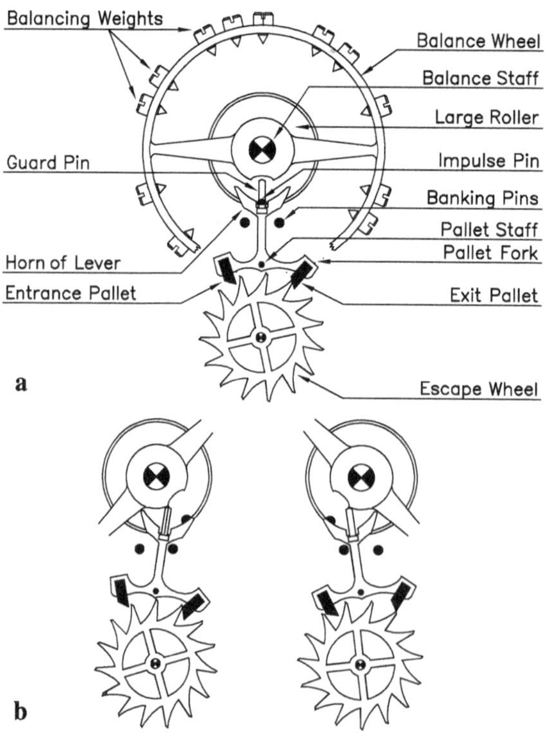

Fig. 3.14. Balance wheel escapement

For temperature compensation, the most simple of a number of possible approaches consists in making the rim of the balance wheel of steel and connect it to the hub with two spokes of brass. When temperature rises, the high thermal expansion of brass relative to steel (19 to 11) distorts the rim into an oval, which lesser moment of inertia compensates for the balance wheel's expansion. The balance screws along the perimeter of the balance wheel allow for fine tuning the oscillations.

To understand why the moment of inertia of an elliptic ring is less than that of a circular one, we consider the extreme case of a ring of radius R and circumference $2\pi R$ being compressed into a straight piece of length L and circumference $2L$. This gives $L = 2\pi R/2 = \pi R$ for the length of the squeezed ring.

The moment of inertia of a ring of mass m is mR^2, and for a bar $(mL)^2/12$, in our case $(\pi^2/12)mR^2$. From this, we get the ratio of the moments of inertia of circular to compressed ring as $1 : \pi^2/12 = 1 : 0.822$, indicating an up to 18% reduction in the ring's moment of inertia.

Spring-driven clockworks

Once the design of the balance wheel escapement had been established, further development went in step with the advance of technology in general. The story of Nuremberg's machinist Peter Henlein (1480–1542) and his first portable timepieces in 1504 is well known. Unlike later periods' pocket watches, which had to be wound every 24 hours, Henlein's kept going for forty hours on one winding. They had only one hand, showing the hour.

Henlein had to forge the mainspring by hand, a job that in later times went to hot and cold rolling mills. In the process, he must have dealt with a host of metallurgical problems, such as slag and sand inclusions and inconsistencies in the steel's chemical composition.

The principal difference between iron and steel is in carbon content, 2% to 5% in cast iron, but only 0.2% in soft steel. Steels with 0.33% C and up are heat temperable, and carbon contents between 0.9% and 1.25% yield spring steel. Nowadays, chemical and spectral analyses allow for controlling the minute changes in carbon content crucial for the steels' properties, but sixteenth-century machinists and locksmiths had to rely on their instinct in selecting, heat treating, and forging the right slabs for their delicate work.

Heat treatment is not the only way of hardening steel. Work hardening occurs by repeatedly cold rolling or persistent hammering of steel and nonferrous metals not temperable otherwise. Spring makers could choose between the risks of heat treatment, including accidental overheating, and cracks and distortion from chilling; or they could rely on work hardening, which then involved long hours of swinging the hammer.

Understandably, Henlein's spiral springs weren't as thin and long as those in modern clocks, and their torque varied according to their state of unwinding. So much so, that a special element, the *Stackfreed*, was developed to compensate for such discrepancies. Most chronicles credit Peter Henlein with the invention of this device, which consists of an eccentric cam and a spring-loaded roller follower. The torque of the mainspring was supposed to ebb down in step with the pressure exerted by the roller and the resultant frictional braking – strongest at the high point of the cam and gradually decreasing with the cam's edging toward the center.

Electric clock

While mechanical clockworks improved with the availability of increasingly purer raw materials and refined methods of metal finishing, the age of electricity bore a new type of clock with the 60 cycles per second period of alternating current as timebase, much like the synchronous electric motor, which in its basic layout turns at $60 \times 60 = 3600$ rotations per minute.

Likewise, electric clockworks are driven by small synchronous gear motors. A 1 : 3600 gear reduction gives the 1 turn per minute for driving the seconds hand, and a subsequent 1 : 60 reduction makes the minutes hand advance by one full turn per hour. And finally, a 1 : 12 reduction controls the hour hand.

An alternative is in the use of four and eight pole motors with 1800 and 900 rpm respectively, combined with gear trains of proportionally lesser rates of reduction.

What made synchronous clocks practical is the close control power plants keep on the frequency of their AC electricity. However, they must be reset after power failures, and American made clocks cannot be used on European standard 50 Hz networks.

Battery powered clocks would be free of such problems, but the speed of rotation of DC motors changes with variations of load and voltage. Thus, it took the invention of the transistor to make motor controllers for network independent clockworks feasible.

Tuning fork ratchet control

Tuning forks, traditionally used for attuning musical instruments, such as pianos, produce sinusoidal sound waves at their highly stable resonance frequency, while most musical instruments blend their fundamental frequency with harmonics.

For instance, striking an "A" on a violin produces sound of 440 Hz (2.5 feet of wavelength) along with overtones of 880, 1320 Hz, etc. The human auditory system hears each of those frequencies separately, but the brain recombines the overtones into what we call timbre – a unique characteristic of each particular category of instrument. On the screen of an oscilloscope, this mélange shows as a complex waveform.

The electrical clockwork in Fig. 3.15 shows a tuning fork with the core of a solenoid mounted to the ends of its legs. Unusual is the cores' tapered shape, selected to make the air gap between core and coil the narrower the deeper the core moves into the coil. This compensates for the loss of the solenoid's drag when the magnetic center of the core approaches the magnetic center of the coil. The solenoid's pull would drop to zero when they coincide.

Further, the narrowing air gap balances the leg's mechanical resistance to bending, which increases with amplitude.

Magnet cores are pressed and subsequently sintered from powdered metals, such as iron, cobalt, nickel, vanadium, and rare metals (niobium), which makes them many times more powerful than conventional horseshoe magnets of soft steel. Associated with the core is the ferromagnetic mantle that guides the magnetic flux lines into enclosing the coil all around.

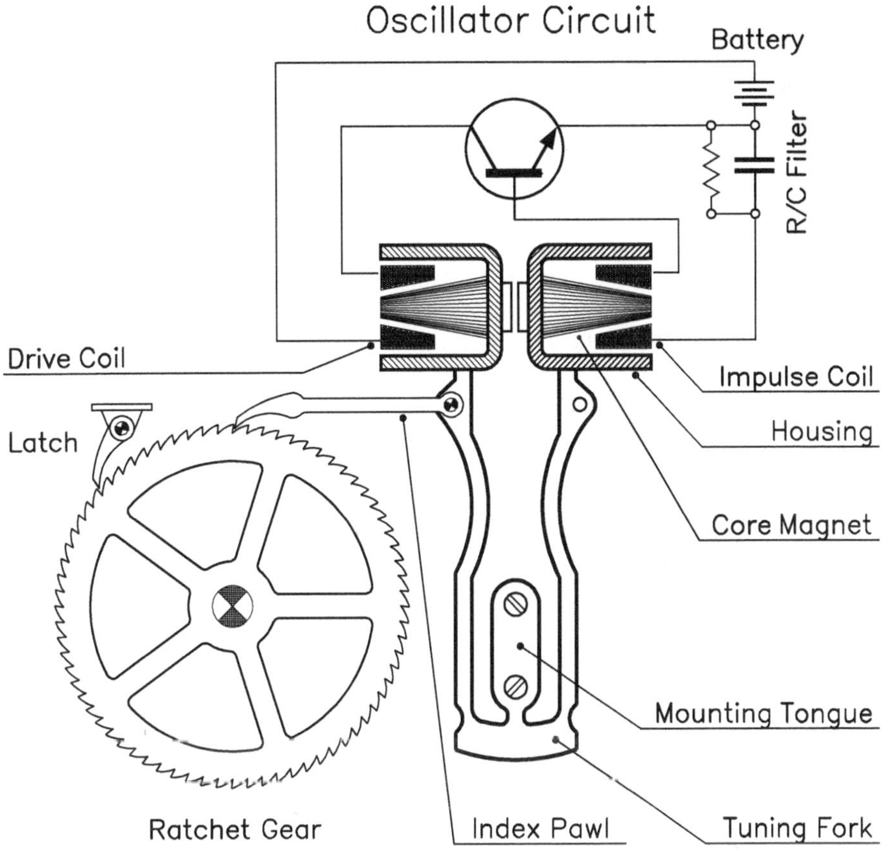

Fig. 3.15. Tuning fork controlled clockwork

The device is powered by a transistorized L/C oscillator circuit with feedback from the coil, shown in Fig. 3.15, right. This leg swings freely at the fork's resonance frequency and induces in its coil the electric pulses that keep the left side leg playing along and drive the ratchet wheel.

The conversion of the fork's vibrations into the motion of the hands happens "the old fashioned way," similar to Fig. 3.12. Not surprisingly, the sawtooth gear is of very fine pitch as to accommodate the minute amplitude of the fork's oscillations. Tuning fork controlled watches have reportedly been made to hold time within two seconds per day.

Crystal-controlled watches

The idea of crystal-controlled oscillator circuits reaches back to the early days of radio technology, when they became the basic elements for stabilizing the frequency of transmitters. If it weren't for their outstanding stability, the noise from our radios and the images on our television screens would fade in and out time again. In watch making, the quest for higher accuracy led to the replacement of the moving elements, such as pendulum, balance wheel, and tuning fork, by a pulsating piezoelectric crystal.

The so-called L/C oscillator circuit includes a capacitance in parallel with an inductance – read coil –, while the piezoelectric crystal makes part of the feed back loop. Oscillations occur as electric energy stored in the capacitor flows through the windings of the coil and builds a magnetic field until the charge of the capacitor is used up. Subsequently, a surge of current generated by the coil's degenerating magnetic field recharges the capacitor.

This flip-flop between electric and magnetic field would go on forever, weren't it for the power drain of the emitted electromagnetic waves and the energy converted to heat in the coil's copper wire. Transistor amplified positive feedback makes up for these losses and maintains oscillations. Radio transmission and time measurement have in common the need for a stabilizing element, which in both cases is a piezoelectric crystal.

Piezoelectricity is a material's characteristic to generate a voltage in response to strain from compression or elongation. Conversely, an applied voltage causes a piezoelectric crystal to shrink or to swell, depending on the applied voltage's polarity.

The natural frequency of a bar of silicon dioxide, SiO_2, vulgo quartz, depends – not surprisingly – on its dimensions, and second, on the direction of its cut relative to the direction of the quartz's crystalline structure. Crystals for quartz watches are usually sized to oscillate at 32 kHz[1], and digital frequency splitters either on discrete chips or as part of clock chips bring that down to better manageable levels. The basic type outputs one electrical pulse for every other input pulse, and thus divides by two. Others divide by ten. For instance, splitting the 32 kHz crystal frequency consecutively five times leaves you with a 1000 Hz oscillation, followed by an electronic counter which emits one output pulse per 1000 inputs.

Digital clocks hitting the market in the early 1970s were expected to become the prototypes of the ultimate timepiece. But when the dust settled, most people found it more convenient to grasp the positions of hour and minute hands rather than interpreting a four-digit luminous or sometimes shadowy readout. This led to a different kind of quartz clock, where a stepping motor, powered by the pulses of the oscillator, drives the hands over one of the usual gear trains.

Stepping motors resemble synchronous motors insofar as their speed of rotation keeps step with the frequency of their power source. As synchronous motors of 2, 4, 6, 8 poles, respectively, run at 3600, 1800, 1200, and 900 rpm, stepping motors come with much higher numbers of poles and run as slow as needed.

Although laboratory quality quartz clocks may keep time within 10^{-5} seconds per day, they still rely on mechanical oscillators, namely, resonating crystals.

1 32,000 oscillations per second.

Clocks and the atom

The common denominator of the classical set of units of measure, such as foot, pound, meter, and kilogram, is their analog definition with the inherent shortfall of leaving room for variations within the no-man's-land between error margins of our instruments of measure.

For instance, the definition of the meter as the 10 millionth part of the meridian quadrant, once hoped to be absolute, was unsettled by every single one of follow up surveys with newer and better instrumentation. Presently, the best figure available is 10,001,965.7293 m. Had the makers of units stuck to their original definition, the meter would have grown by 0.01965% from then to our times. But the facts of life made that the metric length unit remained the man-made platinum iridium bar in Sèvres near Paris, and without much ado, the wish for a nature-based unit faded in face of the realities of ongoing industrialization.

Only the discovery of the quantified structure of the atom resurrected the idea of truly nature-based units of measure. In the course of that development, Albert Einstein proposed in 1904 the proportionality of the energy of the photon (light quanta) and its frequency of radiation, $E = hf$. Herein, f is frequency, and h the proportionality constant, better known as Planck's constant, $h = 6.62618 \times 10^{-34}$ joule seconds. Rewritten into $f = E/h$, this formula allows for the definition of frequency through the related energy of radiation.

That still wouldn't help the search for a nature-based time standard if it weren't for the set of precisely defined energy levels for each of the shells wherein electrons orbit the atomic nucleus. Without that restriction, electrons, the carriers of elementary electric charges, would generate radiation while hurling around the atomic nucleus, much like the flow of electrons from a capacitor into a coil and back in radio transmitters generates electromagnetic waves. Electrons would thus rapidly spend their kinetic energy in the emission of radio waves and end up crashing into the nucleus. In short, matter couldn't exist.

In search for an explanation of this contradiction, Max Bohr, a Danish physicist, postulated the existence of a number of well-defined orbits where electrons do not radiate. As long as electrons are seen as particles, this reads much like an arbitrary supposition, but it makes sense if you attribute them wavelike properties. Here, standing waves would form exclusively in orbits whose circumferences equal a whole number, n, of their wavelength λ. With R for the orbital radius, this leads to $n\lambda = 2R_n\pi$. The radii of nonradiating orbits would thus become $R_n = n\lambda/2\pi$.

The idea of a physical entity, such as the electron, acting in two distinct manners, later known as the *duality principle*, was alien to classical physics. However, it became a necessity when the famous Michelson-Morley interference experiment disproved the existence of the world ether, previously thought of as an ideally thin and elastic substance pervading cosmic space. In the absence of such a wave carrier, light had to be particle radiation as it effortlessly crosses the billions of light-years between Earth and distant galaxies. But conversely, refraction and diffraction are typical wave phenomena.

In 1924, Louis Victor de Broglie published equations for the relation of wavelength and particle size, which gave a wave of 1.67×10^{-10} meters of length as the equivalent of the mass of the electron. The circumference of an electron's orbit must equal this value or a whole number multiple of it for an electron's matter wave to close in itself and not to radiate.

These preferential orbits or shells are customarily referred to by their respective levels of energy, *hf*. For the hydrogen atom, those levels are given as $E_n = -13.6/n^2$ electronvolt, with $n = 1, 2, 3$, etc. The negative sign stems from the definition of potential energy as being zero at the distance $R = \infty$.

As long as a hydrogen atom is left to itself, it remains in the "ground state" with $n = 1$ and $E_n = -13.6$ eV. The other extreme occurs for $n = \infty$ and $E_n = 0$, where the atom is fully ionized: only the nucleus, a proton, is left. The transition from ground state to full ionization of a hydrogen atom could thus be triggered by the impact of a photon of 13.6 eV of energy. With the equivalence of 1 electronvolt $= 1.602176 \times 10^{-19}$ joule and Planck's formula, we get the frequency of such a photon as

$$f = \frac{E}{h} = 13.6 \times \frac{1.602176 \times 10^{-19}}{6.62618 \times 10^{-34}} = 3.288 \times 10^{15}\ \text{Hz}$$

The equivalent wavelength turns out to be

$$\lambda = c/f = 3 \times 10^8/3.288 \times 10^{15} = 91.2 \times 10^{-9}\ \text{meter} = 91.2\ \text{nanometer}\,,$$

far in the ultraviolet.

Likewise, all atoms have a set of characteristic oscillation frequencies, each derived from the energy it takes for moving an electron from one orbit into some other, visualized by the unique set of spectral lines of each chemical element. Among the most conspicuous are the yellow sodium lines of respectively 589.0 and 589.6 nm of wavelength, which – by the way – are easy to produce by dropping grains of table salt (sodium chloride) into a Bunsen flame. From the related frequencies, of 508.8828×10^{12} and 508.3672×10^{12} Hz, the period of one of these oscillation, $T = 1/f$, could in principle become a nature-based unit of time.

A truly nature-based frequency standard

When you irradiate hydrogen atoms with radio waves at successively increasing frequencies, total ionization occurs at 3.288×10^{15} Hz, equivalent to a period of $T = 1/f = 1/3.288 \times 10^{15} = 0.304 \times 10^{-15}$ s. That too could be taken as a nature-based time standard, defined by the frequency where free protons from hydrogen atoms are being generated.

But real world atomic clocks wouldn't employ quantified energies as high as the 13.6 eV afforded for the ionization of hydrogen. Actually, the smaller the threshold energies we use, the more accurate a time standard we get. Best results were obtained with the "hyperfine state" of atoms, discovered by Isaac Isidor Rabi (1898–1988) in 1945 as the minute differences resulting from the orientations of the magnetic fields of the nucleus and the atom's outermost electron. Quantum mechanics allow for only two orientations: equal or opposed.

An atomic clock selects atoms in one hyperfine state and irradiates them at the frequency that causes the transition from one state to the other. This frequency is then used to define the "atomic second".

Unambiguous transitions between hyperfine states occur in elements with only one electron in their outermost shell, which leaves hydrogen, rubidium, and cesium 133 (Cs), a yellow-silvery alkali metal of density 1.873 and 25 °C melting temperature, as possible choices for atomic clocks. Cesium atoms have six electron shells, housing a total of 55 orbiting electrons in numbers of 2, 8, 18, 18, 8, 1. With this distribution, each shell houses the highest possible number of electrons it can support, except for the outermost shell with its single electron. Only the latter will flip in response to radiation at the cesium's critical frequency, while the inner shells are too stable to be affected. This critical frequency turns out to be 9,192,631,770 oscillations per second. And by international agreement, the SI second (atomic second) became the interval of time for 9,192,631,770 oscillations of the cesium 133 atom when exposed to suitable excitation.

NIST assumed in 1990 the operation of the world's most accurate cesium clock, NBS-4, which had been completed in 1968, and in 1999 brought the clock's accuracy to about one second in 20 million years.

The precision of atomic clocks opened the doors for GPS navigation (where the miss of one billionth of a second could cause positioning errors of one foot); synchronization of the internet, and prediction of the positions of celestial bodies with the certainty needed for landing planetary probes on the moon, Venus, and Mars. Precise position measurements further allowed for the design of long-baseline radio telescopes, a network of receiving disks linked to operate like one single parabolic antenna of unprecedented size.

Building an atomic clock

As if the complexities of the principles behind atomic clocks were not enough, actually building one is worse. The components shown in Fig. 3.16 are placed in a Mu metal can for magnetic shielding. Mu metal, a specially heat-treated high-permeability nickel-iron alloy, is capable of blocking magnetic fields as high as a 1000 oersted (0.1 tesla). The entire apparatus is further enclosed in a high-vacuum (10^{-13} atm [ca. 10 nPa]) vessel for free mobility of the cesium atoms.

Initially, the liquefied isotope cesium 133 is vaporized in a furnace from which the atoms exit at about 250 m/s toward a magnetic field for the separation of the ones in the ground state (F3), where only the magnetic fields of nucleus and the outermost electron are determinant. In Fig. 3.16, they are marked with light arrows. The magnet deflects the ones off the ground state toward carbon absorbers. The remainders traverse the Ramsy cavity resonator, where they get exposed to electromagnetic radiation that sweeps backwards and forwards through the cesium's transition frequency of 9,192,631,770 Hz (ca. 9.193 GHz). Every sweep generates a certain number of transitions, so that the resonator's output consists of atoms in the F3 as well the F4 state. A second magnetic field at the exit of the cavity oscillator directs the atoms that have changed into the higher energy state toward the detector. By optimizing the detector readings, the sweep of

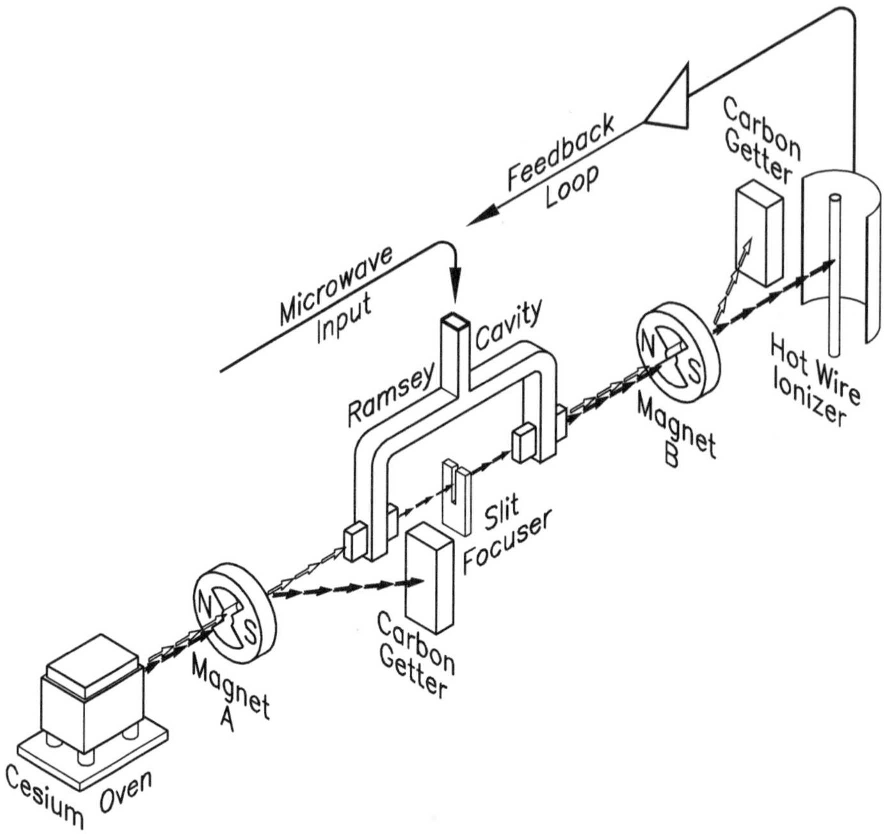

Fig. 3.16. Atomic clock (exploded view, semi-schematic)

the radio frequency is being narrowed until coincidence with the transition frequency of 9,192,631,770 Hz has been reached.

Resonating cavities for ultrahigh radio frequencies are highly critical components. But by coincidence, they had already been developed independently into a well understood technology for the reception of microwave signals from direct broadcast satellites at frequencies close to the cesium transition frequency.

As cesium atoms emerge from the oscillator and the magnetic field, they are exposed to laser light and become fluorescent. The intensity of the fluorescence is highest at the precise transition frequency. Oscillator frequency can thus be fine-tuned by feedback from a light sensor which reacts to the intensity of the fluorescence of the output radiation.

The world's most accurate clock, NIST-F1, started ticking in 1999 with an expected accuracy of one second in 20 million years, proceeded by the officialization of the atomic time standard equaling the second with time for 9,192,631.770 oscillations at the output of the cesium atomic clock.

4 Velocity and acceleration

Achilles, the superman in Hellenistic folklore, couldn't win a race with the turtle, to whom he allowed, say, a 100 yards headstart. Although Achilles runs ten times as fast as his opponent, he reaches the place from were the turtle took off no sooner than the latter has crawled 10 yards ahead, to 110 yards from Achilles' start-line. And when Achilles makes it to 110, he finds the turtle at 111. At 111, the turtle is still ahead of him, at 111.1. Since such reasoning can be carried on indefinitely, one might be tempted to conclude that Achilles, had he survived into our days, would still be racing the turtle. But he, who is no Achilles, should not try to apply such rules to a police cruiser approaching from behind, and better watch his car's speedometer instead.

Although hardly an instrument to make headlines, a speedometer is an analog computer with calculus capabilities; it constantly finds the derivative of distance toward time. In mathematical lingo, it derives the vehicle's speed, v, from the variables of space, s, and time t, as the differential $v = \mathrm{d}s/\mathrm{d}t$.

Tachometer

The key element of a mechanical speedometer as in Fig. 4.1 is an *eddy current slip coupling*, which without much ado converts the continual rotation of the instrument's input shaft into deflection of the needle.

Eddy currents are generated by induction, just like the output of our omnipresent plug-in power supplies, with the difference that eddy currents flow in solid bodies of any size or shape, rather than through wires. Instead of following mechanically predetermined channels, eddy currents freely circulate their domain, where their presence shows up as heat. If transformers had solid cores, eddy currents would heat them up. That's why AC machines, including the popular induction motors, use cores of stacked silicon-steel laminations, insolated from each other by solid layers of paper or varnish or both. Meanwhile, direct current devices, such as DC relays, get away with solid iron cores.

For that same reason, the metal cup in Fig. 4.1, made of electrically conductive material, such as aluminum or copper, tends to follow the rotation of the magnetized disk rotating within: Eddy currents generated by the rotating field of the magnet haul the drag cup along until counterbalanced by the torque of the hairspring, which increases proportional to the pointer shaft's angle of deflection.

The instrument is driven through a flexible shaft (remember the dentist's not so "painless" drill from a few decades ago?) that connects into the vehicle's power train by a set of miniature worm gears.

Not shown in Fig. 4.1 is the *odometer*, a tiny counter inserted into the speed-

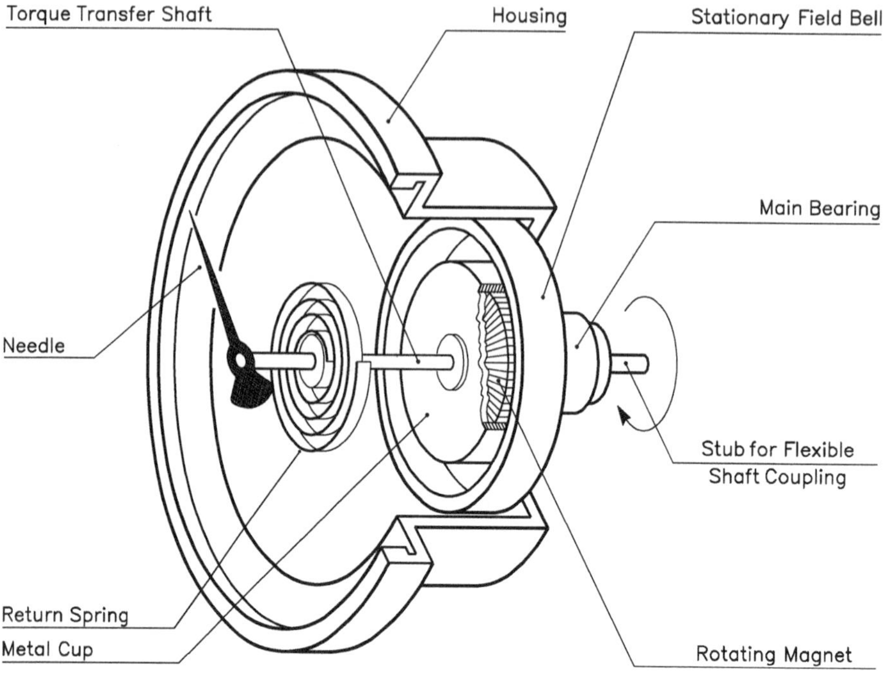

Fig. 4.1. Analog tachometer

ometer-dial. Although the odometer counts the total of rotations of the flexible shaft, the readout comes in kilometers or miles.

The *trip meter* starts counting at the push of a bottom and shows the mileage of each particular trip (presuming the driver remembers to press *reset* no later than at *liftoff*).

Additionally, most cars have a *tachometer* on their dashboard, showing the rotations per minute (rpm) of the engine. In short, speedometers measure the rotational speed of the wheels, converted in mph or km/h, while tachometers show the speed of the engine in rotations per minute (rpm).

Save for some exceptions, the mechanical design of the analog tachometer survived almost a century of automotive development and is still going strong. It meets the legal demands for accuracy in simple and inexpensive ways. However, a few drops of lubricating oil seeping into the flexible shaft housing could cause the device to fail.

Let's keep in mind that your speed readings depend on the rotations of the wheels, rather than the distance covered. With D for the external wheel diameter, the mileage equals the sum of rotations times $D\pi$. That proportionality of speedometer readings and wheel diameter can be misleading in case you decide to replace your good old standard tires with the "cool" types, of greater width and lower profile. Your speed readings could turn out wishful thinking.

An alternative – electrical speed-measuring devices – are popular mostly for the sake of simplicity of their installation. Just like yesterday's friction wheel-

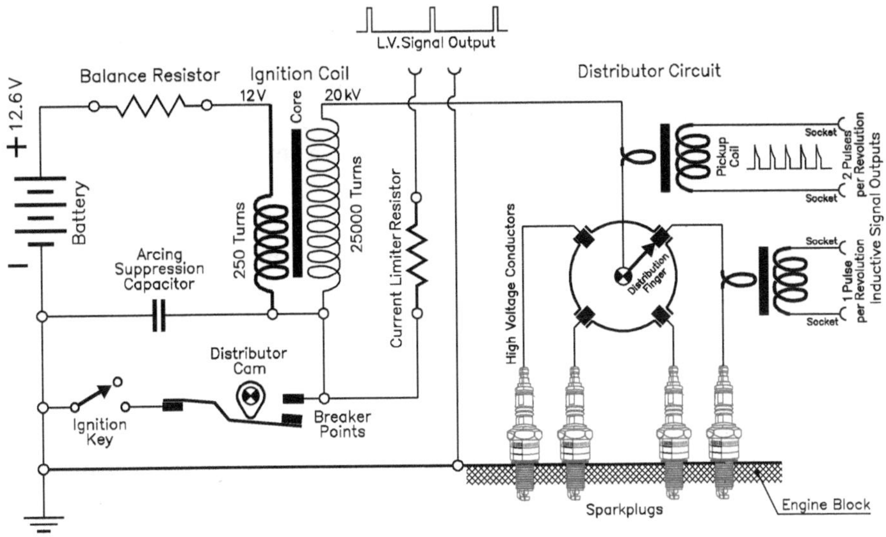

Fig. 4.2. Automotive ignition system with pulse pick up points

driven dynamos for powering bicycle headlights, *electric analog tachometers* are essentially DC dynamos, which voltage is proportional to the speed of the rotor. Therefore, a standard voltmeter can double as readout instrument (if you redraw its scale from volts to rpm), and a pair of electric wires is all you need to connect it.

For the special case of internal combustion engines (except diesel motors), engine speed can be derived from the firing rhythm of the ignition system (Fig. 4.2). Pulses for triggering an electric or electronic counter are taken either directly from the primary circuit of the ignition coil (left in Fig. 4.2), or inductively from the secondary, where voltages as high as 15,000 to 40,000 V are common (right in Fig. 4.2).

But keep in mind that each spark plug in a standard four-stroke engine fires *once* in every *two* rotations of the crankshaft. For instance, at 3000 rpm each spark plug fires only 1500 times per minute.

Pulse count is far more accurate than the readout from analog instruments. For instance, in a V8 engine fire four spark plugs per rotation of the crankshaft, so that the pulse count comes to four times the motor speed, equivalent to $1/4$ rotations per pulse. 10,000 sparks equal $10{,}000/4 = 2500$ rotations, with an accuracy of $0.25/2500 \approx 1/10{,}000$, that is one hundredth of one percent. By contrast, most analog instruments are rated at $\pm 2\%$ precision, save for the laboratory class, of $\pm 1\%$, and high precision instrumentation with $\pm 0.5\%$ precision.

The current-limiting resistor in series with the pulse pickup line in Fig. 4.2 makes sure that an accidental short wouldn't unleash the several hundred amps a car battery is capable of delivering.

While low-tension pulses from the primary portion of the ignition coil have the negative grounded, their inductively coupled counterparts from the high-tension side are decoupled from the rest of the system. That's crucial in circuits where "common ground" may establish unwanted crossover connections.

The voltage to generate a spark within the highly compressed (nowadays typically 1 : 10.5) air-fuel mixture in the cylinder of an internal combustion engine is far higher than under atmospheric conditions, which makes ignition voltages as high as 20,000 V quite common. The conversion of the 12.6 V battery tension to this level by a conventional transformer would make the ratio of primary to secondary windings $20{,}000/12.6 \approx 1600/1$. For every turn on the primary side, the secondary would need a whopping 1600. Such monster transformers were actually built for early TV sets, but turned out prohibitively expensive and a mortal danger for repairmen and experimenters.

By contrast, the high voltage we get from the surprisingly compact ignition coils of car and aircraft motors stems from the instantaneous interruption of the coils primary current supply by the breaker contacts. With L for the coil's inductivity and $\mathrm{d}J$ for the flow of current during the time-span $\mathrm{d}t$, the law of induction, $E = \mathrm{L}\,\mathrm{d}J/\mathrm{d}t$, would give the resulting voltage peak $E \to \infty$ because the abrupt opening of the breaker points ($\mathrm{d}t \approx 0$) interrupts the flow of battery current, J, through the primary windings of the ignition coil. $\mathrm{d}J$ is then the change from J to zero, and the induced tension tends toward

$$E = \lim_{\mathrm{d}t \to 0} \frac{\mathrm{d}J}{\mathrm{d}t} \approx \frac{J}{0} \to \infty \,.$$

In reality, E cannot rise infinitely thanks to the internal DC resistance of the coil windings and, in the case in point, the presence of the *balance resistor* in series with the coil. Most systems also include a spark gap (not shown), which discharges excessive voltage to ground – the engine block.

But the condition $\mathrm{d}E/\mathrm{d}t \to \infty$ alone wouldn't do for the modest 1 : 100 ratio between the ignition coil's primary and secondary to create 20,000 V pulses for firing the spark plugs. Crucial is the dimensioning of the arc suppression capacitor (left in Fig. 4.2) in ways that make it resonate in third harmonic with the primary inductance of the ignition coil – the same old trick that makes the tiny flyback transformer in TV sets generate the horizontal deflection pulses and still the high anode voltage for the picture tube.

If the conversion of the pulse count into a visual readout or into a control signal would be attempted with hard wired individual components, such as resistors, capacitors, transistors, diodes, inductors, etc, the task would be monumental. Some of us may still recall the sad story of the guy who mail-ordered something labeled "computer" from an auctioneer for the sum of USD 1000.00 (a bargain at times when computer usage was sold by the minute), only to receive a boxcar filled to the brim with 6AU7 tubes and all the rest of those times electronic hardware. Since then, the technology of integrated circuits has compressed such carloads of components into chips that sell for pennies (so to say). For our pulse counters, an integrated circuit, comprising a pulse shaper, a counter, a time-base (clock), and the readout circuitry does the trick.

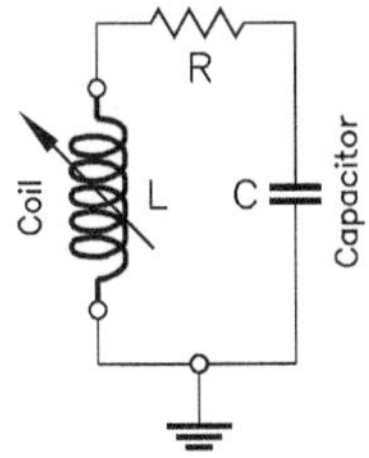

Fig. 4.3. RF oscillator circuit

Pickups

Apart from the special case of internal combustion engines, the *proximity sensor* presents itself as a convenient "non-contact" pulse source in a host of applications. In particular, the outstanding sensitivity of ferrous metal detectors stems from an internal L/C oscillator (Fig. 4.3), fine-tuned to the circuit's resonance frequency (Fig. 4.4). The circuit actually resembles the "local oscillator" at the front end of heterodyne radios. It works by repeatedly exchanging the energy of the coil's magnetic field for the electrical energy stored in the capacitor, and vice versa. In short, the capacitor discharges into the coil, which then recharges the capacitor, and so on. If it weren't for the resistor *R*, symbolizing the DC resistance of the coil-windings, this could go on indefinitely. As it is, *R* announces its presence by converting electrical energy into heat. This loss must be compensated for by feeding trigger pulses into the oscillator at its resonance frequency.

Resonance occurs if the equation $\omega^2 CL = 1$ is fulfilled. Herein, ω is the *radian frequency*, $\omega = 2\pi f$. f is the frequency of oscillation, C and L stand for the capacity of the condenser in farad, and the inductance of the coil in henry, respectively. Here again, we get $E \to \infty$ for the resonance frequency and $R = 0$, but even with the damping effect of R considered, E exceeds by far the voltage of the pulses that keep the oscillation going (Fig. 4.4).

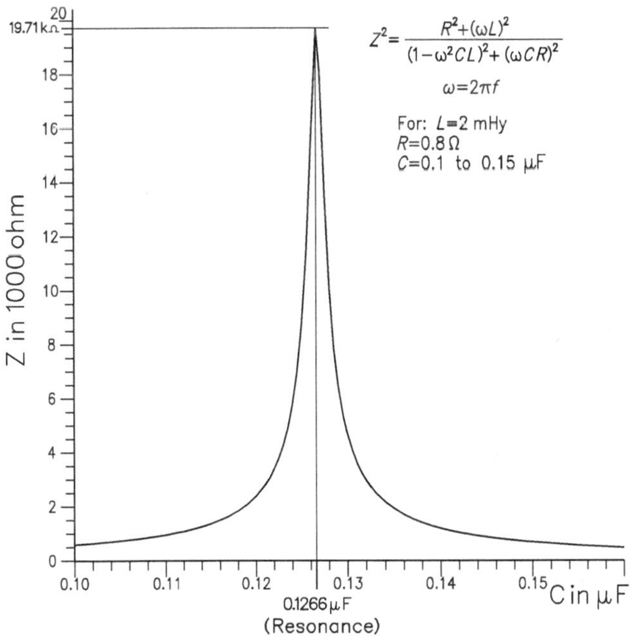

Fig. 4.4. Resonance curve for a parallel L/C circuit

The mechanical counterpart to this phenomenon is a swing, which needs nothing more than a light push every time it swings back in order to lift the child riding it far up into the air. Presuming, that is, that the timing of the pushes coincides with the rhythm of the free swaying swing.

Likewise, high tensions in an oscillating circuit occur only at resonance frequency, and fall steeply if the input pulses aren't synchronized (Fig. 4.4). Since the birth of wireless technology, this characteristic al-

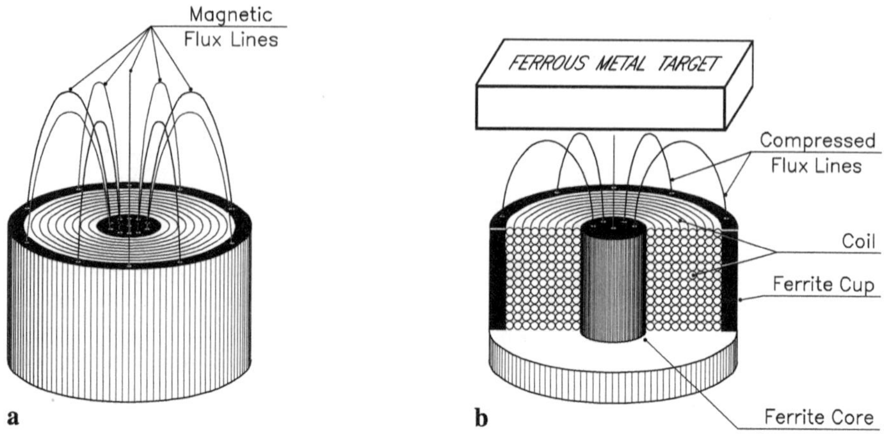

Fig. 4.5. **a** Undisturbed, **b** compressed magnetic flow lines of a coil

lowed radio receivers to distinguish stations rather than receiving a blend of the entire broadcast band at once.

True to this principle, a proximity sensor has a coil such as in Fig. 4.5 wired in parallel with a capacitor, dimensioned to resonate at the frequency of the sustaining pulses from an electronic clock. Undisturbed, the system is synchronized to oscillate at the peak of the curve in Fig. 4.4, but even minute changes in the inductance L cause an abrupt loss of the circuit's overall impedance and herewith the induced voltage. Thereof the nickname "oscillating–nonoscillating" circuit.

Such minute changes happen when a piece of ferrous metal gets into the magnetic field of the coil (Fig. 4.5 b), making the system deviate from the state of resonance with a steep drop in the oscillator's peak voltage. A proximity sensor placed near the periphery of a rotating spur gear will pop out of resonance at the passage of every tooth, while the corresponding voltage drop shows up as a negative pulse at the sensor's output.

Figure 4.6 a shows a pair of proximity sensors controlling the operation of an internal combustion engine by generating a pulse at each passage of one of the 36 straight shoulder teeth on the index wheel. One sensor is being triggered by each passing tooth, the other by a dog once per full rotation, marking the wheel's reset position. With 36 teeth (as in Fig. 4.6), this system measures the engine's crankshaft position within ten degrees of arc and, helped by a microprocessor, allows to fine-tune the spark plugs' firing points with regard to throttle position, cooling fluid temperature, and engine speed.

Figure 4.6 b details the electronics hidden under the skin of a proximity sensor. Shown on the right is the pulse generator, engineered around the venerated 7400 chip, which houses no less than four NAND gates. We recall NAND gates along with AND, OR, and NOR gates as the basic building blocks of computers. Common to all gates are two inputs feeding into a single output. Within this concept, the AND gate conducts if *both* inputs are high (5 V in TTL systems), while the NAND gate turns on only if *none* of the inputs is high. Only two of the four NAND gates

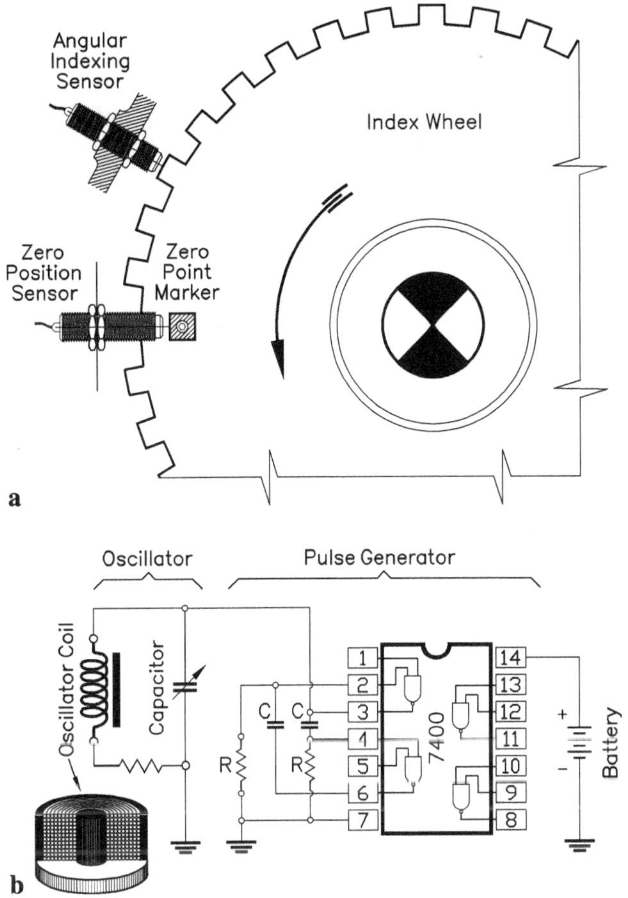

Fig. 4.6. Proximity sensor application

available on the chip get used in the circuit in Fig. 4.6, along with the externally mounted resistors and capacitors that make up an oscillating flip-flop.

The basic flip-flop, also named Eccles–Jordan circuit, has two inputs and two outputs. The latter are mutually exclusive, meaning that one is off if the other is on, and vice versa. Each input pulse inverts the state of the flip–flop. Tube-driven Eccles–Jordan circuits once triggered the horizontal deflection in TV sets. In Fig. 4.6, such pulses come from the discharge of the externally mounted capacitors through their associated resistors. Similar timing circuits are being built with #555 chips.

Proximity sensors can be seen as the forerunners of remote sensing devices, one group using sound waves, and the other Hertzian (radio) waves. In the days when people split logs with the ax, you could tell how far away the lumberjack was working by checking the interval between seeing and hearing the impact of the ax. With 340 m/s for the velocity of sound, a delay of, say, half a second meant $340/2 = 170$ meter of distance. Likewise, we can guess how far a thunderstorm is building up by counting the seconds between seeing a flash and hearing its thunder. Dividing that number by 3 tells the kilometers between you and the bad weather front. Why three? Three times 340 gives 1020 meter, or approximately one kilometer. Hence the relation. Try it for fun next time a thunderstorm traps you at home, but remember that the mile equals 1.609 kilometers.

If that appears crude science, it is yet the guiding principle of some of our brothers and sisters in the animal kingdom, such as bats, flying foxes, shrews, certain cave dwelling birds, and dolphins. What all those species have in common is their inborn system of spatial orientation by sound waves. Bats in particular emit sound at ultrasonic frequencies and envision a scenario of their immediate environment from the echo's intensity, delay, phase shift, and direction, in a process called "mental modeling".

I hate to say (after enjoying all those "Flipper" movies) that dolphins are said to use their acoustic capabilities to no better end than for separating the "grain from the chaff", so to say, which in their habitat means the difference between edible and repulsive fish. Whales transmit infrasonic frequencies, which greater wavelength and herewith superior reach helps their orientation and navigation.

Although the discovery of the bats' remarkable sense of orientation happened as early as 1793 by the Italian biologist L. Spallanzani, only in 1940 succeeded the American researchers D.R. Griffin and R. Galambos in unveiling the underlying principles. Research into how bats convert the acoustic information into appropriate motions of their limbs helped in making cybernetics a separate branch of science.

Sound and ultrasound

In the anthropogenic world of human achievements, the *Galton whistle* is believed the first instrument to reach ultrasonic frequencies of up to 100 kHz. Galton whistles are still being used industrially for purging dust from exhaust fumes.

Sirens, on the other hand, generate oscillations with 200 to 300 watts of sound power, and some have been built powerful enough for igniting a ball of cotton with their sonic beam. But a siren's frequency is determined by the rotational speed of its rotor, which sets limits to how high it can go.

Most loudspeakers (Fig. 4.7) vibrate their diaphragm by forces induced on a lengthwise movable coil within the field of a strong permanent magnet. The system works best in the audio frequency band of 20 to 20,000 Hz, but lacks the power for ultrasound related applications. Only magnetorestrictive speaker devices can be made to reach frequencies of up to 300 kHz.

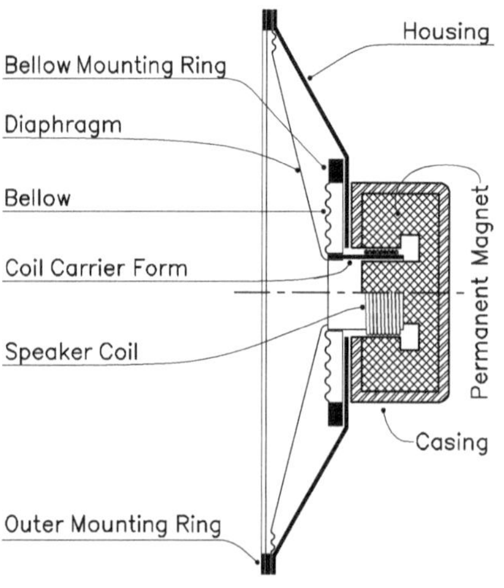

Fig. 4.7. Moving coil speaker

Since the emergence of radio transmission, the piezoelectric properties of quartz crystals were key to frequency stabilization. The proper frequency of a slate of quartz, 1 mm thick, is 3 MHz (3×10^6 Hz). Higher frequencies can be derived as harmonics of such base frequencies.

Since then, the classical piezoelectric materials, such as quartz, Rochelle salt, etc., have been replaced by powder metallurgical ceramic and plastic composites, such as listed in Table 4.1.

1-3 composite is a somewhat flexible piezoelectric material, while all others are rigid.

Table 4.1. Piezoelectric materials (from Robert A. Day, GE)

	Coupling coefficient		Acoustic impedance	Resonant frequency	Max. temp.
	Thickness mode	Radial mode			
Units	k_t	k_p	Acoustic ohm	MHz	°C
Copolymer PVDF	0.2	12	3.9		80
Lead titanate PT	0.51	<0.01	33	<20	350
Lead metaniobate $PbNb_2O_6$	0.30	<0.1	20.5	<30	570
1-3 composite	0.6	≈0.1	9	<10	100

Table 4.2. Transmission and reception efficiencies relative to quartz

Property	Material					
	Quartz	Lithium niobate	PZT-4	PZT-5A	Cadmium sulfite	Zinc oxide
Relative transmission efficiency Y_t	1	2.8	65	70	2.3	3.3
Relative reception efficiency Y_r	1	0.54	0.235	0.21		1.42
Pulse/echo efficiency Y_t/Y_r	1	1.51	15.3	14.7		4.7

The *coupling coefficient* k_t is the ratio of the acoustic energy output of a piezoelectric probe to its consumption of electrical energy. Due to conversion losses, $k_t < 1$. The terms "thickness" and "radial" mode can be interpreted by comparison with the beam of a flashlight. Even if aimed straight forward (thickness mode), some of the beam's rays stray sideways (radial mode). Ideally would be values close to $k_t \to 1$ and $k_p \to 0$.

Transmission and reception efficiencies, Y_t and Y_r respectively, are listed in Table 4.2 relative to quartz. Radial ultrasound emissions cause interference, which makes materials with low radial coupling coefficients mostly preferable.

Piezoelectric speakers (Fig. 4.8) use two thin squares of piezoceramic material, sandwiched together

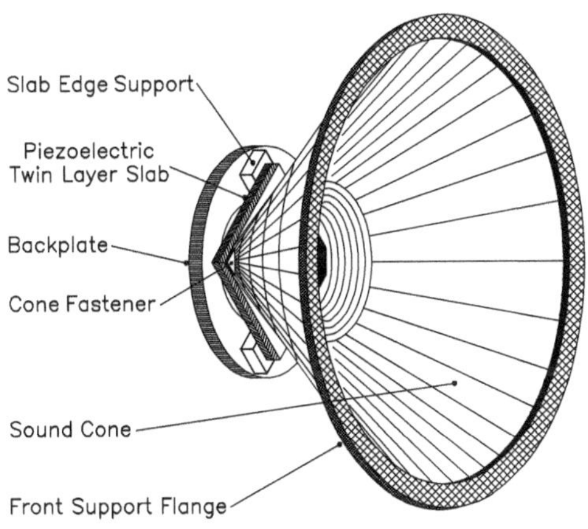

Fig. 4.8. Piezoelectric speaker

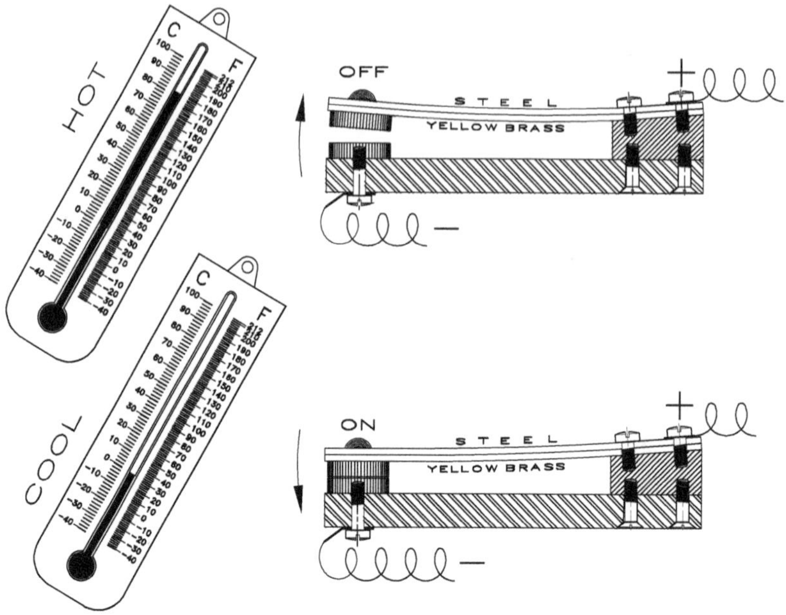

Fig. 4.9. Principle of bimetal controlled switch

at 90° between their effective crystal axes. Only two opposite edges of the piezoelectric plate are supported on the speaker's baseplate. The other two edges support the truncated sound cone, which narrow end is flattened for this purpose. If driven by alternating voltage surges from a push-pull circuit, one square expands while the other remains unaffected, and vice versa, so that the amplitudes of deformations add up.

The functioning of sandwiched piezoelectric plates can be compared with that of the traditional bimetallic temperature switch in Fig. 4.9. Sheets of different metals are laminated together and cut into bimetal strips which warp according to their temperature. Best results give steel and yellow brass, with thermal expansions of 11.6×10^{-6} and 20.3×10^{-6}, respectively. When a, say, 50 mm long bimetal strip is being heated by 100 °C, the steel expands by

$$11.6 \times 10^{-6} \times 50 \times 100 = 0.058 \text{ mm or } 58\ \mu\text{m}\,.$$

For brass, we get

$$20.3 \times 10^{-6} \times 50 \times 100 = 0.1015 \text{ mm or } 101.5\ \mu\text{m}\,.$$

The difference, of $101.5 - 58 = 43.5\ \mu$m, makes that the brass side of the strip buckles, as shown in Fig. 4.9. Snap action is achieved by adding a small magnet, which locks the switch in the closed position until thermal expansion exceeds the magnetic forces and the bimetal tongue breaks away.

Echo sounding

The rise of echo sounding goes back to W.W.II, when ultrasound piling systems, such as Asdic and Sonar, became paramount for locating submerged objects, in particular submarines. Mounted on the underside of a ship's hull, ultrasonic transducers sent sound pulses downward. They were reflected either by the seafloor or by objects in their path. A second transducer was used to receive the echo and convert it into electrical signals. The depth gauges in today's sport fishing boats are the peacetime offsprings of those techniques.

The range of sound waves in a body of water far exceeds that of radio waves, which makes that echo piling is going strong even in today's electromagnetic environment. Sound piling at great wavelengths allows for ocean wide surveillance, but the environmental side effects of such powerful signals caused concerns regarding marine life and oceanic ecology in general.

In seawater of 20 °C, sound propagates with 1520 m/s, so that the echo from the seafloor of a, say, 760 meter deep body of water, will take 1 s to be heard. In precision measurements, variations of the speed of sound due to the water's temperature and salinity must be accounted for.

Ideally, pulses from an ultrasound transmitter should be of short duration and high amplitude. The time delay between pulses emitted and received can be shown on an oscilloscope with its "external sync" terminal connected to the pulse generator, and echo pulses on the "input" terminal. The tube will show two successive spikes, spaced proportional to the separation of transmitter and reflector.

In the early days of echo piling, a rotating disk illuminated by a bright neon lamp was used to display these delays just as stroboscopes make us see an engine's moving elements as if petrified in time. The neon lamp was triggered by the transmitted and the received pulses alike, which made the angular separation of the two flashes in every turn of the disk proportional to the time delay between the pulses, visualized by synchronizing the speed of rotation of the disk with the pulse frequency. If the separation of the in-and-out flashes exceeds 180°, the disk's rotation was halved and readings doubled.

Historically, rotating disks were also the key scanning element in the very first television transmitters and receivers, which had to rotate in unison. Since synchronizing the receiver's disk was done by means of a foot brake, the steadiness of the image depended on the dexterity of the brake operator. Imagine the audience' outrage if the image of a rock dancer slipped out of sync at the very moment when ... but that is a different story.

In any case, all that became history as the event of a diode-controlled circuit, known as the phase discriminator, made the TV brake operators lose their jobs anyway. But an offspring of the rotating disk in the form of rotating mirror assemblies made a comeback as color separator in certain top-of-the line projection TVs.

One of ultrasound's numerous applications is in *nondestructive* detection of internal cracks and other discontinuities in blocks of material, such as ingots and castings in general as hidden flaws become ultrasonic reflectors. Underground mineral, oil, coal, and natural gas deposits can be located with ultrasound, generated by a few sticks of dynamite.

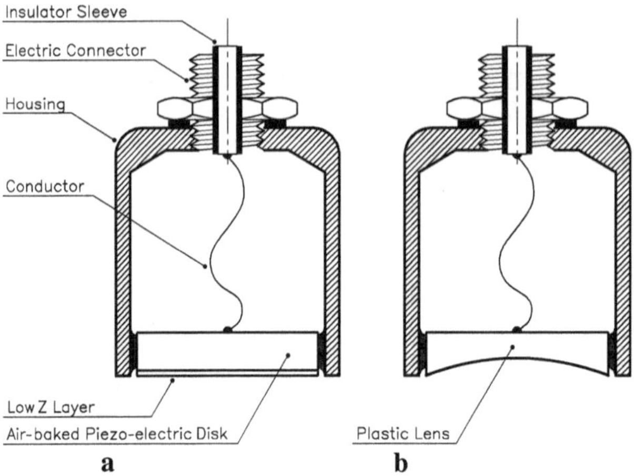

Fig. 4.10. Piezoelectric probes

A breakthrough in medical science came with the introduction of scanning beam ultrasound machines, showing soft tissues rather than the characteristic bone structures of X-ray images.

Piezoelectric probes

For quality control and other industrial applications, piezoelectric probes come in various configurations. Figure 4.10 a shows a straight transducer, which beams its oscillations through a "low z" protective layer to the workpiece. "z", the acoustic impedance, the counterpart to electric impedance, is the sound pressure in 10^5 Pa on a given surface, divided by the sonic flux through that same surface in m^3/s. This unit of acoustic impedance, also borrowed from electric technology, is called the acoustic ohm, $10^5\ \text{Pa}/(\text{m}^3/\text{s})$. Along this line of reasoning, Ohm's law can be expressed for the flow of sound by the rule: "The lower the acoustic impedance, the higher the flux."

The pascal (Pa), unit of pressure in the International System of Units (SI), is defined as one newton per square meter. With 9.80665 newton to the kilogram-force (kgf), that leads to the surprisingly simple conversion factors of:

$$10^5\ \text{Pa} = \frac{10^5}{9.80665} = 1.0197\ \text{kgf}/\text{cm}^2 \approx 1\ \text{kgf}/\text{cm}^2 \text{ or } 1 \text{ at (technical atmosphere)}\,.$$

This makes the acoustic ohm the ratio of sound pressure in technical atmospheres or kgf/cm^2 to the flux in cubic meter per second (m^3/s).

The second variable, acoustic pressure, not to be confused with static pressure, is the *fluctuating mean pressure* in a media agitated by sound waves.

If those metric units sound confusing, take heart. Americans are not obliged to adopt the metric system, but formulas expressed in metric units are simpler to use and understand than their foot/pound counterparts.

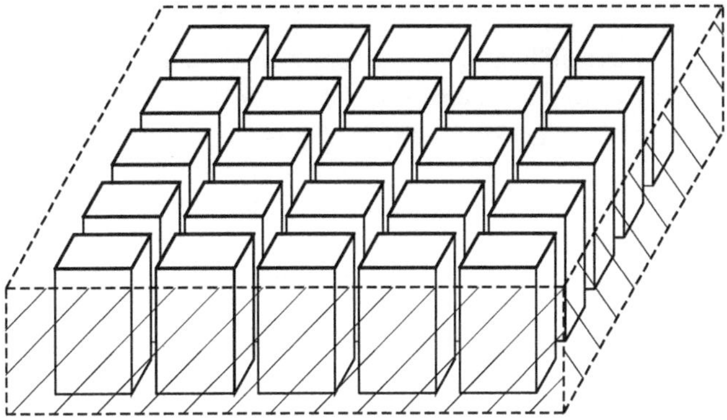

Fig. 4.11. Piezoelectric array of thin rods of ceramics embedded in a polymer matrix

Out of a great variety of ultrasound transducers, Fig. 4.10 shows a pair of industrial pick up units. The probe in Fig. 4.10a employs a cylindrical piezoelectric crystal, protected by a layer of low sonic impedance material. Figure 4.10b shows a transducer with a lens-shaped crystal which produces a focused sound beam, capable of pinpointing flaws and impurities in pipelines and irregularly shaped objects.

Figure 4.11 shows a piezocomposite transducer that can be tailored for the special needs of particular applications. It consists of an array of piezoceramic rods imbedded in a polymer matrix. For a 5 MHz probe, slits must be kept small, typically 0.1 millimeters (0.004 inch) in width, and are filled with castable polymer plastic, such as epoxy. Fabrication of such arrays includes criss-cross slicing of a rectangular piezoceramic slab from the top down to the point where only some 20% of the slab's height is left intact underneath. The slap is then cast into a massive block of polymer until all clearances are tightly filled. After the cast has hardened, the solid part of the array is ground off until all the parallelepipeds are separated from each other, leaving a multi-element transducer with coupling factors of up to 0.65.

Radar

Radar, a radio detection and ranging decive, the electronic counterpart to a sonar device, uses Hertzian (radio) waves in ways sonar uses ultrasound. Like sonar, radar was developed in W.W.II, but that's where the similarities end. Sound waves are mechanical, longitudinal oscillations of matter, such as air, water, and solids of all kind, while radio waves are oscillations of interchanging electric and magnetic fields. Sound waves need a carrier, such as air or water to sustain their propagation, while radio waves, including light, also transit through empty space.

Like sonar, radar works on the principle of deriving an object's distance from the time difference between emission of a signal and reception of its echo. With electromagnetic waves, this includes the measurement of time delays in the range of microseconds and fractions thereof. Light and radio waves travel 300 meters (about 1000 feet) in one millionth of a second (microsecond), which makes such time intervals too short for direct measurement. But take heart. He, who at the risk of his life once climbed the roof of his home with a tool kit and an armful of aluminum extrusions in hopes of assembling them into a hi-tech TV antenna has, unknowingly, taken the first step toward such measurements. How come?

By measuring ghosts. Ghosts – parasitic images caused by stray signals reflected from buildings and other obstacles on the way from the transmitter to the receiver – are less luminous than their directly received "big brothers," because the lion's share of the beam's energy is lost in the process of reflection.

The luminous spot that scans the TV screen and through variation in its intensity generates the image, takes 53.5 μs from the left to the right of the screen. If a secondary image precedes the original by, say, $1/5$ the width of the screen, we can guess the time delay caused by the interfering object, such as a high-rise, as $53.5/5 = 10.7$ μs. With 300 m/μs for the velocity of radio waves, this gives $10.7 \times 300 = 3210$ m or 3.21 km of difference in the length of the direct versus the reflected path from transmitter to receiver. Repeating that test with TV sets in various places would allow to localize the interfering building by triangulation[1].

That much on electromagnetic wave reflection. But unlike broadcast TV, which operates in the frequency range of 54 MHz (channel 2) to 216 MHz (channel 13), radar transmits pulses of typically one microsecond of width at the rate of one per millisecond. Wavelengths vary between one and ten inches (0.0254 to 0.254 meter), which gives for the band of typical radar frequencies

$$\frac{300 \times 10^6}{0.254} \approx 1200\ \text{MHz} \ \text{to} \ \frac{300 \times 10^6}{0.0254} \approx 12{,}000\ \text{MHz}$$

Radar signals can be seen on an oscilloscope. With the horizontal sweep frequency at 50 μs per scan (20 kHz of frequency), one microsecond shows on the screen of a four-inch oscilloscope as a $4/50 = 0.080$ inch displacement. Reflections from a, say, 3 km distant building arrive with $2 \times 3000\ \text{m}/(3 \times 10^8\ \text{m/s}) = 20\ \mu\text{s}$ delay, and thus generate spikes lagging those from the outgoing signals by $0.080 \times 20 = 1.600$ inches. That makes the scale of the display:

$$1.600/3 = 0.533\ \text{inches or}\ 0.533 \times 25.4 = 13.54\ \text{millimeter per kilometer}\,.$$

Doppler shift radar

The radial velocity of a moving target can be found by dividing the difference between two successive readings by the time lag between them. But this rather awkward procedure gave way, in most applications, to the Doppler shift radar, which allows for one-shot speed checks. The system reaches back to the Austrian physicist Christian Doppler, who was first to explain, in 1842, the different pitch of the whistle of a railway engine as it approached the observer and subsequently sped away. It should not surprise that sound, which normally propagates at 340 m/s,

1 deduction of the position of the vertex of a triangle from the length of the baseline and the magnitude of two angles.

travels faster if emitted by a source moving toward the listener. For instance, sound from a train, speeding at 17 m/s toward the observer, must arrive faster by $17/340 = 0.05 = 5\%$, and therefore with 5% shorter wavelength, which makes the pitch more strident. Likewise, one would expect the pitch to lower by 5% when the train passed the observer and moves away. This reasoning is expressed by the formula $f_o/f_s = (v \pm v_s)/v$, where v is the velocity of sound, v_s the velocity of the source (i.e., whistle), f_o the observed frequency, and f_s for the source frequency.

For the above example, we get $f_o/f_s = (340 \pm 17)/340 = 1 \pm 0.05$, a straight five percent up or down, respectively.

But if you keep the source of sound put and let the observer approach or recede at the velocity v_o, the formula becomes $f_o/f_s = v/(v \pm v_o)$. This leads to the slightly different rates of $f_o/f_s = 340/(340 \pm 17) = 1.05263$ and 0.95238 for approaching and receding observers, respectively.

A glimpse at relativity

Up to the 20th century, similar equations were expected to express the propagation of light, while space was imagined home to the "world ether", an ideally tenuous, omnipresent substance, acting as the carrier of light waves. But to most scientists' dismay, this test, famous as the Michelson–Morley experiment, proved the speed of light as invariable, the tenet used by Albert Einstein in the development of his special theory of relativity. Recalculated in relativistic terms, the traditional Doppler formulas melt into the relativistic Doppler equation

$$\frac{f}{f_o} = \frac{\sqrt{1 + v/c}}{\sqrt{1 - v/c}}$$

where $c - 300{,}000$ km/sec stands for the speed of light in vacuum, and v for the relative velocity between light source and light receiver. f_o is the frequency of the light source, and f the frequency of the light at the receiving end. True to the principles of relativity, the equation contains only terms of *relative velocity* of source and receiver with regard to each other, independent of an external frame of reference.

Applied to the change $\Delta\lambda$ of the wavelength of selected emission and absorption lines in the spectra of stars and galaxies, this formula allows for computing the radial (along the line of sight) velocity of such celestial objects. Known as the *redshift*, $\Delta\lambda$ can be figured from $\lambda = c/f$ and the relativistic Doppler equation as $\Delta\lambda/\lambda = (\sqrt{1 + v/c}/\sqrt{1 - v/c}) - 1$. Inverted, this formula reads

$$\frac{v}{c} = \frac{(1 + \Delta\lambda/\lambda)^2 - 1}{(1 + \Delta\lambda/\lambda)^2 + 1},$$

and lets us deduce the radial velocity of celestial objects from their redshift. Thanks to the conspicuous spectral lines of hydrogen and potassium (H and K lines), discernable even in the spectra of the dimmest stars and galaxies, the redshift can be measured with amazingly high precision.

Radial velocities of galaxies have been found approximately proportional to their distances from our own, which led Edwin Powell Hubble (1889–1953) to envision the world as expanding. The Hubble constant, of 73 km/s per megaparsec, is a measure of the degree of that expansion. For instance, the distance of a galaxy speeding away with one tenth of the velocity of light, namely, $c/10 = 300{,}000/10 = 30{,}000$ km/s, should be $30000/73 = 411$ megaparsec. With 3.26×10^6 light-years to the megaparsec, this amounts to 1340 million light-years. Radial velocities as high as $0.8c$ (240,000 km/s) have been observed, originating from objects as far away as 80% of the radius of the universe, regardless of similarities with their brothers closer to home.

Back on Mother Earth, the Doppler principle led to the design of radar detectors with direct readout of the speed of the target, eternalized in street signs such as "Road Construction ahead – Speeds radar controlled – Fines doubled," which some people take as proof that better technology is not always synonymous with greater happiness. But look at the bright side: radar meters are handy tools in a host of peaceful applications. Even over-the-counter models relieve their human masters of such backbreaking jobs as checking the size of a room with a tape measure.

Radar design principles reach back into the early days of radio technology. Electronic enthusiasts with access to vintage radios from the 1920s and early 1930s know the ear piercing inter-station noise of those early music boxes. By FCC rules, AM (broadcast) frequencies are to be 10 kHz apart as not to interfere with each other; and yet, tuning from one station to the next got you superimposed transition signals from both as whistle at the 10 kHz *beat frequency*. If the transmitter kept its signal modulation strictly within the assigned 10 kHz, and the receiver didn't surpass those limits either, the noise subsided once a station was tuned in. But since those times receivers' bandwidth used to vary over the 530 to 1650 kHz broadcast frequencies, higher amplification was fatally linked to equally higher noise levels.

Relief came with the invention of the superheterodyne (or superhet) receiver, which made "a virtue of necessity," so to say, by using beat frequency, the classical noise maker, as the centerpoint of a novel receiver design. The superhet system adds a *local oscillator* to the front end of the receiver which frequency is consistently kept 455 kHz above the receiving frequency, whatever the latter. The signal from a station of, say, 1000 kHz, finds the local oscillator tuned at 1455 kHz. Superimposed, those two frequencies generate beat frequencies of $1455 + 1000 = 2455$ kHz and $1455 - 1000 = 455$ kHz. The next higher station, sending at 1010 kHz, encounters 1465 kHz at the local oscillator, and generates the beat frequencies of 2475 kHz and, – once again – 455 kHz. Filtering out the higher beat frequencies makes that all signals along the entire broadcast band exit the receiver's front end at the "intermediate frequency" of 455 kHz, unified and ready for further amplification. This way, the two or three intermediate frequency (IF) stages that follow operate at 455 kHz throughout and therefore don't generate parasitic beat frequency signals with each other.

Which brings us back to the radar speed detector. Just like the IF in a superhet is the product of superposition of two frequencies (the receiving and the local oscillator frequencies), so comes the output of a Doppler radar as the beat frequency Δf of the frequencies of the transmitted and the reflected beam, respectively. Δf is positive for approaching targets, and negative for receding targets.

As a beam of, say, 5000 MHz (5 gigahertz) hits an automobile traveling at 108 km/h (30 meter per second), the frequency of the reflected wave, $f \pm \Delta f$, is derived from

$$\frac{\Delta f}{f} = 2\frac{v}{c} = 2 \times \frac{30}{3 \times 10^8} = 2 \times 10^{-7},$$

which yields:

$$\Delta f = 5 \times 10^9 \times 2 \times 10^{-7} = 1000 \text{ Hz}.$$

Herein, the multiplier of 2 accounts for the superposition of the Doppler shifts of the incident beam and the reflected beam. From 1000 Hz for the beat frequency related to 108 km/h, we figure the frequency for 100 km/h as $1000 \times 100/108 = 926$ Hz, and subsequently the instrument's scale factor as 926 Hz/100 km/h.

The conversion of frequency counts into digital readouts in miles or kilometers per hour could be talked away with the classical "it's all done by mirrors", rephrased into "it's all done by chips," but that would be a non-explanation. True, the understanding of a chip's operation by looking at its circuit diagram is "insiders' business", but we will analyze the principles of logical circuits in later chapters.

Long before radar beams began scanning the seas, pilots of ocean going vessels figured their position by *dead reckoning*, the plotting of a ship's course on a sea-chart from the ship's speed and direction. Their instrument of choice was the (non-electronic) *log chip*, a bulky triangular board weighted on one edge and tied to the so called log line, which unrolled from the "log-reel" back on board ship.

Today, we would use a stopwatch to check how long it takes the line to unroll to its very end, and get the ship's velocity by dividing the length of the line by the time for its unreeling. But the best chronometers our forebears had to rely on were 15-second sandglasses, hampered by their incapacity of fractional readings. That led to the inverse approach: The sailors let the reel run for the sandglasses fifteen seconds emptying time, and then checked how much of the line had pulled off. To that end, the line was marked with regularly spaced knots along its length, used in those times like the benchmarks on tape measures today. The unit of *knot* for one nautical mile (1.151 statute miles) per hour stands as a reminder of those days knotted ropes.

Velocity and acceleration

Acceleration, the velocity gain per unit of time, is defined by $a = (v_2 - v_1)/\Delta t$. For instance, a car, claimed to reach the speed of 72 km/h (20 m/s) in 5 s, accelerates by an average of $a = (20 - 0)/5 = 4.00\ \text{m/s}^2$.

To figure acceleration from the distance s covered by a moving object in the time span Δt, we use the formula $a = 2s/(\Delta t)^2$. Take the example of a vehicle which speeds up toward a 50 meter mark in 5 seconds. Its acceleration is $a = 2 \times 50/5^2 = 4\ \text{m/s}^2$, incidentally the same as in the foregoing example.

And finally, given the velocity gain Δv of an object traveling the distance s, acceleration is figured from $a = (\Delta v)^2/2s$. Let the car in the present example reach the speed of 20 m/s at a point 50 meters from the start, and you get again $a = (20 - 0)^2/(2 \times 50) = 4\ \mathrm{m/s^2}$.

So far, all cases considered presupposed constant acceleration, like in the case of free fall close to the surface of the Earth. In space, however, where the decline of gravitational force according to the inverse square law must be accounted for, momentary values of acceleration are given by the differential $a = \mathrm{d}v/\mathrm{d}t$.

In the absence of points of reference, such as milestones, acceleration can be derived from the inertial forces it generates. Inertial force and acceleration are proportional to each other. With F for force, m for mass, and a for acceleration, this is expressed by Newton's second law, $F = ma$. The results come in newton (N), the unit for force in the International System of Units. The newton converts to kilogram-force (kgf) by division through gravitational acceleration, $g = 9.80665\ \mathrm{m/s^2}$. For instance, the thrust F needed to accelerate a 1000 kg spacecraft by $a = 10\ \mathrm{m/s^2}$ follows from this formula as $F = 1000 \times 10 = 10{,}000$ N. With gravitational acceleration rounded up to $g \approx 10\ \mathrm{m/s^2}$, F becomes $10{,}000/10 = 1000$ kgf, equal to the spacecraft's weight. That shouldn't surprise, considering our initial setting of $a = g$.

For fast results, we can use the proportionality $F_1/F_2 = a_1/a_2$ to figure inertial forces. In the foregoing examples on speeding up cars, we would set $a_1 = a$ and $a_2 = g \approx 10\ \mathrm{m/s^2}$, and get the inertial force F_1 on the driver relative to her bodyweight W:

$$a/g = F_1/W = 4.00/10 = 0.40\,.$$

As the car speeds up, a 150 lb (ca. 68 kg) person would feel 40% of 150, that is 60 pounds (ca. 270 N) of thrust forcing her backward.

An interesting experiment inside a spacecraft would be the use of a pendulum clock to gage acceleration. According to $T = 2\pi\sqrt{L/g}$ (chap. 3), the time T for a full swing of a mathematical pendulum is inversely proportional to $\sqrt{g}$. In a spacecraft which rocket engines accelerate it consistently at $a\ \mathrm{m/s^2}$, we get $T \propto \sqrt{1/a}$. For instance, a pendulum clock ticking away in a spaceship that accelerates at the rate of $g/3$ would go $\sqrt{3} = 1.732$ times slower than on Earth; it would stop altogether once the craft settles at cruising velocity, whatever breakneck speed that might be. But don't forget to flip the clock upside down when you stop and head for home.

The identical effects of gravitational and inertial forces make it possible to use similar measuring instruments for both. Even a kitchen scale would show linear and centripetal acceleration through deflection of its internal spring. *Strain gages* double as accelerometers in applications such as triggering an airbag. *Gravimeters*, which deduce gravitational force with high precision by timing the free fall of metal rods inside an evacuated container, actually determine acceleration.

Passengers in a windowless airliner would have no way of guessing the speed they are traveling with, but would readily recognize speed changes that force them forwards and backwards in their seats. Glued to our planet as we are, we all ignore that we house on a planet speeding at 30 km/s (108,000 km/h) around the Sun. Neither does it bother us that we travel at 20 km/s along with the local star cluster the Sun makes part of. Not even the 250 km/s we cover as part of the Milky

Way are being felt. If astronomers wouldn't tell, we would never have guessed. In a system of two moving objects, A and B, in space, it is impossible to determine if A is in motion and B stands still, or vice versa; or still if A and B are both moving, each at its own pace, unless we have access to external points of reference.

Take for instance a stream flowing at the velocity v_A, and a boat plowing the waters at the velocity v_B, and you get $v = v_A \pm v_B$ for the velocity of the boat relative to the river banks. The plus sign is for downstream travel, and the minus for the tiresome journey upstream. So far, we discussed cases with $v_B = 0$, that is, the laws of motion for a moving object in stationary environment. The scenario changes if $v_B > 0$, which brings us to the question of flow measurements. Devices such as the log chip could measure the liquid surface velocity in open channels, but would fail to obtain other significant data, such as a waterway's potential for power generation, where velocity differences with depth get into the picture.

Liquid flow metering

The classical sensor here and in similar applications is the *Pitot static tube*[2], basically an angled glass tube with one end stretched into the shape of a nozzle (Fig. 4.12).

In a body of stagnant water (or any other fluid), the liquid level in the Pitot tube coincides with the liquid surface level, H_S. In moving water, with the nozzle facing the flow, the fluid's *dynamic pressure* raises the level in the Pitot tube by H_D. The work (in joule) for lifting a quantity m of fluid up to this level is mgH_D, and the energy for performing that work is the kinetic flow energy of m, $mv^2/2$. Equaling the two terms leads to the equation $mgH_D = mv^2/2$, where m cancels out and we get $H_D = v^2/2g$.

H_D is the dynamic *water head*, the height of the water column in the Pitot tube above the water level. In metric units, 10 meters of water column equal the pressure of 1 kg/cm^2. This because we have 100 cm to the meter, which makes the volume of a ten meter high water column, one square centimeter across, $10 \times 100 \times 1 = 1000$ cm^3. And since *one gram* is the weight of *one cubic centimeter* of pure water (at 4 °C), 1000 cm^3 weigh 1000 g or 1 kg. In short, the pressure of 1 kg/cm^2 equals 10 m of *water head*. From this, we can figure the pressure at the faucet at the base of a, say, 50 meter high water tower, as $50/10 = 5$ kg/cm^2 or 5 at (technical atmospheres) (ca. 490 kPa).

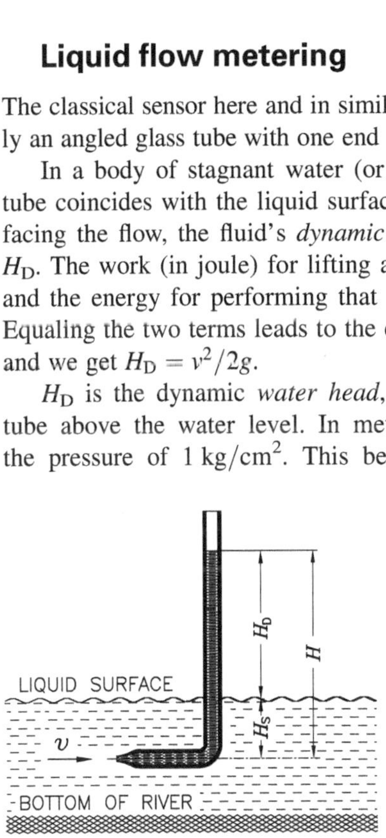

Fig. 4.12. Flow measurement with pitot tube

For liquids other than water, pressure is proportional to the fluid's specific gravity, γ.

2 Henri Pitot, 1695–1771.

For instance, gasoline weighs 0.68 kilogram per liter, or 0.68 g/cm^3. A ten-meter column of gasoline thus weighs 0.68 kg and the aforementioned water tower, used with gasoline, would deliver its contents at $50 \times 0.68/10 = 3.40$ at (ca. 334 kPa) of pressure at the tap.

Summing it up in general terms leads to the hydrostatic formula $p = \gamma g H$. With $H_D = v^2/2g$, this term yields the *dynamic pressure* $p_D = \gamma H_D = \gamma v^2/2g$ or inverted, the flow velocity, $v = \sqrt{2gp_D/\gamma}$.

In a setup like Fig. 4.12, H_D can be read directly on the vertical leg of the Pitot tube, regardless of the static pressure (p_S) at the depth of the nozzle. In a piped systems, however, a Pitot tube reads the sum of static and dynamic pressure. The former must be measured separately at right angles to the direction of flow and be subtracted from the readings of pressure, p, from the Pitot tube, $p_D = p - p_S$. This brings the equation for flow velocity into the form:

$$v = \sqrt{\frac{2g(p - p_S)}{\gamma}} .$$

To get v in m/s, this formula must be used with metric units, that is g in m/s^2, p and p_S in N/m^2, and γ in kg/m^3.

The consistency in the choice of units is essential. If we used technical atmospheres (kg/cm^2) for the pressure differential and expressed specific gravity, γ, in kg/dm^3, and used m/s^2 for terrestrial acceleration, the above equation would get us

$$v = \sqrt{\frac{(m/s^2)(kgf/cm^2)}{kgf/dm^3}} = \sqrt{\frac{m \times kgf \times dm^3}{kgf \times s^2 \times cm^2}} = \sqrt{\frac{m \times dm^3}{s^2 \times cm^2}} = \frac{\sqrt{m \times dm^3}}{s \times cm} .$$

Oops. The unit for velocity is m/s, not $\sqrt{m \times dm^3}/(s \times cm)$!

So, let's use the units of the International System rigorously. Take for example an airliner, flying at the altitude of 5000 m, where ambient pressure is 5500 kg/m^2 and the air weighs 0.736 kg/m^3. Readings of Pitot pressure of 6225 kg/m^2 would tell the pilot the cruising velocity of the plane in terms of

$$v = \sqrt{\frac{2 \times 9.807 \times (6225 - 5500)}{0.736}} = 139 \text{ m/s or } 139 \times 3600/1000 = 500 \text{ km/h} .$$

Pitot tubes of typically 1/2 inch diameter and 10 inches length are standard in aeronautics. Like in most industry grade Pitot tubes, the barrel's central bore is the admission port for total pressure, p, flanked by a pair of longitudinal blind holes with lateral openings for the admittance of static pressure, p_S.

The need to compensate for the decrease of atmospheric pressure with altitude led to the design of the *altitude-compensated aircraft speedometer*. This is a twin bellow pressure gauge with a lever mechanism that lets dynamic pressure deflect the needle, while static pressure causes a displacement of the needle's swivel point in ways that counteract the effect of atmospheric pressure fluctuations.

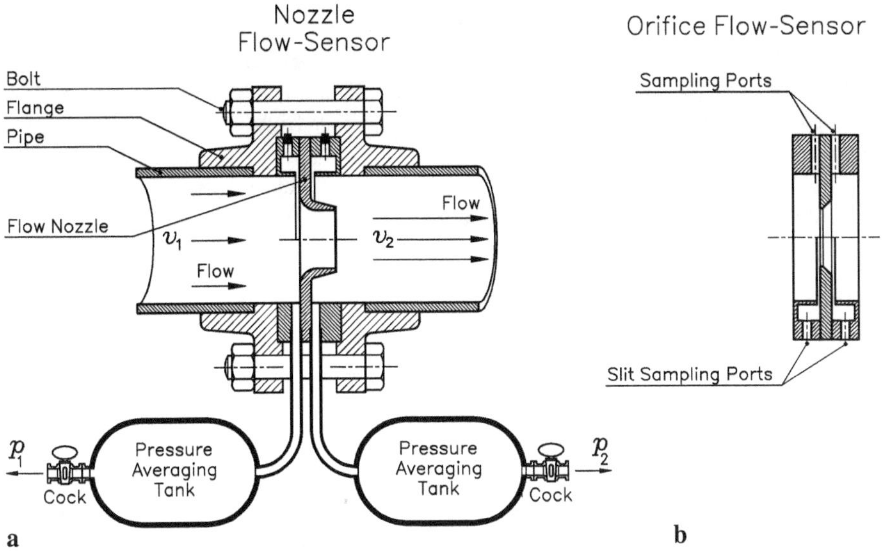

Fig. 4.13. In-line flowmeter in piping and pipelines

Orifice flowmeters

In the chemical and petrochemical industry, industrial fluids tend to clog the thin openings of Pitot tubes. Here, it is better to derive flow rates in pipe lines from the pressure differential before and after an inserted orifice plate or nozzle (Fig. 4.13). Since the principles of measurement with Pitot tubes and orifice plates are mutually converse, the flow velocity in pipes can be derived from the formula $v = \sqrt{2g(p - p_S)/\gamma}$ by replacing the term v by the speed of the fluid, v_2, and $(p - p_S)$ by the pressure difference $(p_1 - p_2)$. This gets us $v_2 = \sqrt{2g(p_1 - p_2)/\gamma}$.

Unlike Pitot tubes, which measure *dynamic pressure*, flow nozzles (Fig. 4.13 a) and orifice plates (Fig. 4.13 b) derive the flow velocity from the *static-pressure* differential between upstream and downstream from the restriction. The energy for accelerating the fluid while passing through the restriction is drawn from the potential energy of the fluid's static pressure, which makes $p_2 < p_1$.

Orifice plates designed to check p_2 as close as possible to the restriction are shown in Fig. 4.13 b in two versions, one being traced above and the other below the center line. The former has direct ports for p_1 and p_2, the latter employs narrow circular slots leading into circumferential averaging channels that feed into the gauge ports.

Not surprisingly, the flow pattern through a nozzle (Fig. 4.13 a) is far more regular than that of a simple opening and matches theory more closely. Nevertheless, the formula for v_2 still needs a correction factor, α, in order to yield usable flow data, which makes it into

$$v_2 = \alpha \sqrt{\frac{2g(p_1 - p_2)}{\gamma}}.$$

α depends on the ratio d/D between the size of the restriction and the internal pipe diameter. For nozzle type restrictions (Fig. 4.13 a), we have $\alpha = 1$ for $d/D = 0.46$, but the factor becomes 1.10 for $d/D = 0.71$, and 0.99 for $d/D = 0.32$. For orifice flowmeters (Fig. 4.13 b), α deviates widely from unity. For instance, we have $\alpha = 0.625$ for $d/D = 0.50$.

The pressure averaging tanks between output ports and pressure gages are there to equalize the input to the pressure gauges and keep their needles from vibrating in response to eddy formation and occasional pressure surges.

Variable-area flowmeter

For lower flow rates and smaller installations, such as laboratories, an eminently practical instrument is the variable-area flowmeter, often called Rotameter (Fig. 4.14). It consists of a beveled tube of transparent plastic, such as acrylic or polysulfone, with a spherical or cylindrical float inside. As liquids or gases stream from the tube's narrow lower end up through the annular passage between inner tube and float, they drag the float along. In a tapered tube, that passage gets wider the higher the float comes up, whilst the fluid's velocity around the float decreases. The float settles in the position of equilibrium between lift and weight.

Empirically made up scales for these instruments, usually engraved on the proper tube, come out surprisingly regular (Fig. 4.14). Minor aberrations from linearity are compensated for by making the taper slightly bulged.

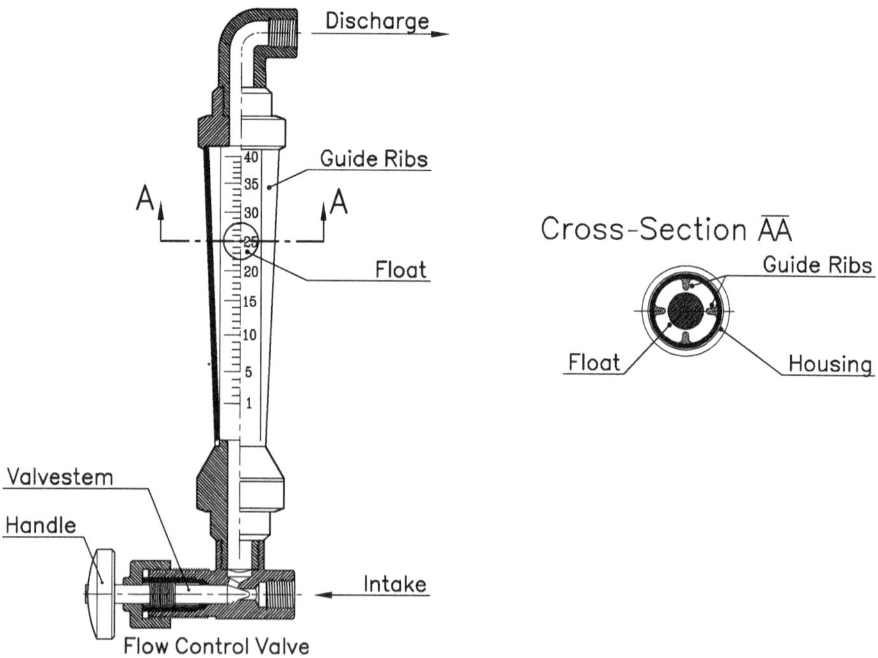

Fig. 4.14. Variable area flowmeter

Positive displacement meters

Where flow rates convert into monetary values, principally at the gas pump, positive-displacement meters are the rule. The term "positive displacement" is best envisioned by the simile of gear drive versus belt drive. The hands of a clock are linked through a 1 : 12 gear train to make sure that, for the lifetime of the clock, the small hand will point accurately at one of the numbered marks on the clock-face every time the big hand stands on 12. That's positive engagement. If the hands were belt driven rather than positively engaged, the slightest slips of the belt in each turn of the pulley would sum up to unacceptable levels. Imagine a 20 mm diameter pulley oversized by a mere $1/100$ millimeter, and you promptly get after a mere 2000 rotations an error of $2000 \times 0.01 \times \pi = 62.8$ mm, the equivalent of $62.8/20\pi \approx 1.0$ full rotation of the pulley. The catchword here is "error accumulation," absent in gear drives, regardless of their state of conservation, but inherently present in friction and belt drives, as precise as they might be.

In hydraulics, the classical piston pump is the prototype of a positive displacement machine, as opposed to centrifugal pumps, which allow for the passage of fluid even if the rotor has stalled.

Positive displacement meters encompass the liquid flow in its entirety, which makes linearity inherent to their principles of operation. Flowmeters of that type with 0.02% precision are not uncommon.

Their design resembles inversely operating pumps, with some exceptions: For instance, the principle of the venerable piston pump is not readily reversible. Gear pumps, such as the oil pump encased in the motor block of a car engine (Fig. 4.15), could in principle be made to turn by forcing oil through them, but still would perform poorly as flowmeters.

On the other hand, gear pumps with three- or four-lobe impellers are quite common as volumetric measuring instruments with typically $\pm 0.1\%$ accuracy. This because a three-lobe impeller, such as in Fig. 4.16, provides far bigger pockets than standard gear pumps to trap and discharge the fluid. The involute profile of their teeth makes them roll rather than slip on each other at the points of engagement, which allows for mutual matching to very small tolerances. Figure 4.17 pro-

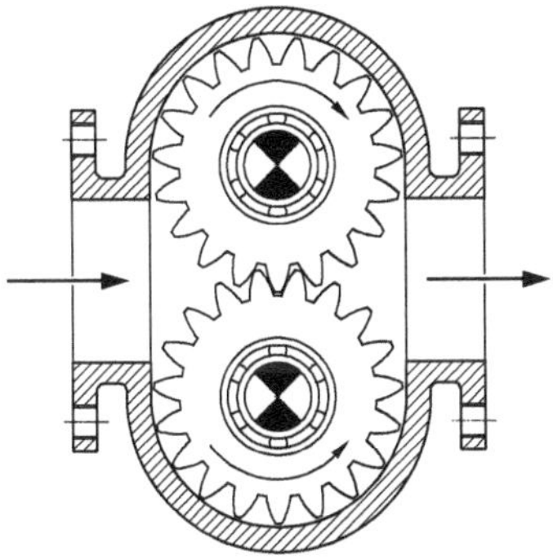

Fig. 4.15. Gear pump

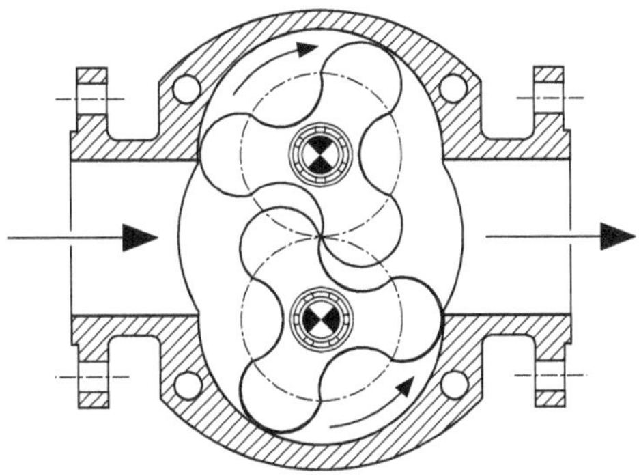

Fig. 4.16. Positive-displacement flowmeter

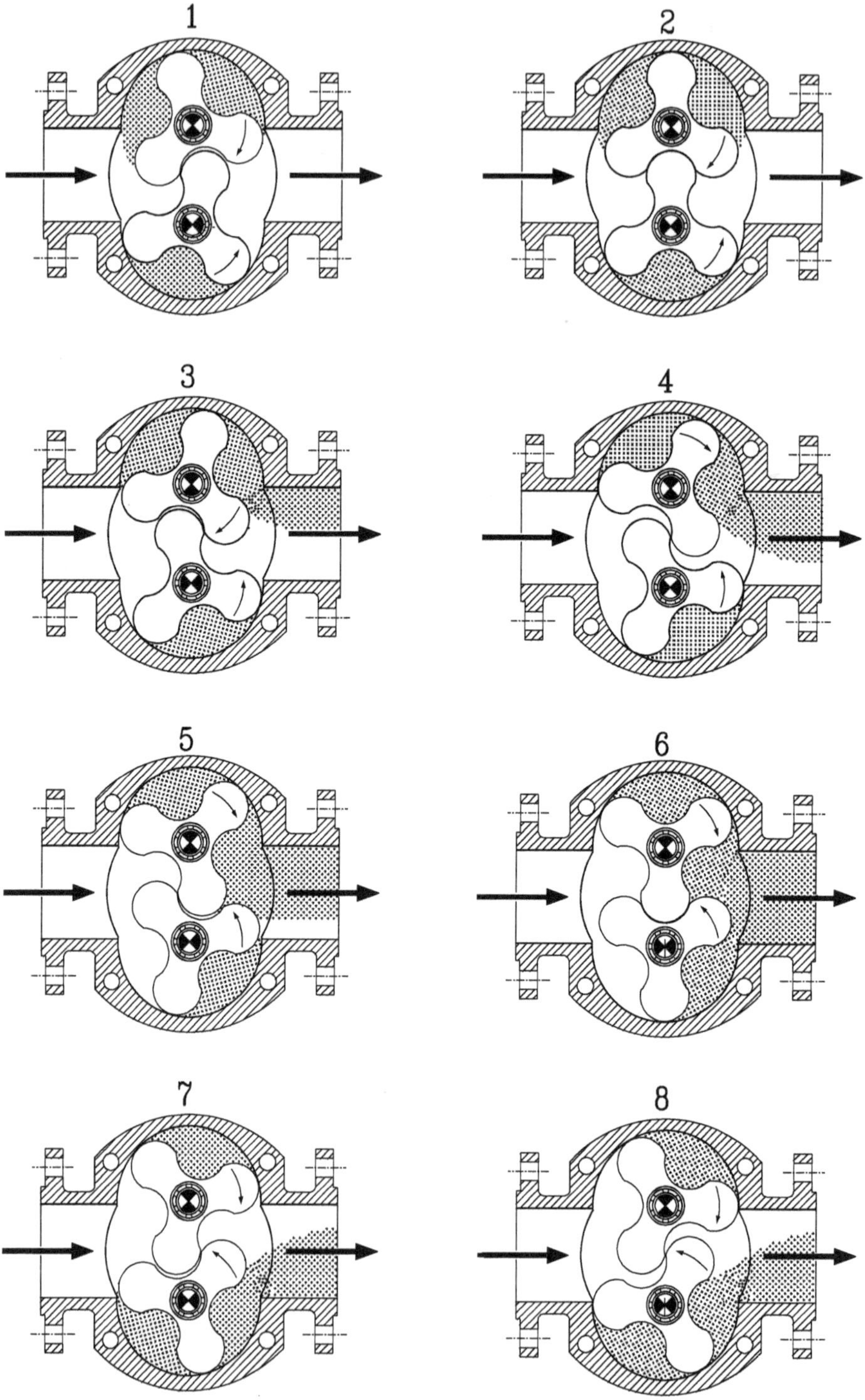

Fig. 4.17. Operating sequence of positive-displacement flowmeter

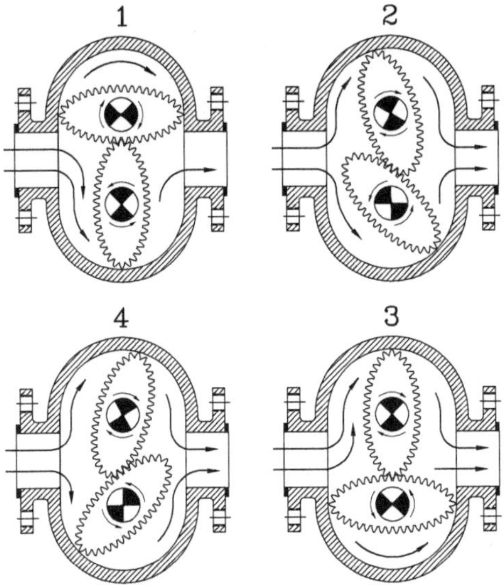

Fig. 4.18. Ellipsoidal-gear positive-displacement flowmeter

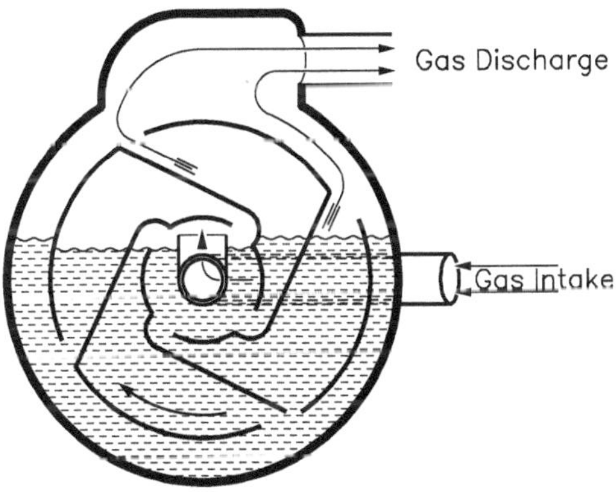

Fig. 4.19. Rotary fluid piston gas meter

gressively displays the operating sequence of a three-lobe positive-displacement meter.

For quiet operation, gears with helical twist are common. Their operating principle equals that of their spur gear counterparts, yet with less noise and vibrations.

Several other options for gear pump-derived flowmeters are available. An elegant solution (Fig. 4.18) uses a pair of ellipsoidal spur gears encased in a twin semicircular housing. Gears and the inner surface of the housing are precision-machined for mutual sealing as to prevent backflow at the point of tooth engagement. At stage 1 in Fig. 4.18, pressure from the incoming fluid acts on the left flank of the upper gear and makes it rotate clockwise, while the gear below is subjected to equal forward and backward turning pressure and does not produce torque.

At stage 2 in Fig. 4.18 a pocket of fluid is about to be trapped underneath the lower gear, while the upper gear begins to discharge. This process repeats itself at stages 3 and 4, yet with the functions of upper and lower gears inverted.

The demand for inherently leak-free meters led to the development of *wet-gas meters*, such as shown in Fig. 4.19. They are less critical to manufacture than their toothed counterparts, as fluids such as glycerin and light oils provide for friction-free sealing. They became the instruments of choice in the early days of domestic heating gas supply.

In Fig. 4.19, the four-chamber impeller, rotating around a hollow shaft that doubles as gas supply port, admits the gas into the chamber at left while its exit port is still below fluid level. Gas entering that chamber makes that the impeller rotates clockwise until the chamber's discharge port clears the liquid level and allows the accumulated gas to exhaust, while the following compartment is being charged. All that happens gradually and keeps the impeller in consistent, regular rotation.

However, rotor speed must be kept below the point where the fluid would eddy in circles along with the rotor. The resulting limits to the instrument's capacity and the need for periodic refills made that this meter type became one of the lesser used in our days.

Gasometers

Gasometers, the classical industrial gas storing units, allow for the measurement of the amounts of gas withdrawn, but do not provide flow rates directly. Telescoping bells extend the storage capacity of industrial gas tanks to several times the volume of the base vessel, while laboratory gasometers (Fig. 4.20) are of the single bell construction.

Gas pressure is given by the difference of the weight of the bell minus that of the counterweight, divided by the bell's cross-sectional area, $D^2\pi/4$. It can be set to the desired values by selecting the counterweight appropriately.

In installations where the consistency of gas pressure is critical, the buoyant force on the submerged part of the hull must be accounted for. It causes pressure to be highest when the gasometer is full, and decreasing in step with the hood's descent. This hydraulic lift can be counterbalanced by selecting the chain that

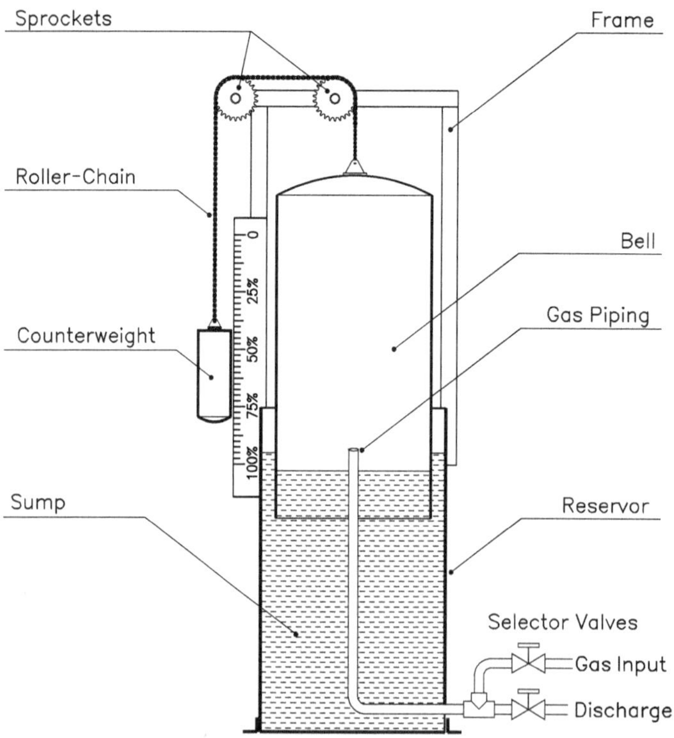

Fig. 4.20. Laboratory gasometer

holds the counterweight so that the weight of the descending portion of the chain equals the hydraulic lift of the bell's mantle.

The more of the length of the mantle gets submerged, the longer becomes the portion of the chain weighing down the bell. For instance, a 5/8 inch roller chain, which weighs 2.06 lb/ft, or approximately 3 kg/m, adds 3 kg to the weight of the bell for every meter of descend. A bell of $D = 500$ mm (0.500 m) diameter, made of, 0.0747 inch, or 1.90 mm, thick (#14 MSG) steel sheet, displaces per meter of height of the mantle the volume $0.500 \times \pi \times 0.0019 \approx 0.003\ \mathrm{m^3/m}$. Since the specific gravity of water is $1000\ \mathrm{kg/m^3}$, the resulting lift is $0.003 \times 1000 = 3\ \mathrm{kg/m}$, equaling (by lucky coincidence?) the weight per meter of the suspension chain.

For other dimensions, this coincidence is bound to happen if you make the diameter of the bell

$$D = \frac{\text{weight per meter of mantle}}{\pi \times \gamma \times \text{wall thickness}} \,.$$

γ is here the specific gravity of the fluid in the gasometer, $1000\ \mathrm{kg/m^3}$ for water, and $1261\ \mathrm{kg/m^3}$ for glycerin. Double checking this formula with our former results gets us

$$D = \frac{3\ \mathrm{kg/m}}{\pi \times 1000\ \mathrm{kg/m^3} \times 0.0019\ \mathrm{kg/m^2}} \approx 0.500\ \mathrm{m}$$

as one of the possible combinations of the size of the hood and the type of chain capable to stabilize the gas pressure along the entire descend of the bell. The gas holding capacity of the gasometer per meter of height of the bell is

$$V = 0.500^2 \times \pi/4 = 0.196\ \mathrm{m^3} \approx 200 \text{ liter per meter of descend of the bell}\,.$$

Let's take this example of a simple device, such as the gasometer, as a reminder that direct measurement is usually the most reliable and the most rewarding approach, because each of the components in a measuring mechanism introduces its own set of uncertainties.

5 Force, mass, weight, and torque

Physical forces make part of our daily live, and yet, the concept of force in itself remains an abstraction. Forces cannot be seen or touched, but fit by definition consistently into the laws of physics due to a set of characteristics that we, humans, have assigned to them.

Forces are conceived as capable of the deformation of solids, a feature that allows for their physical measurement. Herein, the bridge between the philosophical concept of force and the real world phenomenon of deformation finds its expression in Hook's law, which states the proportionality between an applied force and the elastic deformation it causes, or more specifically, the linear relation between stress and strain.

Stress is the force per unit area, i.e., force divided by the area it is acting upon. Hence, stress is expressed in units which intrinsically divide force by area, such as the *pound per square inch* (psi), *kilogram per square centimeter* (kgf/cm^2), or in the International System, the *newton per square meter*, also called *pascal* (N/m^2 or Pa).

Strain, on the other hand, refers to deformation. It is the ratio of compression or elongation to the cross sectional dimension of the object under stress.

Hook's law, stating the linear relation between stress and strain, can be understood intuitively: The stronger you pull on a rubber band, the longer it gets, period. As a formula, Hook's law reads

$$\text{strain} = \frac{\Delta L}{L} = \frac{1}{E} \times \text{stress}\,,$$

where L stands for the length of the specimen under test, and ΔL for its elongation (or compression); $1/E$ is the proportionality constant, and its reciprocal, E, is called the *elastic modulus*. For commercial steel, E equals 30×10^6 psi, or 21,000 kilogram per square millimeter (kg/mm^2).

For instance, $\Delta L/L = 0.001$ stands for elongation or compression of a 1 meter long bar by 1 mm, or 0.1%. For the case of a steel bar, the required stress follows from Hook's law as $S = 0.001 \times 30 \times 10^6 = 30{,}000$ psi, which can be visualized as a load of 30,000 pounds stretching a bar of 1 square inch cross-sectional area and 100 inch of length by $0.001 \times 100 = 0.1$ inches.

Note that the use of the inverse of E as the proportionality constant in Hook's law is merely a question of convenience. After all, *30 million* is easier to remember than 0.0333×10^{-6}.

Permanent deformation (material flow) of steel sets in at about 40,000 psi, which represents the *elastic limits*, the boundaries of Hook's law. Stress beyond the elastic limits causes the material irreversible changes, and the stress/strain relation becomes unpredictable.

Recapitulating, Hook's law brings the problem of force measurement down to the simple task of checking length differences. Tensile stress testing machines submit the sample to successively increasing loads and show the related elongation on a readout dial, numerical display, or as a graph.

Strain gages and the Wheatstone bridge

Strain gages are sensors which derive the change of length of a test bar from changes in its electric resistance. Most load cells employ resistive strain elements for being rugged, simple to install, and inexpensive.

The resistance of a piece of wire is – not surprisingly – proportional to its length L and inversely proportional to its cross-sectional area A. This is another of the intuitively comprehensible laws of physics: A short length of thick wire conducts better than a long, slender one. To express that in mathematical terms, we define the resistance of 1 meter of wire of 1 mm^2 cross-sectional area as the *resistivity* ρ of the wire's material, which allows for postulating a conductor's electrical resistance as

$$R = \rho L/A\,.$$

At 20 °C, the resistivity of steel wire is $\rho = 0.12\,\Omega/\text{m} \times \text{mm}^2$, of aluminum wire 0.0283, and of electrolytic copper wire 0.0172. Resistance wires (as used in hair dryers, ranges and room heaters) rate much higher. The specific resistance of 20/80% chromium–nickel alloy is 1.1, of Cekas and Kanthal filaments 1.4 and 1.45, respectively, and for the temperature-stabilized alloy Constantan 0.4965. Note that these values for ρ are for L expressed in meter, and for A in square millimeter.

A wire being stretched gains length at the expense of its diameter; that is, it gets longer and slimmer. Since the amount of material contained in a piece of wire and herewith the wire's volume cannot change, stretching a wire lengthwise by, say, 1%, causes it to constrict likewise by 1%, to 0.99 of its original cross-sectional area. This doesn't change the wire's volume, because $1.01 \times 0.99 \approx 1$. However, the resistance of the wire, $R = \rho L/A$, rises by a factor of $1.01/0.99 \approx 1.02$, a 2% gain. In other words, the degree of elongation of a wire shows up double in the wire's electrical resistance. That made resistive strain sensors the most widely used elements in force-measuring equipment, such as strain gauges and load cells.

Electrical resistance is best measured with the Wheatstone bridge (Fig. 5.1), a circuit for the comparison of an unknown resistor R_x with a resistor of known resistance, R_3. The bridge is "balanced" when the flow of electric current through the galvanometer is zero. That's obviously the case if we use identical resistors throughout, namely $R_1 = R_2 = R_3 = R_x$. Otherwise, the equation $R_1/R_2 = R_x/R_3$ must be satisfied.

Initially, R_2 is adjusted for zero meter reading with R_x in "stress-free condition". Subsequent changes of R_x show by the deflection of the meter needle. The

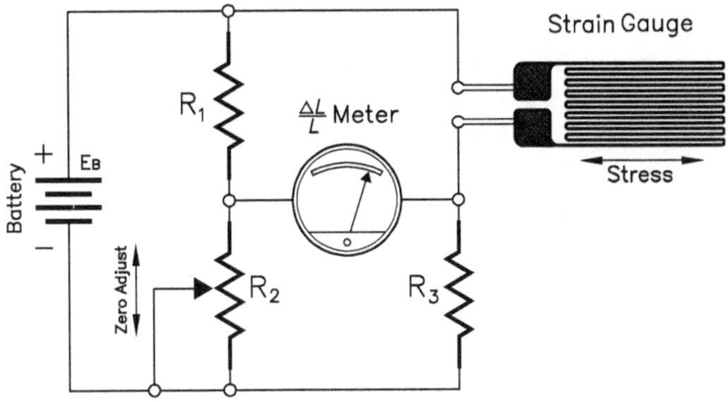

Fig. 5.1. Bridge circuit with strain gage

degree of change is found by replacing R_3 with appropriately selected resistors until zero meter reading is regained. With that new value for R_3, we use the relation $R_1/R_2 = R_x/R_3$ to get

$$R_x = R_3(R_1/R_2)\,.$$

A wire-wound potentiometer will work fine for R_3, but a resistance box, consisting of an array of precision resistors, which you plug together in series or in parallel, is more accurate.

In Fig. 5.1, R_x, the resistance of the strain gage, is secondary in importance to its change of resistance, ΔR_x, when stress is applied. Rather than derive that change from the difference of R_3 before and after application of the load, direct measurement of this difference yields results of much higher accuracy. To this end, R_3 is made up of a fixed resistor in series with a resistance box. If the latter is shorted during zero adjustment of the meter, it will later show the variation, ΔR_3, rather than the total of R_3. From this, the variation of R_x can be figured as

$$\Delta R_x = \Delta R_3(R_1/R_2)\,.$$

For instance, measuring the change of R_x from 100 Ω to 101 Ω with a 0.1% accuracy meter would make errors of up to $100 \times 0.001 = \pm 0.1$ Ω possible. But that's 10% of the 1 Ω change of R_x, while direct measurement of the difference would yield $10/0.1 = 100$ times better accuracy.

For fast and fairly accurate results, Fig. 5.2 is an immensely practical variety of

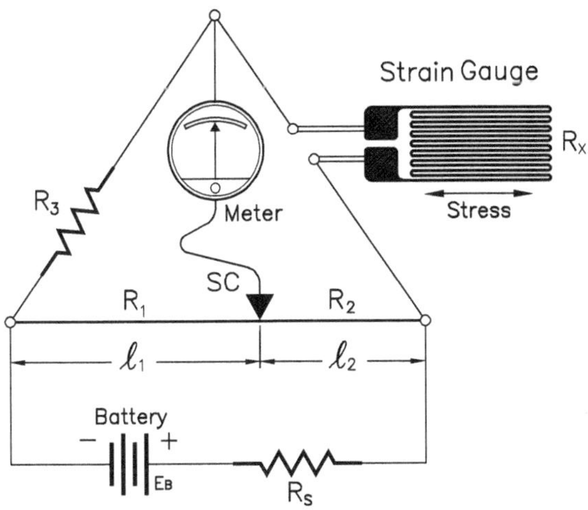

Fig. 5.2. Slide wire resistance bridge

the Wheatstone bridge circuit. It combines R_1 and R_2 into a single resistance wire of length ℓ. This wire is tapped by a sliding contact or cursor (SC in Fig. 5.2), which position determines the values of R_1 and R_2 as proportional to the length of wire left and right from the cursor's point of contact. At center, we have the equivalent of $R_1 = R_2$ and $R_x = R_3$. For the cursor positioned at ℓ_1, the ratio $R_1/R_2 = \ell_1/(\ell - \ell_1) = R_3/R_x$ allows for figuring the unknown resistance as

$$R_x = R_3(\ell - \ell_1)/\ell_1 \,.$$

This circuit needs a current-limiting resistor R_S in series with the power supply, because the combined resistance of R_1 and R_2 alone would be too low to keep the resistance wire from heating up and draining the power supply.

The secret of the Wheatstone bridge is its independence of the supply voltage E. A used battery in reasonably good shape will work as well as a brand new one. That's a virtue few other instruments can match. But in cases where manual balancing would be too cumbersome a procedure and direct readouts are required, this feature must be given off and a well regulated power supply becomes a must. Though the accuracy of Zener diodes as voltage regulators was not always enough, present days' precision voltage regulator chips fit most applications.

The gauge resistance R_x in Fig. 5.1 can be read from the meter deflection if we leave the bridge zeroed in and draw the dial to the appropriate scale. Under these conditions, the voltage on the left meter contact becomes $E_1 = E \times R_2/(R_1 + R_2)$ and – if the meter is of the high impedance type (from 10 MΩ up) – remains virtually immune to changes of R_x. The voltage on the right meter contact, $E_2 = E \times R_3/(R_x + R_3)$, makes the meter reading of $E_1 - E_2$ indicative for the sensor resistance.

A formula for $E_1 - E_2$ is expressed best with the introduction of an auxiliary variable:

$$\varepsilon = \frac{E_1 - E_2}{E} \quad \text{or} \quad \varepsilon = \frac{R_2}{R_1 + R_2} - \frac{R_3}{R_3 + R_x} \,.$$

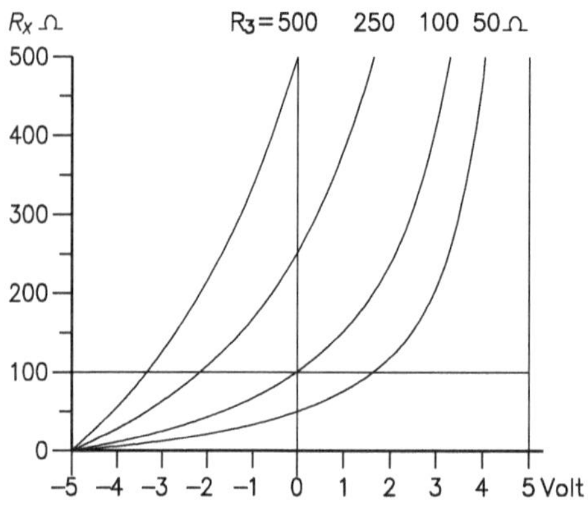

Fig. 5.3. Meter readings for Wheatstone bridge circuit

Summing up those fractions and extracting R_x yields

$$\frac{R_x}{R_3} = \frac{R_1 + R_2}{R_2(1 - \varepsilon) - \varepsilon R_1} - 1 \,.$$

The curves in Fig. 5.3 and the scale in Fig. 5.4 are plotted from this equation for identical values, of 100 Ω, for R_1 and R_2, and a 10 V power supply. This locks the left contact of the meter at half the supply voltage (5 V in this case) leaving equal space for negative and positive outputs alike.

Fig. 5.4. Scale for the meter in Fig. 5.3

As long as R_x is the ohmic resistance of a strain gage, only the positive arms of the curves must be considered, but with other types of sensors, such as temperature-sensitive resistance bulbs, negative readouts are as likely as positive ones.

The formula's inverse proportionality between ε and the supply voltage E makes a well-regulated power supply paramount. Precision voltage regulator chips are widely available.

Nonlinear scales are common in electric instruments, as for instance in hot wire volt- and wattmeters, and the popular analog multimeters. The latter show resistance on a left-side compressed scale, with usually 0.2 ohm per division on its right end, 1 ohm at center, and 500 ohm in the scale's left-hand (kilo-ohm) region. Like in the case of the slide rule (may he rest in peace), such scales must be read with an eye on the varying significance of their divisions. The scale, based on $R_3 = 100\,\Omega$ in Fig. 5.4 comprises 10 ohm in its first division, 20 ohm around the middle, and 100 ohm for right-side divisions.

If the sensing resistor R_x were remotely positioned, temperature differences between the latter and the rest of the bridge circuit would show up in the meter reading, because any material's electrical resistance changes with temperature. The resistance of steel wire increases by 0.50% per degree Celsius (1.8 °F), of Nichrome by 0.04%, of Manganin by 0.0001%, and of Constantan by 0.0028%. Although the proper selection of the strain gauge's resistive material can minimize the negative effects of temperature fluctuations, better still is to etch the entire bridge circuit (safe for the power source and the meter) onto the same printed circuit board to keep them at identical temperatures.

R_x in Figs. 5.1 and 5.2 is a bonded metal strain gauge, consisting of a conducting grid, etched out of a substrate of resistive material on a flexible, insulating carrier, such as printed circuit board, plastic sheet or ribbon. Since the parallel sections of the grid are serial connected, their resistance sums up to that of a long flat wire. R_3 is usually a "dummy load", resembling R_x, yet positioned at 90° to the latter. While the sensor resistor stretches longitudinally under load, the dummy resistor's transversal deflection does not significally change its proper resistance.

Notwithstanding that nonlinear outputs introduce an extra variable in process control systems, *digital signal processing* (DSP) permits virtually unlimited ways to manipulate data extracted from analog inputs. Combined with microprocessors, these chips operate much like computers, using algorithms applied to digital input data signals in order to obtain the output in the desired format. They can be programmed to "stretch" or reshape nonlinear functions. The respective algorithms are stored in a permanent programmable memory (EPROM) or, more recently, flash memories, and if programmable signal processor chips are used, their internal memory.

Measuring deflection

Figure 5.5 shows a strain gauge set up for measurements of the deflection of an American Standard 8 by 4 inch double T beam with a length of $\ell = 5$ meters between support points. Load is generated by a 3 inch inner-diameter hydraulic jack,

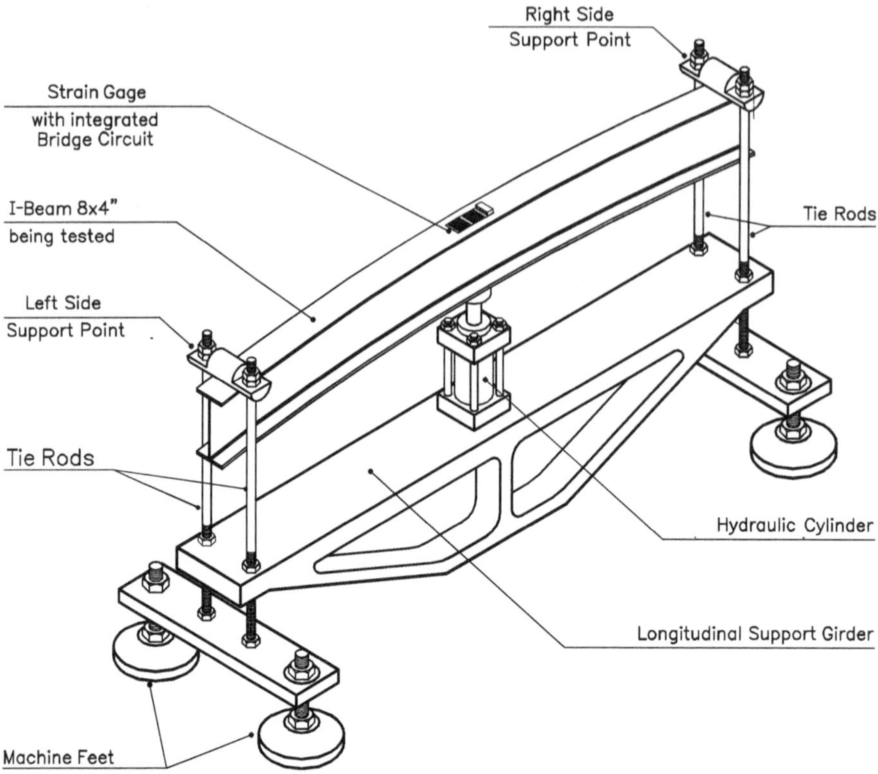

Fig. 5.5. Test rack for measurements of stress and deflection in beams and girders

which flexes the beam upwards and thus stretches its upper flange while the lower flange gets compressed. The centerline of the beam remains unaffected. With a hydraulic cylinder of, say, 3 inch inner diameter and 1000 psi oil pressure, the load F on the beam becomes

$$F = (\pi/4) \times 3^2 \times 1000 = 7069\ \ell\text{b} \text{ or } 7069 \times 0.4536 = 3206\ \text{kgf}\,.$$

Reaction forces on the beam's support points amount to half of the load each, that is, $F/2 = 3206/2 = 1603$ kg.

For $\ell = 5$ meter between support points, the *bending moment* M at the beam's center becomes

$$M = (F/2) \times (\ell/2) = 1603 \times 5/2 = 4007.5\ \text{m} \times \text{kg} = 400750\ \text{cm} \times \text{kg}\,.$$

Stress, σ, follows by dividing this bending moment, M, by the beam's *section-modulus* Z, which is listed in reference books for the 8 by 4 inch standard I-beam as 14.2 inch^3. With 2.54 cm to the inch, that converts into $14.2 \times 2.54^3 = 232.7\ \text{cm}^3$, and stress becomes

$$\sigma = \frac{M}{Z} = \frac{400750}{232.7} = 1722\ \text{kg/cm}^2\,.$$

For a safety factor of 4.6, the *ultimate tensile strength* of the steel the beam under test must be made from becomes $\sigma_o = 4.6 \times 1722 = 7920\ \mathrm{kg/cm^2}$.

Herein, the safety factor of 4.6 is the product of four material characteristics, *a, b, c,* and *d*. *a* is the ratio of yield strength to ultimate tensile strength. *b* equals 1 for stable loads, 2 for variable loads, and 3 for cycling loads. *c* is 1 for slowly applied loads, 2 for tossed-on loads, and 3 for impacting loads. And finally, *d* takes care of contingencies, such as accidental overloads and the effects of material imperfections.

In our example of a beam of commercial steel with typically $3500\ \mathrm{kg/cm^2}$ of yield strength, the constant *a* becomes $8000/3500 \approx 2.3$. Since the force of a hydraulic cylinder rises slowly and consistently, we can set $b = 2$. Likewise, gradual application of the load makes $c = 1$, and, with accidental overloads unlikely, we set $d = 1$ too. Altogether, we get for the safety factor $F_S = 2.3 \times 2 \times 1 = 4.6$.

In the present example, strain gauge readings check fairly well with the computed stress values. Stress readings above the mathematically predicted would indicate a beam of lower-strength steel.

The beam's deflection, *d*, at center, can be predicted with the formula

$$d = F\ell^3/48EI\,,$$

where *E*, the elastic modulus for commercial steel, is 30×10^6 psi $= 2{,}100{,}000\ \mathrm{kgf/cm^2}$; *I*, the beam's *area moment of inertia,* can be found in reference books as $59.9\ \mathrm{in^4}$ that is, $59.9 \times 2.54^4 = 2493\ \mathrm{cm^4}$. With $F = 3206$ kgf and $\ell = 500$ cm, the formula gets us the rise at the beam's midpoint as 16 mm.

Deflections of this amplitude could be measured directly and used to figure stress. But field checks on elements of huge structures, such as bridges and towers, would be formidable jobs, which a cemented-on strain gage solves effortlessly and accurately.

Crystal strain gauges

An alternative to strain gauges are piezoelectric force sensors. They compare to their resistive counterparts much like vacuum tubes compare to transistors, insofar as a tube's basic function is voltage amplification, while transistors are current amplifiers. Likewise, piezoelectric sensors generate *voltage*, while the output of resistance bridges is *electric current.*

The piezoelectric effect, discovered by Pierre and Jacques Curie in 1880 in certain anisotropic crystals, is understood as a crystal's electric polarization when stressed along one of its electrical axes. Application of typically $500\ \mathrm{kgf/cm^2}$ (7000 psi) of pressure may generate 5000–15,000 V/cm. Conversely, application of an electric field makes a piezoelectric crystal undergo mechanical deformation.

Although quartz and tourmaline are the classical piezoelectric materials, they are joined by barium titanate, lead titanate–zirconate ceramics, and polyvinylidene fluoride. Rochelle salt, which we remember from pickups of long outdated record players, is 10,000 times more effective, provided we keep it dry and at temperatures between -18 and $+24$ °C. All these materials have in common that they react to forced distortion along their electrical axes by accumulating positive and negative electric charges on opposite surfaces.

Unlike resistance bridge sensors, which need a power supply, piezoelectric sensors generate their proper voltage. If you were to connect a neon lamp between opposite surfaces of a piece of Rochelle salt, the lamp would flash every time you hit the crystal with a lightweight hammer. Likewise, gas lighters ignite a stream of flammable gas upon impact of a spring loaded bolt on a crystal set in the neck of the lighter.

The inverse effect – the deformation of a crystal by an applied voltage – can be observed by connecting the output of an audio generator to a large piece of Rochelle salt: The crystal functions like a speaker – sort of, that is. The audio can be heard distinctly, if not loudly. Nonetheless, the crystal microphone excels by its fidelity in the reproduction of the harmonic and sometimes not so harmonic sound commonly known as music.

Piezoelectric crystals' dimensional changes in response to an applied voltage made them the classical frequency stabilizing elements in radio transmitters and ultrasound apparatus. Best known applications in the field of consumer products are quartz watches. Quartz crystals are also at the core of ultrasonic machinery in a multitude of applications, such as nondestructive material testing, emulsifying liquids that otherwise wouldn't mix, welding of plastics, and, in combination with the action of abrasive powders, for metal drilling.

From levers to scales and balances

"Give me a lever long enough and a place to stand, and I will move the Earth!" exulted the Grecian philosopher, scientist, and inventor Archimedes (287–212 BC) in a letter to his sponsor, King Hieron II of Syracuse. Regardless that Archimedes spoke overwhelmed by a sudden insight into the theoretically infinite load carrying

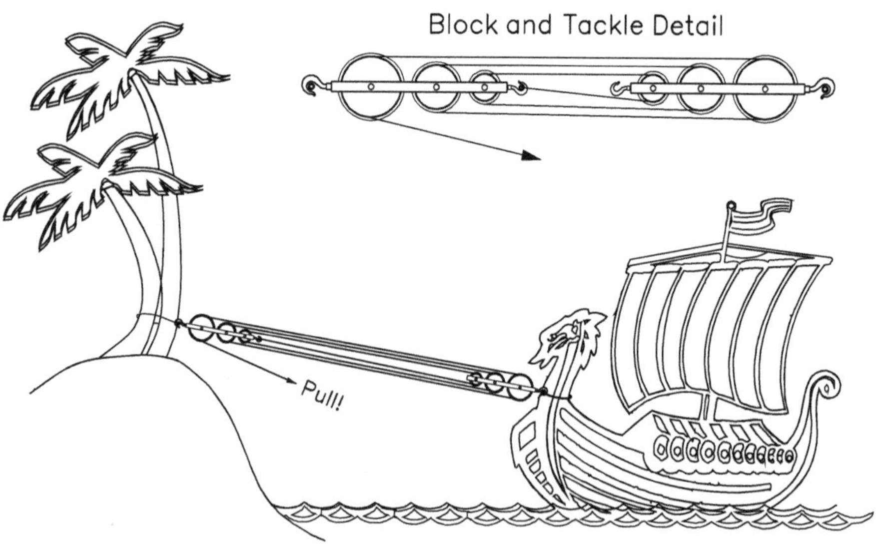

Fig. 5.6. Archimedes ways of homing a ship

potential of levers, the king took him by his word and demanded tangible proof. Short of lifting the Earth, Archimedes agreed to tow a cargo ship, loaded with passengers and goods, out of its moorings – a task that so far had afforded the combined efforts of hundreds of slaves. But Archimedes, the story goes, "holding the head of the pulley in his hand and drawing the cords by degrees, drew the ship in a straight line, as smoothly and evenly as if she had been in the sea".

Block and tackles (Fig. 5.6, top) of n movable pulleys multiply the pull applied on the tie rope by 2^n, so that a device with, say, a total of three pairs of pulleys, powered by one man, could in theory emulate the power of $2^6 = 64$ dock workers. And if Archimedes hauled the cord with approximately 50 kgf (110 lb) of force, the resulting traction, of $50 \times 64 = 3200$ kgf minus frictional losses, might well have sufficed to move a ship. But how far would the ship have traveled in the process? To tow it over 100 meters (about 300 feet), Archimedes would have had to pass $100 \times 64 = 6400$ m $= 6.4$ kilometers (4 miles) of cord between his hands with unwavering force of 110 lb.

So far, so good. That the chronicles miss to report the philosopher's grazed hands and, ultimately, his heart attack from exhaustion, makes it likely that the benign king gave himself satisfied with seeing his vessel creep by as little as two foot or so. But even then, Archimedes would have pulled 128 feet of tightly tensioned rope between his hands. Let's hope that his days had more effective blister ointments than ours.

René Descartes (1596–1650), embodied by his maxim "Cogito – ergo sum!" (*If I wouldn't be around, I couldn't doubt*), would have doubted the royal scribe, which quill might have been guided by his loyalty to the court, along with the specter of a noose on his neck or the proverbial *sword of Damocles* dangling overhead.

For Archimedes, that sword was of a Roman legionnaire after the philosopher's devices, such as heavy catapults and mighty cranes that lifted boats of the Roman fleet straight out of the water, had helped the city of Syracuse to withstand the Roman onslaught for many months. But Archimedes, engrossed in scratching geometrical figures in a bed of sand, even failed to realize when the enemy had overrun the city's barricades and harshly reminded an intruding Roman legionnaire to mind the sand-drawings, which rendered him the death penalty from the soldier's hands.

But here again: Could a common soldier have dared to decide whether a worldwide celebrity should live or die? Or had he been sent by the Roman commander Marcellus to do away with that pest of a tinkerer who had given the proud and far superior Roman forces such a hard time in the conquest of the city? We shall never know, but let's rejoice that Archimedes discoveries were not erased from the books of history, regardless that he fought and died on the wrong side of the fence.

Scales and weights

Findings of prehistoric scales might be interpreted in the sense that humans instinctively understood the laws of levers long before Archimedes, but it was he who cast them into equations. A seesaw (Fig. 5.7) loaded with a child of weight W_2 on one side and an adult of weight W_1 on the other, can be balanced if the

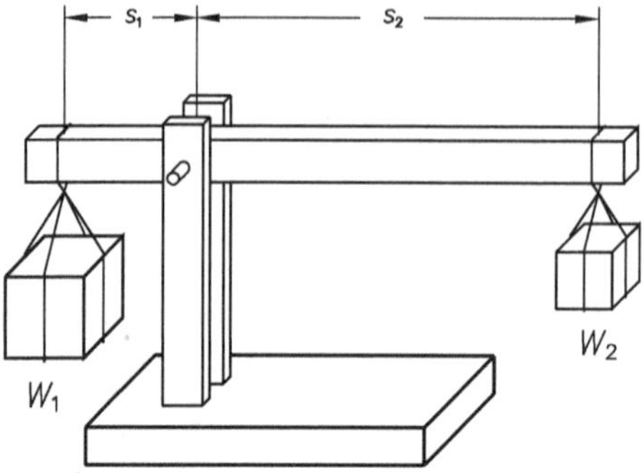

Fig. 5.7. Archimedes' law of the lever

child gets seated farther from the fulcrum than the (heavier) grown-up. We can write that as the equation $W_1/W_2 = s_2/s_1$, showing the length of the right and left levers inversely proportional to the respective loads. Cross multiplication leads to the equation of Archimedes' law of the lever,

$$W_1 s_1 = W_2 s_2 ,$$

which became the working principle of scales and balances ever since.

It also led to the concept of *torque* T for the product of force times s, the distance to the axis of rotation. For the special case of weighing, force is given by the weight of the load and, respectively, the weights, so that the formula for torque becomes $T = Ws$, which, with the law of levers, gets us $T_1 = T_2$: Left and right side torques on a balanced seesaw are equal. This checks with the law of conservation of energy insofar as tilting the beam by an angle α from the horizontal affords the work $W_1 s_1 \sin\alpha$ on the left side and $W_2 s_2 \sin\alpha$ on the right side of the lever. Since α is the same on both sides, $\sin\alpha$ cancels out and, here again, the work-derived equation blends into the static formula $T_1 = T_2$. Rewritten into $T_1 - T_2 = 0$, this becomes a special case of the general law of statics, $\Sigma T = 0$.

Seesaw-like devices cannot be used as such for beam scales because even the slightest displacements of the weights or the load would change the outcome of the weighing. Hanging pan scales (Fig. 5.8) are free of such shortcomings because

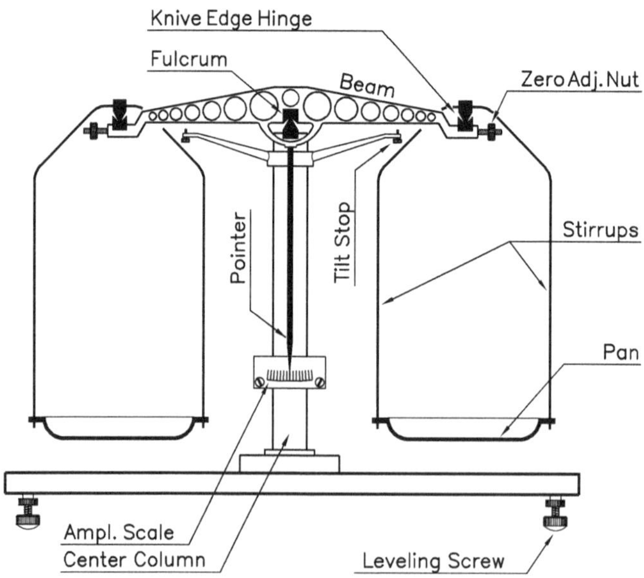

Fig. 5.8. Hanging pan scale

the load and the weights, respectively, act always on the pans' suspension points, no matter where the loads and weights rest on the pans.

Beam scales and equal-arms scales find the mass of an object by comparison with calibrated standard masses, such as a "set of weights", usually including weights of 1, 2 × 2, 5, 2 × 10, 20, 50, 2 × 100, 200, and 500 gram.

Remnants of beam scales have been found in Egyptian burial sites from 5000 BC, some with allegedly 1% of accuracy – quite an achievement for a device made from a wooden beam with three holes and strings for pivot and pan suspensions.

A different concept, the *sliding weight scale*, goes back to 1400 BC in Egypt, but likewise to the Etruscans, who populated the Apennine peninsula previous to the Romans. In a different field, Etruscans are known as the earliest quantity producers of iron. Also in antiquity happened the invention of the *unequal-arms scale*, which balances load and weights by sliding the fulcrum. Spring scales show up no sooner than toward the end of the 17th century.

Scales in ancient Roman had their center pivot and points of suspension marked by holes drilled through the beam. Although some finds of Roman scales have a knife-edge support-bearing at the fulcrum, only in the 18th century became the knife edge support, as in Fig. 5.9, the rule.

Since the spacing between pinholes cannot be changed once they are drilled, Egyptians rather rounded the ends of the beam and slung the strings over them. The scale could subsequently be fine-tuned by carefully filing the ends of the beam to the appropriate length.

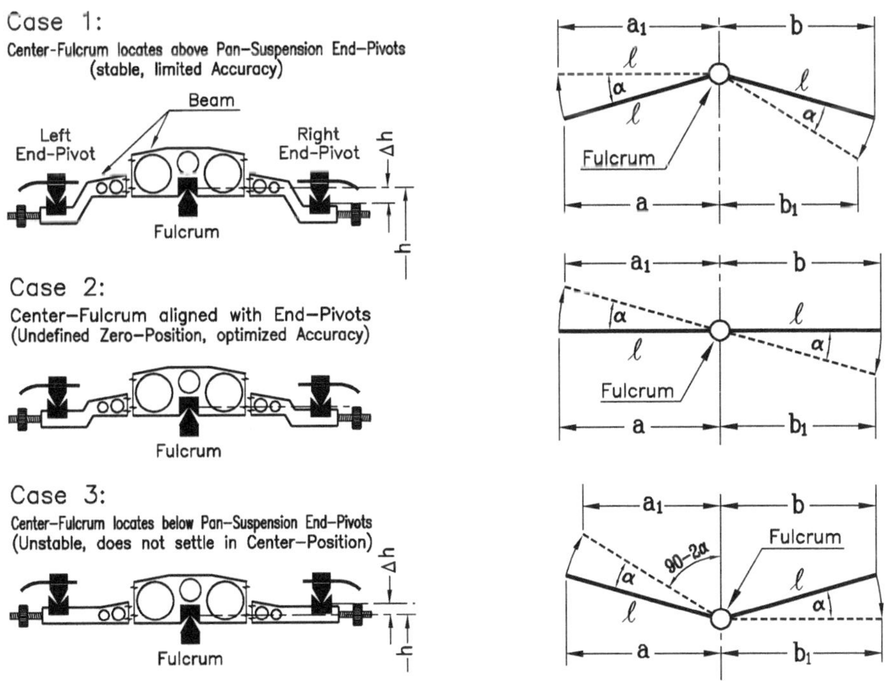

Fig. 5.9. Stability of beam scales

A scale's sensitivity depends on the vertical positioning of the beam's center pivot and the pan supports. In Fig. 5.9, three mutual placements of those cardinal points are shown: case 1, fulcrum above pan supports; case 2, fulcrum in line with pan supports; case 3, fulcrum below pan supports. The geometry of each configuration is shown on the right with thick black lines for the beam in balanced condition, and in thinner dashed lines if tilted by an angle α. Since gravitational forces act exclusively in the vertical, the effective (torque-generating) force at the ends of a beam of length 2ℓ is proportional to $\ell \cos \alpha$.

As long as fulcrum and swivel points are aligned (case 2), the values for α are identical on the left and on the right for whatever position of the beam, and so is $\ell \cos \alpha$. Lacking a preferential position, the beam swings freely without a predetermined point to level off.

With the pan supports above the fulcrum (case 3), tilting the beam by α makes the effective length of the right arm *increase* from $\ell \cos \alpha$ to ℓ, while that of the other arm *decreases* from $\ell \cos \alpha$ to $\ell \cos(2\alpha)$, hereby adding to the disparity of the arms' effective lengths. Such a scale wouldn't settle at midpoint but end up locked into either one of the extreme positions.

Placing the fulcrum above the pivot points (case 1) favors the midpoint position. For instance, tilting the beam clockwise by α makes the left arm's effective length ℓ, and that of the right arm $\ell \cos(2\alpha)$. With the cosinus function always <1 and declining with rising α, we have

$$\ell \cos(2\alpha) < \ell \cos \alpha ,$$

which makes the effective length of the beam at right less than that on the left and causes it to swing back. Likewise, it swings back toward center if the initial tilt happened counterclockwise.

With equal loads on both pans, the scale settles at midpoint after a few oscillations, which makes this the design of choice. Since sensitivity depends on the relative positioning of fulcrum and pan support points, such scales can be designed for a predetermined degree of sensitivity. However, what we gain in sensitivity is lost in stability and vice versa. That's why weighing on a chemist's balance takes so much longer than on a kitchen scale.

Perfect centering of the pivot is determinant for the performance of a scale. The knurled nuts, marked "zero adjust counterweights" in Fig. 5.8, are meant to compensate minor differences in the mass of the semi-arms of the beam, the attached pans, and the stirrups; but they cannot be used to make up for an out-of-center pivot.

If, for instance, the center were 0.1% off to the right, the right side semi-arm would measure 99.9% of its nominal length, and the left side semi-arm 100.1% To correct that for a load of, say, 100 gram on the left pan, $100(100.1/99.9) = 100.2$ gram would be needed on the right pan – an addition of 0.2 gram. In contraposition a load of only 50 gram would need only a 0.1 gram correction. In other words, the amount of correction would be load dependent. That's why the only way to right the wrong is by resetting the beam's knife-edge bearings.

If the worst comes to the worst, we may ignore the problem and weigh as al-

Table 5.1. Estimation of equilibrium position

Amplitude left	Average	Amplitude right	Average
−10			
		7	
−9			
	−45/6 = −7.5	6	25/5 = 5
−8			
		5	
−7			
		4	
−6			
		3	
−5			

ways, but follow up with a second weighing where we leave the weights in place but replace the load by weights from a spare set. When the scale settles again, those weights equal the true weight of the load, regardless of the beam's symmetry.

Modern scales have knife-edge hinges of synthetic sapphire and a long pointer made to display the amplitude of the beam's oscillations on a dial. Its scale will usually show 0.1 milligram (1 mg = 0.001 g) per division. For instance, if the pointer comes to rest at 3 divisions left of center, an additional 0.3 mg should be put on the left pan. That way, 0.1 mg precision can be achieved without physically using such tiny weights.

As not to wait through the intrinsically long settling period of precision balances, operators often deduce the weight of the load from the position of the midpoint between left and right side deflections. Since we deal with a *damped* oscillation, which loses energy with every stroke, an uneven number of semi-swings must be used for averaging, as demonstrated in Table 5.1 for a total of 11 semi-swings. Damping is assumed as one unit of amplitude per full swing.

The expected position of equilibrium is the average of −7.5 and +5.0, that is: $(-7.5 + 5)/2 = -2.5/2 = -1.25$, meaning that we have 0.125 mg more weight on the right pan than on the left.

For highest accuracy, atmospheric buoyant forces on weights and load must be considered. For instance, if we use a set of brass weights, of density 8.6, to weigh a piece of aluminum, of density 2.7 g/cm^3, we get unequal lift on left and right side. The volume of 1 g of brass is $1/8.6 = 0.116\ cm^3$, and the volume of 1 g of aluminum $1/2.7 = 0.370\ cm^3$. According to Archimedes' law of buoyancy, the lift of a body equals the weight of the surrounding medium – air in our case – displaced by the body. With 1.20 mg for the weight of 1 cm^3 of air at 20 °C, buoyant force on the brass weights becomes 0.116×1.20 mg, and on the aluminum load 0.370×1.20 mg. The difference $(0.370 - 0.116) \times 1.20 = 0.30$ mg, must be added per gram of load in order to compensate for the latter's greater lift. If the apparent weight of the aluminum bar was 100 g, its true weight will be $100 + (100 \times 0.00030) = 100.030$ g.

The *equal-arms scale* (Fig. 5.10) we remember as one of grandmother's kitchen gadgets, was designed in 1669 by Gilles Persone de Roberval, a member and professor at the French Academy of Science. It employs a lever mechanism shaped after a parallelogram, which maintains the pans horizontal and accessible from above. The mechanism keeps the pan support mounts vertical and the pans themselves horizontal, emulating the role of gravitation in the hanging pan scale. Therefore, neither the location of the weights nor that of the load has any bearing

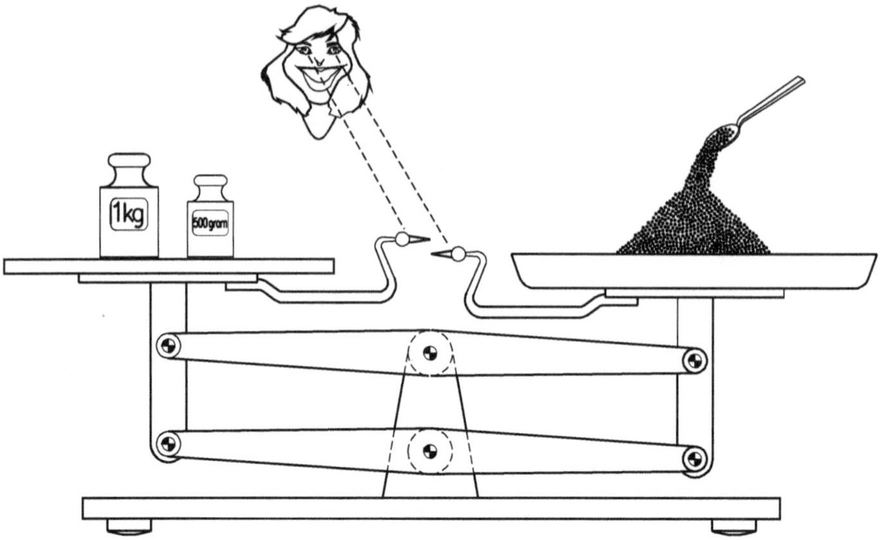

Fig. 5.10. Classical Roberval kitchen scale

on the results of the weighing. For best precision, the swivel pins shown in Fig. 5.10 are replaced by knife-edge supports.

Though equal-arms scales still survived for many years as traditional kitchen outfit, dial scales (Fig. 5.11) edged in during the mid 20th century. Grocery stores jumped at the chance to get rid of the chore of handling weights, except for the mighty Austrian *Meinl* chain, which popularity among the country's homemakers rested on some free extra weight of what the customer was buying, sufficient to tilt the scale past the point of equilibrium without bankrupting the company. That additional spoonful of sugar, flower, etc., made that the Austrian housewives left the Meinl stores with happy faces and that certain "free lunch" feeling in their hearts.

Fig. 5.11. Dial scale

By contrast, dial scales (Fig. 5.11) showed those little extras as what they actually were – if at all. While Roberval's equal-arms scales (Fig. 5.10) show the *difference* between weights and load, dial scales show the *total*. Like we saw in the case of the Wheatstone bridge, direct measurement of a difference is many times more accurate than figuring a difference by subtraction of two totals. And besides, the old kitchen scales (Fig. 5.10) showed the point of equilibrium through the coincidence of *two* indicators, one of them going up while the other went down, thus doubling the visual effects of extra weight. Whether Mr. Meinl was aware of such mechanical intricacies or just followed his sound business instinct remains an open question, but his deep knowledge of human nature is for sure.

Innovation has many faces, and an innovation of the calendar created the constellation *Libra, the Scales,* seventh in the Zodiac, and for that matter, one of the least conspicuous zodiacal constellations. Unlike the privileged group of asterisms, which resemble what their names imply, such as Cygnus, the swan, Libra graces the night sky as a simple trapezoid. It takes creativity to make two of the figure's opposing sides symbolize the strings suspending the pans, and to insert an imaginary pivot into the stellar void at upper midpoint. Unlike most other constellations, which originate in Grecian legend, Libra is a latecomer, introduced into the Zodiac in 46 BC in the wake of Julius Caesar's calendar reform. Neighboring Scorpius, the scorpion, paid for it with the loss of a pair of stars that, previous to the event, had correctly marked the tips of his claws.

But Libra holds more than what meets the eye. Eratosthenes, the Grecian philosopher who measured the circumference of Earth with an accuracy never equaled until the dawn of the metric system, left us tantalizing data on certain stars of the Libra constellation. He listed Zuben Eschamali, the 2.61 magnitude northernmost star in Libra, as brighter than Antares, the reddish tip at the nose of Scorpio, of first magnitude in our times. Unless Eratosthenes got it all wrong, Antares must have gained brightness since his days, which could signal the star's expansion into a "red giant", something that happens when a star has exhausted its stock of hydrogen fuel and expands. Pessimistic cosmologists expect such a fate for our Sun in about five billion years. If you find that remote, think twice. The notorious Dr. Stephen Hawking, Lucasian Professor of Mathematics at the University of Cambridge, suggests in a speech on this subject matter that we should start building interstellar spacecraft right here and now.

Torque

The tragic of labor bereft of the instant gratification of accomplishment is symbolized – in Grecian mythology – by the person of Sisyphus, founder of the town of Ephyra, which later developed into the sprawling city Corinth. But what most endeared him to the people was his outstanding smartness in solving the problems of every day.

When he suspected his neighbor, Autolykos, of luring cattle from Sisyphus herds into his own, Sisyphus engraved the words "stolen by Autolykos" into the hooves of his livestock and subsequently followed the cattle's coded footprints right into his neighbor's stables. This little trick paid off twofold. Not enough that Sisyphus got his animals back, he also found time to seduce his rueful neighbor's daughter, Antikleia, whose offspring included in later generations Homer's seafaring hero, Odysseus.

When Sisyphus' days in this world were to end, Hades, master of the kingdom of the dead, appeared in person to lead him away. Sisyphus however had no longings for his visitor's nebulous promises of an undefined afterlife and, noticing a pair of shackles in his guest's hands, feigned vivid interest in that innovative tool. Question after question on the device's making and functioning made that Hades, flattered by the other's admiration, let Sisyphus hold the keys while he, Hades, demonstrated the use of the shackles on his own wrists. Had the cuffs been man-made, the divine Hades would have shaken them off at a moment's notice, but those gadgets from beyond resisted all and every spell he tried to get them off.

Suddenly at the mercy of his host, Hades found himself locked up in a closet somewhere in Sisyphus' home, where he would have remained prisoner forever, hadn't it been that, Sisyphus aside, nobody else died anymore in the ancient world – not even the soldiers hacked to pieces in the rage of battle. No sooner than the masters of Olympus got wind of the explosive population growth below did they venture to Earth, freed their peer, and unceremoniously dispatched Sisyphus into the underworld.

There, his penance for playing mightier than one of the ruling class was set at hauling a boulder single-handed up to the peak of an underworldly mountain, but with a catch: At the very moment the task seemed complete, the rock invariably escaped his keeper's grip and rolled back down all the way, leaving Sisyphus with no choice other than to start his toil times over again. We lack information if he is still doing that, but times have changed and what once was known as Sisyphean labor has been renamed into "eight to five".

Meanwhile, we may analyze how gruesome Sisyphus' punishment really was, and whether his employers violated labor legislation then and now. Humans have been found capable of doing one seventh of the work of a horse, and for lack of convincing evidence on a Hades-specific system of weights and measures, we use our earthly unit of horsepower, of 75 kilogram-meter per second (kgf × m/s), to estimate Sisyphus' capacity as $75/7 \approx 10$ kgf × m/s.

On the other hand, the work to haul the weight W of the boulder, and of course, Sisyphus' own weight, W_S, up a hill of height H, amounts to $(W + W_S)H$. And under the assumption that he had one eight-hour day, of $t = 28{,}800$ s, to accomplish that work, the necessary power amounted to $P = (W + W_S)H/t$.

Let's estimate the weight of the boulder as $W = 500$ kg, and Sisyphus' personal weight at 80 kg, and we get for the power required for one climb per day as $P = 580H/28800$. Since we already know the available manpower as 10 kg × m/s, we set $P = 580H/28800 = 10$ kgf × m/s, and get $H = 10 \times 28800/580 = 496.6 \approx 500$ m for the height of the hill Sisyphus scaled with no end in sight.

If Sisyphus had $F = 40$ kgf of muscle power to roll a rock of radius R uphill, the torque he applied relative to the point of contact between rock and ground was $T = F \times 2R$. Likewise, an equivalent force F_o, acting on the rock's midpoint, would generate the torque $T = F_oR$. Equating the two formulas gives $T = F_oR = F \times 2R$, from which we derive $F/F_o = R/2R = 1/2$.

Does that mean that Sisyphus outsmarted his masters again by cutting the severity of his sentence in half?

Whatever. In any case, knowledge of $F_o = 2F = 80$ kg allows for computing the uphill inclination α of the track from the formula $F_o = W \sin\alpha$, as shown in Fig. 5.12, which yields $80 = 500 \sin\alpha$, and $\alpha = \arcsin(80/500) = 9.20°$, equivalent to an incline of 1:6.17. For comparison, the slope of interstate highways is usually kept at less than 1:25. Poor old Sisyphus, having a lousy 1/7 HP to beat our roadsters!

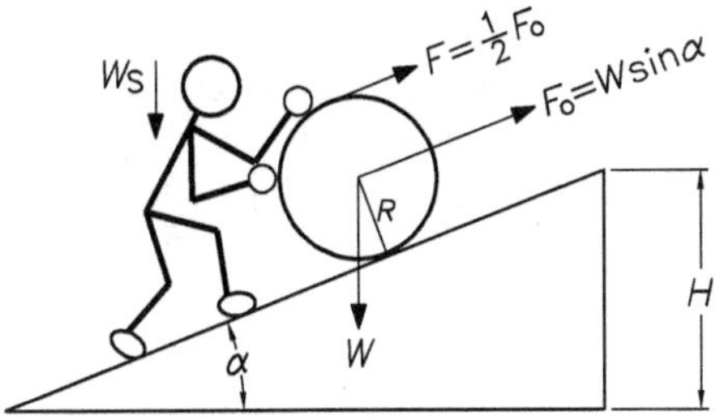

Fig. 5.12. Up and up the hill it goes ...

As to the size of Sisyphus' rock, it seems save to assume that the divine strife for perfection shaped it as a cylinder of equal height and diameter, the most compact among all possible cylinder proportions. With R for the radius of such a cylinder, its height becomes $2R$, and its volume $V = R^2\pi \times 2R = 2\pi R^3$. From that, the radius follows as $R = \sqrt[3]{V/2\pi}$.

The rock's volume V, required to use this formula, can be derived from its weight, of $W = 500$ kg, and the specific gravity of stone, which we find in reference books as $\gamma = 2900\ \text{kg/m}^3$. Since weight equals volume times specific gravity, $W = V\gamma$, the volume of the boulder must have been $V = 500/2900 = 0.172\ \text{m}^3$, and its radius $R = \sqrt[3]{0.172/2\pi} = 0.30$ m.

Thus, Sisyphus' rock measured a puny $2 \times 0.30 = 0.60\ \text{m} = 60$ cm (two feet) in diameter. If you find that disappointingly small, just try to roll it uphill! Or check that, on a 9.20° incline, the boulder tends to roll back down with the force of (not surprisingly) $F_o = 500 \times \sin 9.20^\circ = 80$ kg.

That much on the lifestyle in the world of Hades. For more details, consult http://www.Stynxferryrides.com. Meanwhile, as we prepare to leave the Grecian underworld, let's bag what we found out there on torque, work, and power, and apply it to present day machinery.

Motor torque and de Prony's brake

Take for instance a DC dynamo with a pulley of radius R, driven by a flat belt that's running at the speed of v meters per second. With F for the pull of the belt, we get the power of the drive as $P = Fv$.

The value of v can be figured from n, the rotations per second of the machine, and $2\pi R$, the circumference of the pulley, as $v = 2\pi Rn$. Thus we get the power of the machine as $P = 2\pi nFR$. But 2π is the circumference of the unit circle, and $2\pi n$ is the angle subtended on the unit circle by n rotations per second, known as the *angular velocity* ω of the pulley.

Introducing $F \times R$ for the *torque* T of the drive along with the terms above into the formula for power, we get $P = \omega T$.

Note that n stands for rotations per second, contrary to the customary rotations per minute (rpm). The conversion $n \to n/60$ changes n into the accustomed unit of rpm. With this, the formula for power becomes

$$P = \frac{2\pi n}{60} T = \frac{\pi n}{30} T\,.$$

Herein, the term $\pi n/30$ is once again the drive's angular velocity ω, but this time with n in rotations per minute (rpm).

Hence, the power of an engine or motor can be found by measuring *torque* and *rotations per minute*. The *de Prony* brake (Fig. 5.13) was designed on this principle in 1821 by Gaspard Riche de Prony (1755–1839), Napoleon Bonaparte's choice for supervising the draining of the Pontine Marshes and projects that sheltered the province of Ferrara in Northern Italy from inundations by the river *Po*.

In another field, de Prony worked with Legendre, Carnot and other contemporary mathematicians, and some 70 to 80 assistants, on tables for logarithms and trigonometric functions to at least fourteen and sometimes twenty-nine decimals. In spite of deProny's vast labor resources, it took him several years to complete those tables in 1801. Tragically, the mere volume combined with the low sales potential of this work made that his tables never got published in full (take solace, fellow writers!).

The core of the de Prony brake (Fig. 5.13) is – as the name suggests – a conventional friction brake which force is regulated by a spring-loaded wingnut.

De Prony's brake can be balanced in two different modes. We can tighten the brake shoes until the motor slows down to the desired rpm and subsequently shift the counterweight to the point of equilibrium, where the lever lingers between upper and lower stop pin, touching neither. Or else, we may start with positioning the counterweight for the desired torque and then tighten the brake until the lever idles – once again – between the stop pins.

The motor's rotations per minute are checked with a conventional rpm meter, pressing the beveled (and sometimes slotted) end of the meter shaft-end into the center-hole of the motor shaft end.

In both cases, the torque of the machine is proportional to the distance of the

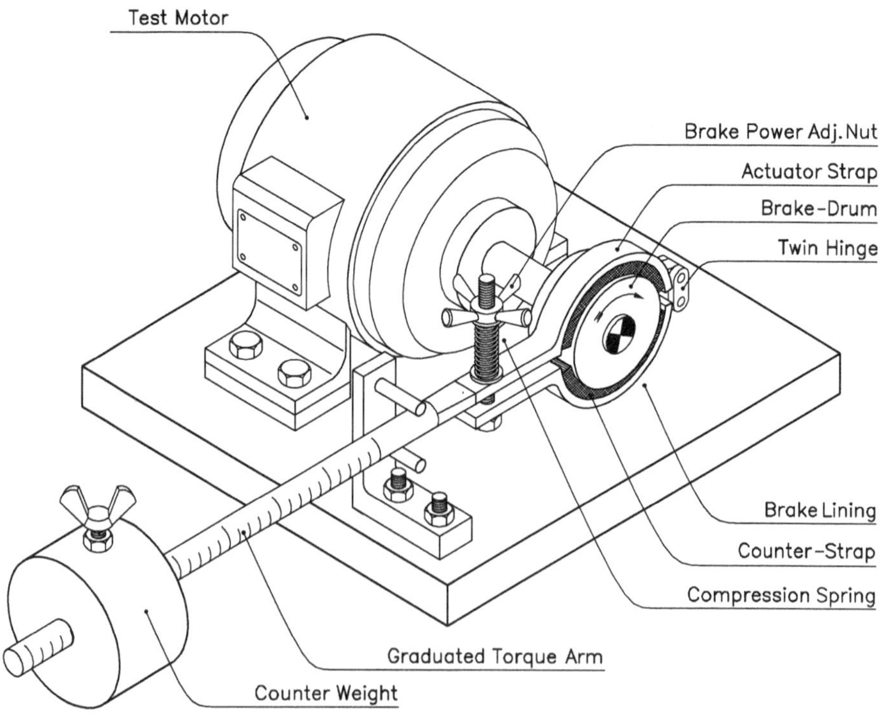

Fig. 5.13. De Prony's brake

center of mass of the counterweight from the centerline of the motor shaft, and can be read from the position of the counterweight on a scale engraved on the lever. The scale should include the torque of the proper weight of the lever.

Assume that we measure the torque of an electric motor at 1800 rpm as 5.41 m $\times$ kgf (meter-kilogram-force) and get from the formula $P = T\pi n/30$ the power of the motor as $P = 5.41 \times 1800 \times \pi/30 = 1020$ m $\times$ kgf/s.

Division by 75 (1HP = 75 m $\times$ kgf/s) converts this into $1020/75 = 13.6$ HP. Although the unit of horsepower doesn't make part of the International System of Units, it is still frequently used to express the power output of primary movers, such as engines and motors. On the other hand, the electricity a motor consumes is measured exclusively in units of the International System, namely, kilowatt. We are no strangers to kilowatts, thanks to our monthly electricity bills, which charge by the kilowatt-hour. If HP-hour were the utilities' unit of choice, the price would be another, but the amount on the foot of the utility bill would still be the same.

To convert the result of 1020 m $\times$ kgf/s into metric units, we multiply by $g = 9.807$ (gravitational acceleration in m/s^2), which enters the picture because the newton (N), unit of force in the International System of Units, relates to the kilogram-force by 1 kgf = 9.807 N. Thus, we get the output power of the machine under test as:

$$1020 \times 9.807 = 10,000 \text{ watt} = 10 \text{ kilowatt (kW)} .$$

If that same motor consumes, say, 12 kW of electricity, its efficiency is $10/12 = 0.83$ or 83%. Should a competitor's offer include a machine with 86% at the same price, which one would you buy? That's what de Prony's brake helps to decide. Scores of dynamos and alternators, admired as great achievements at the times they had been built, ended up on the scrapyard by the magic of this percentage.

Eddy current dynamometer

The eddy current dynamometer (Fig. 5.14) is the electrical equivalent of de Prony's brake. As the name implies, it uses electrical rather than mechanical braking. While the risk of overheating limits de Prony brakes' range of usefulness to within 50 HP and 200 rpm, eddy current brakes and dynamometers have no such restrictions. This because mechanical brakes generate frictional heat and rely on cooling by the surrounding air, while the elements of electrical brakes don't touch and allow for heat dissipation by forced air streams and water-cooling of the brake housing.

We remember eddy currents from discussions of the analog speedometer, where eddy current generated torque converts the rotation of the input shaft into deflection of the instrument's pointer. The eddy current brake in Fig. 5.14 functions on identical principles but is built for much heavier loads. In the Fig. 5.14, eight pairs of electromagnets are shown with the eddy current brass disk rotating between their pole shoes, but industrial models often cram in as many magnets as they can fit into the given circular space. The disk has radial slots, meant to add to the path the eddy currents within must take (Fig. 5.14 b). The polarization of the magnets is such that the north of one magnet faces the south of the corresponding one on the other side of the disk.

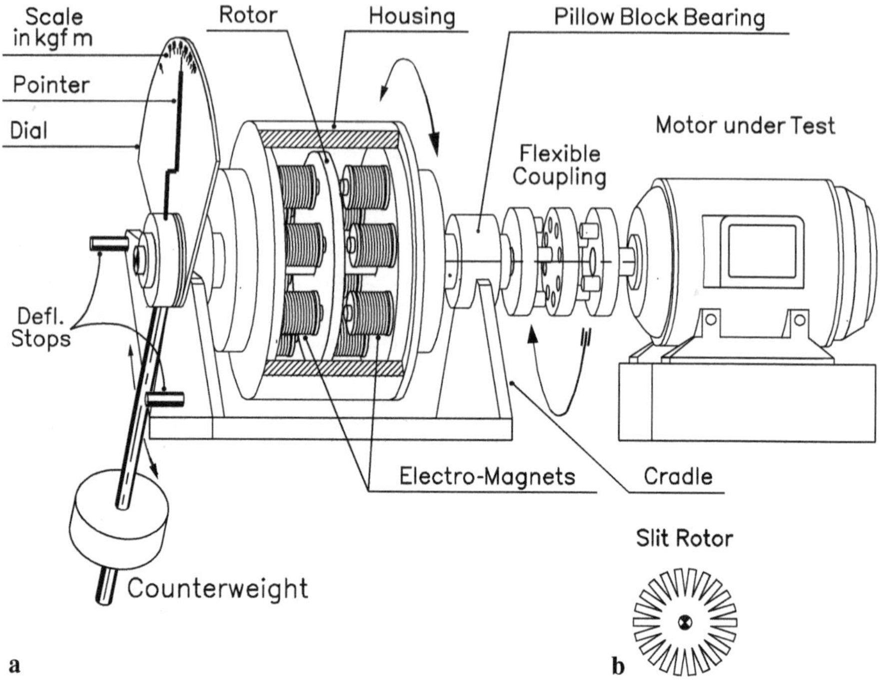

Fig. 5.14. Eddy current dynamometer

The housing, made of high-permeability material, vulgo steel, closes the path of the magnetic flux lines on the rear of the electromagnets.

A flexible coupling, shown in exploded view in Fig. 5.14 a, links the motor under test with the eddy current unit while allowing for a certain degree of linear and angular misalignment. The coupling's center disk is made of rubber or wear-resistant flexible plastics.

The entire brake assembly rests in a cradle in which it tilts until the point where the torque from the counterweight equals the torque of the motor. We could measure that torque like in de Prony's machine by sliding the counterweight into the balanced position. More convenient however is to keep the weight fixed and, like in Fig. 5.14, derive torque from the angular swing of the brake-housing. The scale, which divisions contain the conversion of tilt angle to torque will not be regular. After all, very small torque is enough to swing the counterweight out of its vertical (zero) position, while times again higher torque is needed the farther the weight is up. However, the weight's counter-torque rises only up to 90° of deflection and would decrease from then on up. That's the raison d'être for the stop-pin that keeps the brake assembly from accidentally overshooting that critical point and swirl around.

The dynamometer can be fitted for digital readout if we replace the pointer by a rotary encoder (Chap. 2) with signal processor. The additional investment pays off as higher accuracy, limited only by frictional losses in the cradle bearings. Such losses can be minimized if we support the brake assembly in a set of cam followers, which produce only rolling friction instead of the far higher gliding friction of journal bearings.

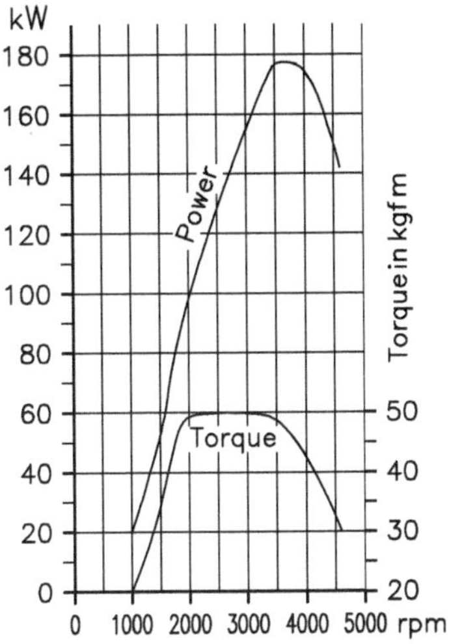

Fig. 5.15. Torque and power of an internal combustion engine

The torque and power curves in Fig. 5.15 are typical dynamometer readings on an internal combustion engine. If the pressure generated by combustion of the air–fuel mixture in the cylinders were constant throughout the work cycle, torque too would be constant and power would rise proportional to the engine's rpm, according to the definition that power is equal to force times velocity. In Fig. 5.15, this situation exists between 2000 and 3500 rpm, where torque remains fairly constant. If we throttle the engine into lower rpm, the density of the aspirated air–fuel mixture sinks and so does the engine's power and efficiency. At the high end, friction between piston and cylinder walls makes that the engine's power decreases the more the faster it turns.

Figure 5.15 also shows why hybrid vehicles are often seen as the cars of the future. Their idea is to keep the engine at all time at its speed of highest efficiency and cleanest exhaust. For an engine with the characteristics shown in Fig. 5.15, that would be somewhere around the middle of the constant torque domain, say, between 2500 and 3000 rpm. In this range, the engine develops enough power for both: open highway driving and trickle charging the battery. In crawling city traffic, power is provided by a battery-powered high-torque electric motor, while the gasoline engine continues at highest efficiency, but this time driving the alternator that recharges the battery.

Further fuel savings are obtained by regenerative braking. Far from being a novelty, this had been the way of slowing down the good old trolley cars with their serial connected DC machines, capable of operating either as motors or as dynamos, depending whether you feed them electricity or drive their rotor by external power sources, such as the braking torque of a rolling wagon. In the latter case, power for turning the motor uses up the momentum of the vehicle and brings it close to a halt. In a wasteful foretime, electricity generated in the process was used up in a bank of resistors, but hybrid vehicles recycle it for charging their batteries.

Engine, alternator, and electric motor operate in unison over a differential gearbox, a glorified version of the classical differential. Digital control systems for fine tuning the interaction of all elements have been introduced to various degrees.

Torque pops up all the time in every day's life. Next time we admire the trains in movies of the wild, wild West and their steam engines' whirling piston and connecting rods, let's think *torque*. And likewise, when we hear that foreboding screech of tires on the street, preceding the deafening bang. But most important, it is thanks to the torque of the motor in your air conditioner that you can read this book, reclining comfortably in your chair, rather than battling sweltering heat or suffering the chill of those long winter evenings.

And next time you see a guy pushing a boulder up a mountain, say "Hi!" for me.

6 Vibrations

The constellation of the *Southern Cross*, flanked by *Alpha Centauri*, the third brightest star in the sky, is as familiar to stargazers south of the equator like the Big Dipper and Polaris to their brethren at northern latitudes. Incidentally, Alpha Centauri is the star closest to the Sun, and consists of two components, Alpha Centauri A and B. Both resemble the Sun in size and surface temperature.

Thus we could venture that conditions on a hypothetical Alpha Centauri planet the size of Earth could resemble home, insofar as the grass would likely be green, and white clouds would float in the blue sky. Déjà vu, except for one thing – silence, an all encompassing, ghostly silence. No rumbling from the interstate, no roaring aircraft overhead. No crackling lawnmowers, no whining police sirens, and principally, no ear-piercing boom boxes. Not even a child's laughter reverberating in the air. Silence so deep that you could perceive the impact of a falling leave on the ground, the rustling of a gentle breeze skimming the meadows, and the morning dew rolling off the leaves. And here and there, the cry of an alien bird of prey, too distant to be seen.

Once upon a time, you needn't travel through deep space to experience such a veil of silence. It endured – if we believe in Grecian mythology – until Hermes, the heel-winged son of Zeus, found himself involved by melodic tunes that seemed to emanate from the windswept carcass of an animal cadaver, whose dried-out sinews still tautly stretched from bone to bone. The magic of these sounds inspired quick-minded Hermes to build a harp worth bestowing to Apollo, the divine sponsor of music, in hopes to make him forget the loss of his herds, which Baby-Hermes had stolen at the record-setting age of one day.

The harp was named after Aeolus, the master of the winds in Grecian folklore, while Hermes, like so many inventors after him, was left to linger in the sidelines. Later instruments designed to the principles of Hermes' musical skeleton became known as Aeolian harps, unique insofar as they are played by a breeze of wind rather than the hands of musicians.

Aeolus had made it into the headlines of those times' newspapers (known by their mouth-to-mouth communication technology) by assisting Odysseus – the errant celebrity seafarer – on his way home from the Trojan war. In a gesture of solidarity without borders, Aeolus gathered all the unruly winds in a bag to guarantee fair sailing for his beloved guest. But once within view of their homeland, the ship's crew raided the bag for hidden treasures and the nasty winds escaped, hurling the vessel back into the raging seas toward new, life-threatening adventures.

Rabbinical records – on the other hand – offer a story linked to King David's enjoyment of the sounds elicited by the midnight breeze from a harp, or Kinnor, above his bedstead. "Other kings slumber soundly through the entire night . . . but I wake up at dawn . . . the north wind, blowing through my chambers, causes the harp to sound by itself. These sounds awaken me and I spend the rest of the night singing psalms and hymns to praise God."

Aeolian harp – the dawn of music

In a different setting, the term *Aeolian* applies to a Grecian musical key, a precursor of present-day minor keys. What sets the Aeolian harp – the sole instrument played by nature rather than by man – apart from the rest of stringed instruments is the identical length of its four to twelve chords of various thickness, all tuned to the same note. By contrast, the effective length of a violin's cord depends on where the violinist holds it down, wherein the purity of the instrument's notes reveals a player's capacity to guess the right spot; an ability that takes many years of arduous training to acquire. Plucking instruments, on the other hand, sport a row of logarithmically spaced bridges, marking the correct length of cord for the notes of the harmonic scale.

The active length of the cords on the Aeolian harp in Fig. 6.1 is the free space between the two hardwood bridges of triangular cross section, on the instrument's soundboard. The chords, which can be gut, steel, or nylon, are pinned down on one end and tied to tuning pegs at the other.

Apart from Hermes' harp-building efforts, harps like that in Fig. 6.1 go back to the Jesuit Athanasius Kircher, born in 1602 in Fulda, Germany. In the early eighteen hundreds, Kaufmann of Dresden and Heinrich Christoph Koch used the names of Windfang or Windflilgel for their line of harps that sported a funneled wind canal placed over the soundboard (Fig. 6.1). It guides and equalizes the airstream perpendicular to the strings and can be sized to fit the width of the window sash encasing the harp.

The omnipresent vortex

Wind induced tunes in parallel airflow stem from the formation of vortexes (rapidly swirling motions around a center) on the downstream side of the cords. At certain well-defined flow velocities, eddy formation occurs in step with the chord's resonance frequency, locking it into *transversal* oscillations, which in turn excite the *longitudinal* wave-motion in air, called sound. The velocity of propagation, c, of the transversal oscillation of a cord is given by

$$c = \sqrt{S/m}\,,$$

where m stands for the cord's mass per unit of length, and S for the applied tension.

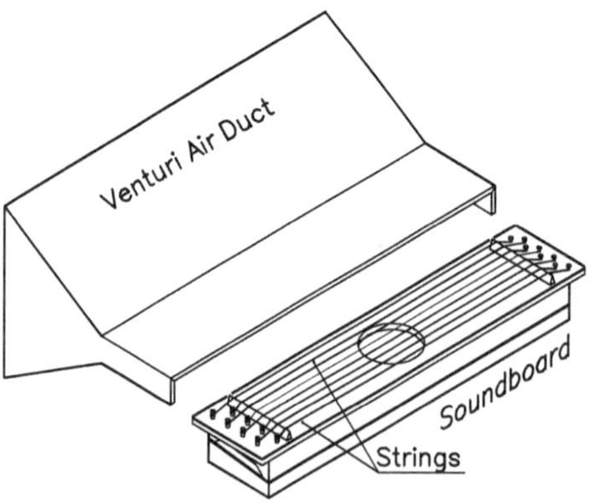

Fig. 6.1. Aeolian harp

The mass per meter or foot of a string of diameter $D = 2R$ and length ℓ can be figured by multiplying its volume, $R^2\pi\ell$, with the specific gravity γ of the string's material, such as metal or gut. Herewith, we get

$$m = \gamma\ell\frac{D^2\pi}{4} = 0.25\pi\gamma\ell D^2 \, .$$

Eddy formation, the cause of the so-called aeolian vibrations, happens downstream from an obstacle, called bluff in this context. It starts with the transition from laminar to turbulent flow and endures into higher flow velocities until the flow patterns turn chaotic. Figure 6.2 illustrates that process. On the upper left is a typical field of *laminar flow* streamlines, prevalent at flow velocities low enough for adjacent layers of fluid not to mix. Here, flow patterns up- and downstream are still mirror images of each other.

The next two panels in the upper row show the onset of eddy formation. In the second row of flow patterns, the eddies grow to the point of instability and periodic inflating and deflating sets in as higher flow velocities make the combined size of the vortexes outgrow the boundaries of the stern wash behind the bluff. The eddies, which so far had been fed with fluid from the stream, start drawing fluid from each other. Either one can trigger this grows at the expense of its counter-

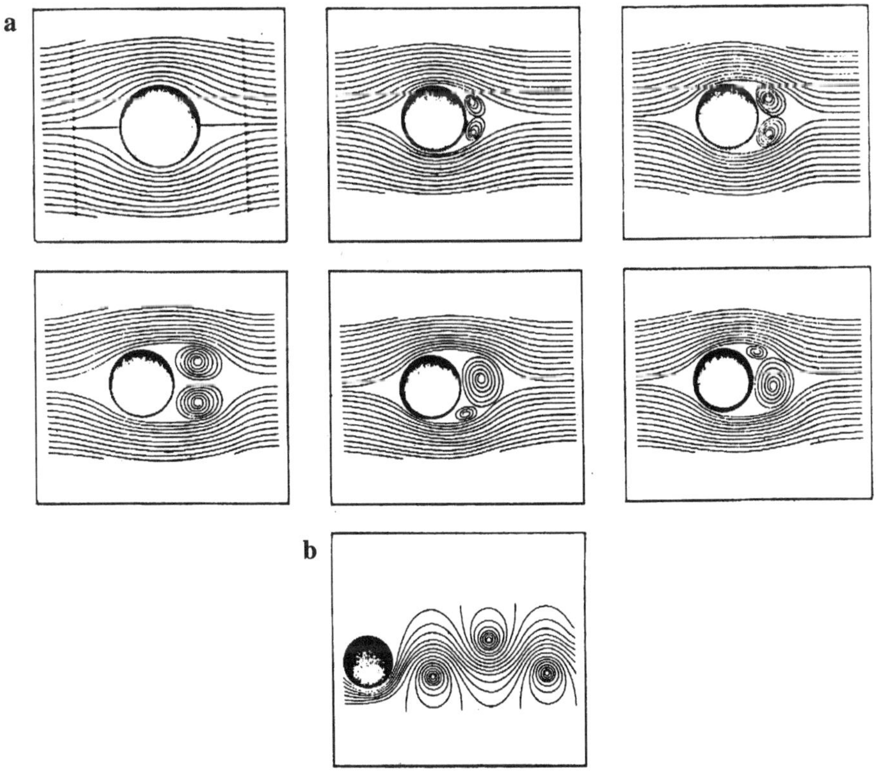

Fig. 6.2. **a** Vortex formation and **b** Karman street

part, until the latter's near depletion makes the scene shift into reverse.

Dynamic pressure of the fluid's motion makes the bluff (cord or string in the case in point) swing contrary to the vortexes while its reactive elastic forces add to the amplitude of the shedding vortexes, boosting their pulsation.

Flow patterns in Fig. 6.2a refer to an observer at rest relative to the bluff, such as somebody on a bridge eyeing the waters below. Figure 6.2b shows the pattern of streamlines seen by an observer riding along with the mainstream. For him or her, the bridge wanders upstream, while the whirling eddies behind the pillars fall into a pattern named *Karman Vortex Street.*

V. Strouhal measured f, the frequency of vortex shedding in air in the wake of a cylindrical bluff, such as a string, as 0.185 times the velocity v of the airflow, divided by the bluff's diameter D:

$$f = 0.185v/D\,. \tag{6.1}$$

Accordingly, a stiff breeze of, say, 0.357 meter per second (1.285 km/h), would cause a 1 millimeter (0.001 meter) diameter cord to oscillate at $f = 0.185(0.357/\ 0.001) = 66$ Hz, incidentally the frequency of "C" in the harmonic scale. That wouldn't pass a hum if it weren't for the harmonics of this and other low frequencies that make the Aeolian harp sing. Moreover, Eq. (6.1) rearranged into

$$v = Df/0.185$$

suggests an interesting method of measuring flow velocity in tubes and ducts, where D is constant, so that v and f become the sole variables. This allows to derive flow velocity from the time-span between the passages of eddies in the Karman vortex street (Fig. 6.2b). To this end, the *vortex flowmeter* sends a light-beam through the flow to a photocell on the other side. As the passing eddies modulate the intensity of the beam, the photocell generates pulsed DC, triggering an electronic pulse counter. Vortex flowmeters must be calibrated according to the fluid they are used with, because Strouhal constants, and with them the frequency–speed relation, are material specific.

Waves and the sine line

Waves in a vibrating string use to resemble the sine line, plotted in Fig. 6.3 for the period of 2π, matching the circumference of the unit circle. The highest value y can have is unity $\left(\sin\frac{1}{2}\pi = 1\right)$. Worth remembering is the area of exactly 2 length-units squared under each lobe of a sinusoid.

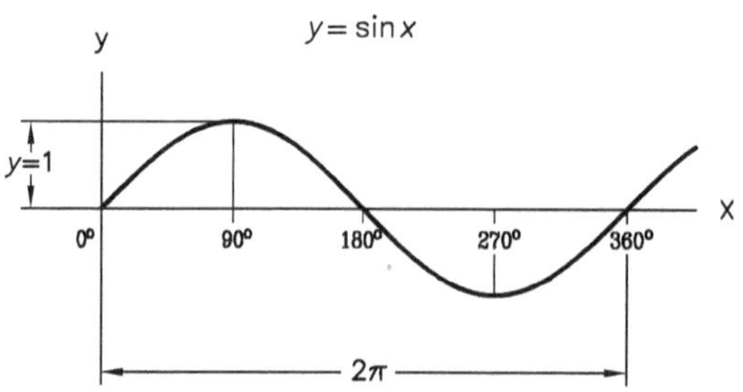

Fig. 6.3. Sine line

Conversely, the wave in Fig. 6.4, curved like a sine line, is the instant image of a string or cord, oscillating with the

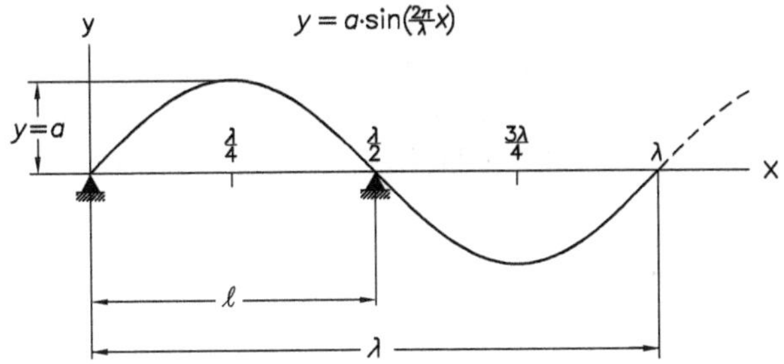

Fig. 6.4. Sine wave

wavelength λ and amplitude a. It is derived from the sine curve by stretching the x values by a factor of $\lambda/2\pi$, and the y values by $a/1 = a$, in the transformation

$$y = a \sin \frac{2\pi}{\lambda} x.$$

From this, we derive the equation of a traveling wave of velocity c (Fig. 6.5), by replacing x by $x - ct$, and get

$$y = a \sin \left[\frac{2\pi}{\lambda} (x - ct) \right].$$

The distance ℓ between the points of suspension of a swinging cord equals the spacing of the nodes of the sinusoid, and herewith $\lambda/2$, the length of the half-wave. Thus, the sound from an oscillating cord is actually the *second harmonic* of a virtual wave of length λ. Figure 6.4 shows that by marking the nodes at $x = 0$ and $x = \lambda/2$ as support points.

An oscillation's velocity of propagation can be derived by comparison with a paddle boat, which speed (in the absence of losses) would be the product of λ, the stroke of the paddle, and f, the number of strokes per second: $c = \lambda f$. Inverted, the equation gives the frequency of the oscillation as $f = c/\lambda$. Oscillations at this base-frequency may trigger others at the frequencies of $2f$, $3f$, $4f$, etc., called *harmonics*. Generation of harmonics in musical instruments depends on the physical properties of their sound board and is recognizable as the characteristic timbre of each category. For instance, the sounds from a guitar are different from those of a

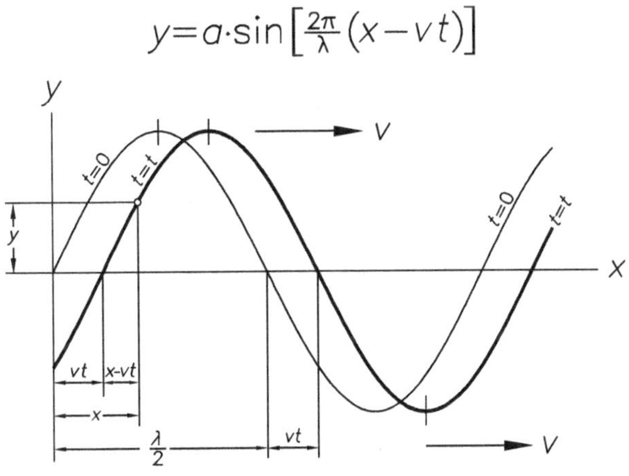

Fig. 6.5. Traveling sinusoidal wave

trumpet, even with both instruments played at identical frequencies. And the owners of a microphone and an oscilloscope will know from experience that tuning forks and certain whistles produce the purest sinusoidal waves, while most other musical instruments emit waveforms of higher complexity.

Since the frequencies of harmonics are multiples of the proper frequency of their source, their wavelength can never exceed that of the source. Formation of harmonics is, so to say, a one-way street. If the distance ℓ between suspension points of a vibrating string equals a whole-number multiple of the half-wavelength $\lambda/2$, the outgoing and the returning (reflected) waves overlay. Figure 6.6 depicts this process in eight "snapshots," separated timewise by $T = 1/f$ seconds. In Fig. 6.6 a, positive crests of the outgoing wave (fine line) match with negative crests of the reflected wave (medium wide line) and mutually annul (coarse straight line). $T/8$ seconds later (Fig. 6.6 b), the outgoing wave shifted to the right, and the arriving wave at the same rate to the left; the resultant is of somewhat lower amplitude than its originators. In Fig. 6.6 c for $\lambda/4$ phase lag, the amplitude of the superposition of in- and outgoing waves has grown to exceed the initial amplitude of the oscillation.

And finally, in Fig. 6.6 e, the in- and outgoing wavetrains overlap und the amplitude of the resultant reaches its peak of $2a$.

In Fig. 6.6 f–h, phase differences augment in in-

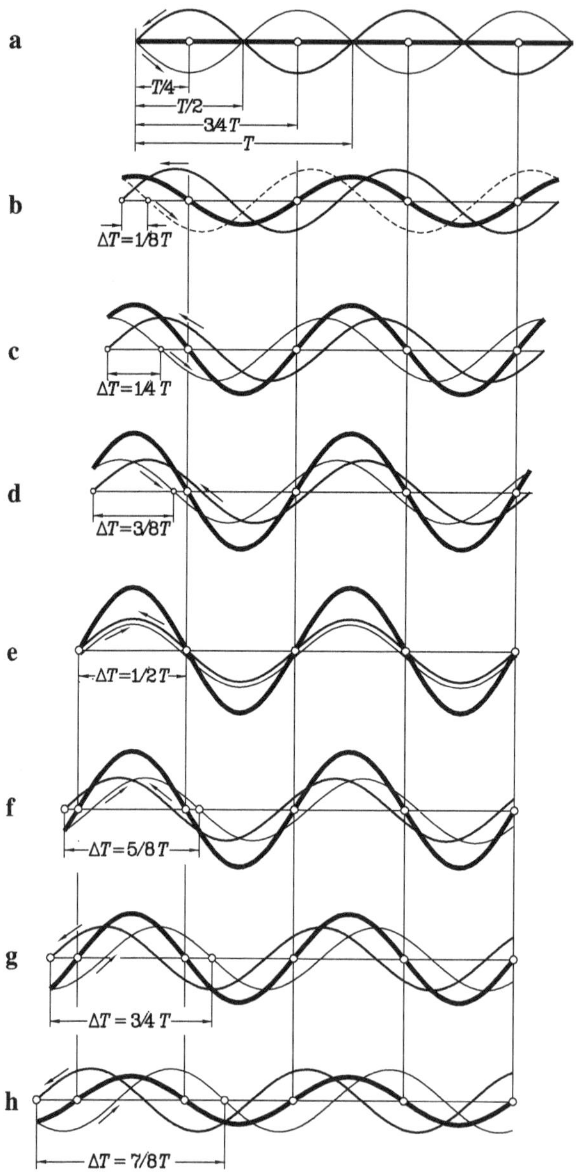

Fig. 6.6. Wave superposition

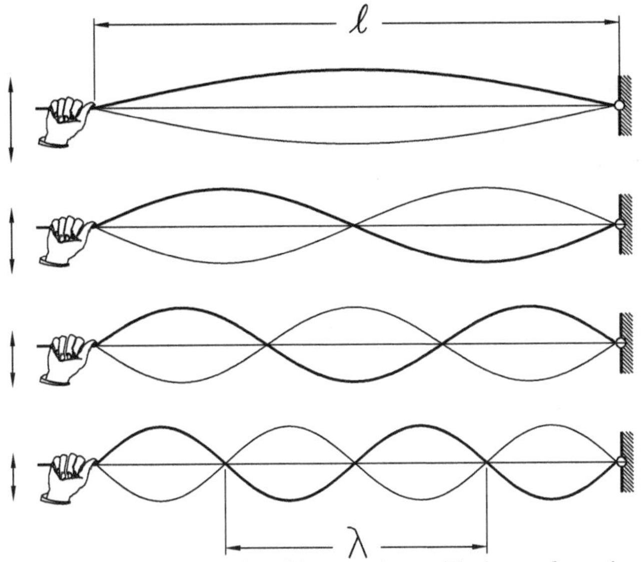

Fig. 6.7. Fundamental and harmonic oscillations of a string

verted sequence until we are back to the straight line of Fig. 6.6a.

In all that, the nodes, points where the resultant waves cross the baseline, never change position. This generates a *standing wave*, similar to Fig. 6.9, which pictures the waveforms at 10 identical time intervals on a greatly expanded y scale.

The fundamental standing wave[1] of a string with $\ell = \lambda/2$ consists of a simple up and down at the middle of the string (Fig. 6.7, top). For $\ell = \lambda$, we get a two-lobe standing wave, and for $\ell = 3\lambda/2$ a three-lobe wave, etc.

Standing waves (Fig. 6.7) arise when the quotient f/f_o goes off without remainder. Only then sustain the outgoing and the returning waves each other. If standing waves from several sources mutually interfere on their way from a musical instrument to the listener, they build up into complex waveforms, as we see on the screen of an oscilloscope connected to a microphone. Joseph Fourier (1768–1830), son of a tailor in Auxerre, France, came up with mathematical proof that any kind of waveform could be created by superposition of a number of sinusoidal waves of consecutively higher frequencies and lower amplitudes. Following that up, we may understand the timbre of the sound of a violin by envisaging the instrument's body as the sum of a number of domains, each in a state of sinusoidal oscillation at its natural frequency. If only Stradivarius had had an oscilloscope!

He, who is not blessed with the possession of such an instrument, can still observe higher order harmonics on a string, such as a clothesline with one end fixed while the other is jiggled. With some degree of dexterity, you can make the line swing at one or the other of its many resonant frequencies with the more periods the faster you swing.

1 wave that remains in an unchanging position.

String theories

The frequency of a string of mass m, tensioned to S, can be derived from the equations $c = \sqrt{S/m}$ and $c = \lambda f$, as

$$f = \frac{c}{\lambda} = \frac{1}{\lambda}\sqrt{\frac{S}{m}}.$$

Since d and m are invariable for a given instrument, the formula can be expressed by $S \propto f^2$, or $f \propto \sqrt{S}$, which means that the pitch of a cord's sound is proportional to the square root of the applied tension.

For instance, the tensioning of a Nylon string of 25 cm effective length and $D = 1$ mm (0.1 centimeter) diameter, meant to sound the note A, of 440 Hz of frequency, can be found as follows.

From $\ell = 25$ cm between supports, we get the full wavelength $\lambda = 2\ell$ as 50 cm or 0.50 meter. The wave speed is $c = \lambda f = 50 \times 440 = 22{,}000$ cm/s.

The mass per centimeter of a Nylon 6 string with specific gravity of $\gamma = 1.1\ \text{g/cm}^3$, is $m = (\pi/4)D^2\gamma = (\pi/4) \times 0.1^2 \times 1.1 = 0.00864$ gram per centimeter of length.

Now, the formula $c = \sqrt{S/m}$, reshuffled into $S = mc^2$, gets us

$$S = 0.00864 \times 22{,}000^2 = 4.18 \times 10^6\ \text{dyn} = 41.8\ \text{newton} = 4.26\ \text{kgf}\,.$$

(Like the newton, the dyne too is a metric unit of force, yet in the original centimeter-gram-second system. The conversion factors are 100,000 dyne to the newton, and 9.807 newton to the kgf).

To find if the string will be up to the task, we take the ultimate tensile strength of Nylon 6, typically 800 kgf/cm^2, to get the strength of a 1 mm (0.1 cm) diameter string as $S_{\max} = 0.1^2 \times 800 \times \pi/4 = 6.28$ kgf. That makes the ratio of applied tension to the string's ultimate tensile strength: $100 \times 4.26/6.28 \approx 68\%$. Small wonder that delirious musicians make their instrument's cords snap.

Aeolian vibrations

An economically important application of the laws of oscillatory motion is in wind-induced (aeolian) vibrations of aluminum electric cables with or without steel core. It all began early in the 20th century with the familiar "song" of telegraph poles that became quite loud with your ear pressed on the pole. What then didn't pass an odd phenomenon became a problem when aluminum cables became the basic building blocks of widespread networks linking hydroelectric power plants with their consumers.

Hydroelectric generation of kinetic energy happens without consumption of any raw material other than sunlight, insofar as radiant heat from the Sun causes oceanic surface water to vaporize and raise into the higher layers of the troposphere, from where it returns as rain. Rain feeds the springs which fill the big reservoirs of plants like Lake Mead with the Hoover Dam; the Foz do Iguaçu in western Brazil; the Austrian Kaprun works, and the Yangtze Three Gorges Dam in modern China – to name only a few. Water admitted at the high end of the utility is recovered at ground level, the place where it would end up anyway, with or without utilization of its potential energy.

However, before you think "Free Lunch", think "Distribution." While coal- and oil-fired plants can be placed close to the centers of consumption, locations of hydroelectric units are Mother Nature's choice. That accounts for the huge distribution networks of hydroelectric utilities which, summed up, often exceed the investment in the actual power generation hardware by a factor of 3 : 1. Conversely, thermoelectricity can be generated right at the consumers' doorsteps.

In the "good old times" of ongoing electrification, distribution lines were fanned out radially from each power plant to substations and consumers, with the result that accidental wrecks of towers and breakage of cables left the users of the affected branch stranded until the damage was repaired in loco. Things got better with the crosswise interconnection of power lines in ways that every center of use could be fed through a variety of pathways. The resulting network, known as *power grid*, virtually eliminated outages, but optimizing its operation became a major headache to mathematicians until computers got into the picture.

While wooden poles and light gauge wires suffice for modest demands, the networks of the world's great hydroelectric plants depend on heavy aluminum cables, which in bundles of two, three, or four, conduct electricity of sometimes one million volt and beyond. Aluminum weighs only 30% of copper, the classical conductor material, and yet retains 60% of the copper's conductivity. This means that one pound of aluminum wire conveys twice the electricity of one pound of copper wire. Add the higher price of copper and you get the average cost of aluminum cables per kilowatt as four to five times lower than for comparable copper cables.

Long spans, such as crossings of Norwegian fjords, became viable with the introduction of the ACSR cable, consisting of a core of high-strength steel cable, embedded in a mantle of spiraled aluminum wires. As the steel cable takes care of mechanical loads, the mantle is the carrier of electricity. Meanwhile, ACSR cables have become the conductors of choice for power transmission lines in general.

However, aluminum's susceptibility to fatigue makes that such cables cannot be tensioned in excess of 18 to 20% of the their ultimate tensile strength. Above that limit, wind-induced vibrations were found to inflict damage within the first few years of use. One of the first cases was an ACSR 2 AWG[2] distribution line in the Midwest of the United States in 1917, which cables were sustained between wooden poles, spaced 55 meter from each other. Half-wave vibrations in the spans became so strong that the tops of the poles wagged lengthwise to and fro until rupture set in.

In 1923, cracks in California's Big Creek transmission line after only ten years of service triggered extensive research into the phenomenon.

Wide spaced cables are too heavy to vibrate full span, save for incidents, such as short circuit induced forces, or the sudden drop of heavy icing. Random vibrations (Fig. 6.8) don't endure, because the induced wave (dark shaded line) and its reflection (dark line) use up their energy to build up the superposition (fine line)

2 American wire gauge.

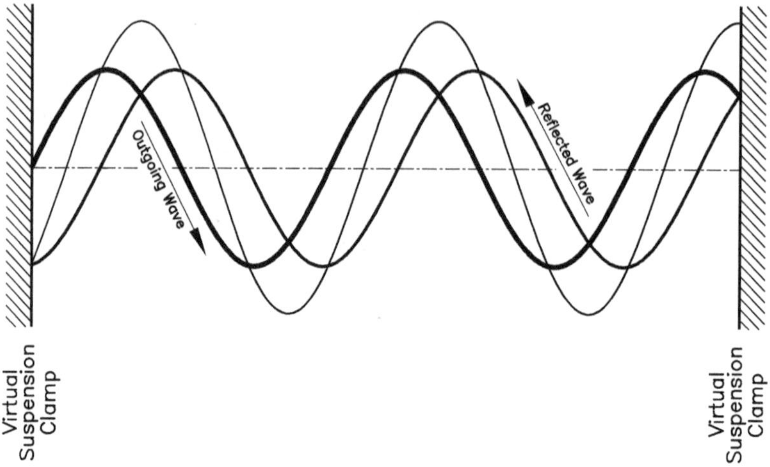

Fig. 6.8. Random wave motions

which happens out of step with the rhythm of vortex shedding and thus is not being sustained.

Therefore, wind-induced vibrations build up only in harmonic frequencies and become stationary through the superposition of the outgoing wave and the wave reflected back from the next support point. Both waves are of equal frequencies, but wouldn't survive unless the space between support points housed an integer number of half-waves, which is 1 for musical cords, while long spans of cable can hold harmonics in great numbers. Standing waves (Fig. 6.9) build up at the frequency of vortex shedding, fed consistently on wind energy.

Interesting phenomena were observed on a 650 meter wide Norwegian fjords crossing, and in a second case, on a transmission line of 965 meter of length with cables of 29 mm diameter, weighing 2.45 kg/m.

Although stress had been held at merely 13% of the conductors' ultimate tensile strength, vibrations occurred at the 250th harmonic of the fundamental frequency of 0.1 Hz, which made the frequency of the 250th harmonic $0.1 \times 250 = 25$ Hz, and the adjacent frequencies $249 \times 0.1 = 24.9$ Hz and $251 \times 0.1 = 25.1$ Hz. These frequencies are so close to each other, that minor changes in wind speed often caused the vibration pattern to jump from one frequency to the next and back, alternately leading to destructive or constructive interference. The result was the

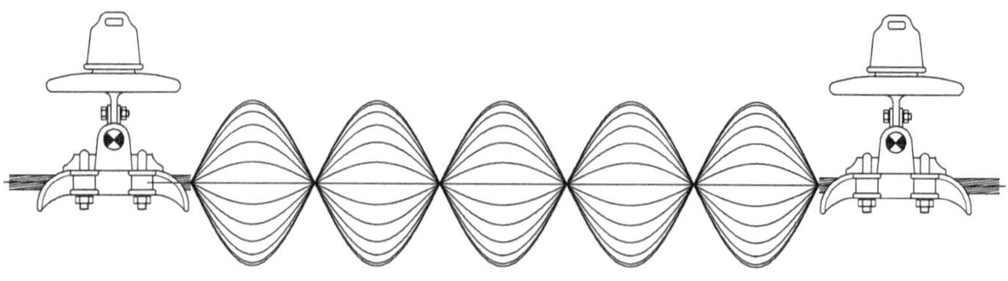

Fig. 6.9. Standing wave pattern in an electric cable

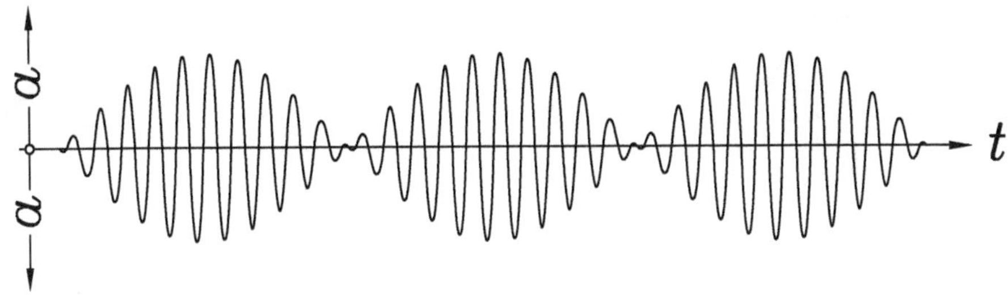

Fig. 6.10. Pulsating waveforms, the result of superposition of two adjoining oscillations

pulsating wave pattern shown in Fig. 6.10, resembling an amplitude modulated radio wave, though the latter has its amplitude commanded by electrical signals from a microphone rather than by wind-swept wave interference.

The phenomenon stems from the superposition of a fundamental oscillation with amplitude a_o and frequency f with oscillations of the adjacent frequencies of $(f+1)$ and $(f-1)$. With t for time, and a_1 for the amplitude of the harmonics, the wave equations read

$$a = a_o \cos(2\pi ft) \qquad \text{for the fundamental oscillation,}$$
$$a = a_1 \cos[2\pi t(f+1)] \qquad \text{for the upper harmonics,}$$
$$a = a_1 \cos[2\pi t(f-1)] \qquad \text{for the lower harmonics.}$$

Summing up those three equations and using the identity

$$\cos(\alpha+\beta) + \cos(\alpha-\beta) = 2\cos\alpha \times \cos\beta\,,$$

gets us

$$a = a_o \cos 2\pi ft\left(1 + 2\frac{a_1}{a_0} \times \cos 2\pi t\right),$$

which indeed is the equation of a modulated wave.

Practice meets theory

Vibrations offer an easy way to measure the sag of a cable between points of suspension, such as poles and towers. All you have to do is hitting the cable with a wooden bat and then firmly grip it so your hand can feel if it vibrates. Count the seconds t until the induced pulse has traveled the distance ℓ to the opposite suspension point, gets reflected, and comes back. Since the wave traveled the free length of the cable, ℓ_c, two and fro at the velocity c, we can set $2\ell_c = ct = t\sqrt{S/m}$ from which we deduce

$$\ell_c^2 = \frac{t^2}{4} \times \frac{S}{m}\,.$$

From here, we get the *sag* of the cable from the formula of a catenary, the basic curve of a hanging rope, which reads

$$\text{sag} = \frac{\ell_c^2 mg}{8S} \, .$$

Introducing the term for ℓ_c^2 from the previous equation yields the simple term

$$\text{sag} = \frac{t^2 Smg}{4 \times 8 \times mS} = \frac{g}{32} \times t^2 = 0.306 t^2 \, .$$

Thus, neither the weight of the cable nor its tensioning matter. Sag is derived uniquely from the time for a bout of vibration to propagate from pole to pole and back. Assuming for instance that the time-lag between hit and return pulse is 6 seconds, you can figure the sag as $0.306 \times 6^2 = 11$ meter. Counting seconds did it all. Compare that with a surveyor's workload in measuring a cable's sag on a river-crossing or other hard-to-reach regions, even if helped by theodolite and laser-gun.

Ill winds

The scope of the damages inflicted by aeolian vibrations is not limited to electric lines, small or tall. One of the most publicized cases in other fields is the collapse of the Tacoma Narrows Suspension bridge on 7 November 1940 at 11:00 a.m. This bridge over the Puget Sound linked Tacoma, a seaport in the state of Washington, to the Olympic peninsula. Its length, 7400 feet overall, included a 2800 feet single span.

Right from its opening on 1 July 1940, the bridge, later nicknamed "Twisting Gertie", became known for its propensity to swing up and down and its likelihood to twist. Thrill seekers often crossed the bridge by foot to no other end than to experience its strange behavior, which however – four months later – led to disaster. But it didn't take a hurricane to bring the bridge down. Rather, it tumbled amidst a light morning breeze, which suggested aeolian vibrations as the culprit.

The case of the Tacoma Narrows Bridge is not unique. Back in the late 1930s, a number of other bridges, the Golden Gate Bridge among them, needed stiffening by extra trusses to prevent unpredicted damage from wind-induced vibrations. So much so that the Tacoma bridge failure marked a turning point in structural design engineering: Rather than considering wind forces alone, vibrational forces and fatigue entered the computations.

Wind power of the nasty kind

The potential of hurricanes and tornados to tear down most kinds of construction by sheer dynamic air pressure is obvious, but the fact that light breezes can have similar effects needed explaining. The dynamic pressure p from wind-forces is given by Bernoulli's equation:

$$p = \gamma v^2 / 2g \, ,$$

where γ is the specific gravity of air, of $1.29\,\text{kg/m}^3$ at $0\,°\text{C}$ ($32\,°\text{F}$), and $1.20\,\text{kg/m}^3$ at $20\,°\text{C}$ ($68\,°\text{F}$).

The wind-force per unit of length (1 meter) on a cable of diameter D is the product of wind pressure p times the area of impact, $D \times 1$ m, namely,

$$F = \frac{\gamma v^2}{2g} \times D = \frac{1.20v^2}{2 \times 9.807} \times D = 0.061Dv^2 .$$

Experimental evidence has shown that Karman forces, F_K, caused by vortex shedding, exceed the straight wind impact forces by a factor of 1.04, which gets us

$$F_K = 1.04\frac{1.20v^2}{2 \times 9.807} \times D = 0.0636Dv^2$$

Take for instance a conductor of 22.4 mm diameter, such as ACSR 477-30/7, code-named *Hen*, which weighs 1.111 kilogram per meter, exposed to $v = 5$ m/s winds, and you get

$$F_K = 0.0636Dv^2 = 0.0636 \times 0.0224 \times 5^2 = 0.0356 \text{ kgf/m} .$$

Compared to the weight of the conductor, of 1.111 kg/m, this force amounts to a mere $0.0356/1.111 = 0.032$ or 3.2%. If that were a constant load, we would have nothing to worry about. But for a force alternating between 0 and a, the question is vibrational energy. We start by estimating the average wind force as $F_A = F_K/2$ and get

$$F_A = 0.0636Dv^2/2 = 0.0318Dv^2 . \qquad (6.2)$$

Then, we reshuffle Eq. (6.1) $f = 0.185 \times v/D$ into $v = fD/0.185$ and get with Eq. (6.2)

$$v^2 = \frac{f^2D^2}{0.03423} = 29.22f^2D^2 ,$$

and from there $F_A = 0.0318Dv^2 = 0.0318 \times 29.22Df^2D^2$, or

$$F_A - 0.929f^2D^3 . \qquad (6.3)$$

Taking it from here, we compute the energy of aeolian vibrations as the product of F_A, the average force per cycle, and the summed up displacements any point of the cable passes through in each cycle.

With a for the amplitude of the vibration, these displacements include the cable's transversal motion from zero to $+a$, back to zero, on to $-a$, and back to zero again, a total of $4a$. With that value, Eq. (6.3) yields the energy of the aeolian oscillations as

$$W = 4 \times 0.929af^2D^3 = 3.716af^2D^3 \text{ m·kgf per cycle} ,$$

or converted into electrical units:

$$W = 3.716 \times 9.807af^2D^3 = 36.44af^2D^3 \text{ watt seconds per cycle} . \qquad (6.4)$$

With $D =$ constant, vibration energy is proportional to the product of amplitude times frequency squared.

Controlling aeolian vibrations

Equation (6.4) can be used to find whether or not vibration dampers are needed. The fatigue resistance[3] of aluminum cables, which we recall from direct measurements by Edwards and Boyd et al. as $s = \Delta\ell/\ell = 75 \times 10^{-6}$ for the critical strain s or likewise, the material's relative elongation under critical stress. Stress higher than s is likely to cause fatigue-related cracking.

With Hook's law $\Delta\ell/\ell = \sigma/E$ and $E = 7000\ \text{kgf/mm}^2$ for the elastic modulus of aluminum, we get $\sigma = E \times \Delta\ell/\ell = 7000 \times 75 \times 10^{-6} = 0.525\ \text{kgf/mm}^2$. Therefore, cables should not be stressed beyond approximately $0.50\ \text{kgf/mm}^2$.

On the other hand, the strain next to the support points of a catenary is given by the formula $s = af/406$, which yields with the above values for stress:

$$af = 406s = 406 \times 75 \times 10^{-6} = 0.0304\ \text{m} = 30.4\ \text{mm}\,.$$

For the boundaries of aeolian vibrations, reaching from 2 Hz to 30 Hz, this places the amplitudes of permissible vibrations between 1 mm and 15 mm, respectively. Transmission lines that exceed those limits must be equipped with vibration dampers.

Brought into Eq. (6.4), this term gets us

$$W = 36.44af^2D^3 = 36.44 \times 0.0304fD^3 = 1.11fD^3 \qquad (6.5)$$

for the highest oscillatory energy a cable can handle without help from vibration dampers.

Vibration dampers

The equation $W \leq 1.11fD^3$ shows the energy a cable can dispose of without structural damage as directly proportional to the frequency of wind-induced vibrations. An ideal vibration damper should thus act in linear relation with frequency, something encountered so far only in types resembling automotive models.

A simple damping element consists of spirally twisted hard drawn aluminum wires, called armor rods, which firmly wrap around the conductor for about one meter in both directions outward from suspension clamps. Where armor rods don't suffice, the devices in Figs. 6.11–6.13 get into the picture. Each has its particular ways of damping. The festoon (Fig. 6.11) shifts the outgoing and the reflected oscillations *out of sync*. It consists of a piece of cable of a gauge lighter than that of the main conductor, clamped with deep sag to the latter. Spacing increases from the suspension points outwards.

Standing waves, entering the space between the festoon's points of support, find themselves trapped in sections which length is *not* a whole-number fraction of the length of the vibration's half-wave, and die out because the festoons out-of-phase

3 endurance against cyclic-loading-induced structural damage.

Fig. 6.11. Festoon damper

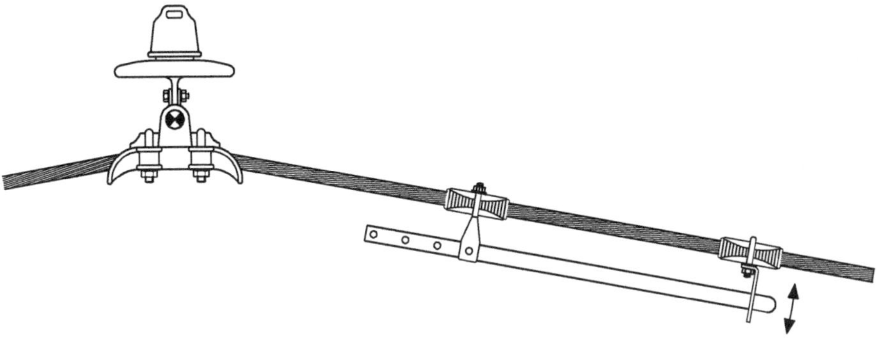

Fig. 6.12. Cantilever damper

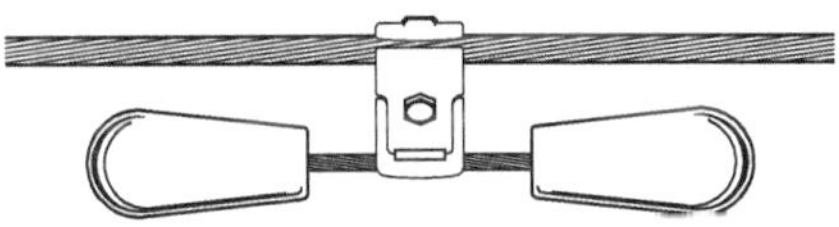

Fig. 6.13. Stockbridge damper

vibration misses out on energy feedback from external sources, such as wind power. The longest waves die away in the wider spaced regions of the festoon cable, while shorter waves get caught between the narrower spaced clamps closer to the cable's suspension points.

The cantilever damper (Fig. 6.12) works on the principle of *energy dispersion.* It consists of a hinged lever which far end swings freely in the vertical slot of a fixed bracket. In the presence of vibrations, the tip of the lever bangs alternately against the upper and lower edges of the slot, hereby dispersing vibrational energy. The frequencies the device will suppress depend on the spacing between the fulcrum and the slot bracket.

Figure 6.13 shows the popular Stockbridge damper, interesting insofar as it draws power from the cable's vibrations for feeding its own particular vibrations, which are twofold: The bell shaped weights vibrate in turn of their mounting points on the ends of a stiff steel cable, called the messenger cable (first peak in damping curve in Fig. 6.14). And second, the weights swing up and down in their entirety, gull-winging the messenger cable in turn of its centerpoint (second peak in damping curve). The eigenfrequencies of the device depend on the elastic properties of the messenger cable and likewise, the spacing and the mass of the weights.

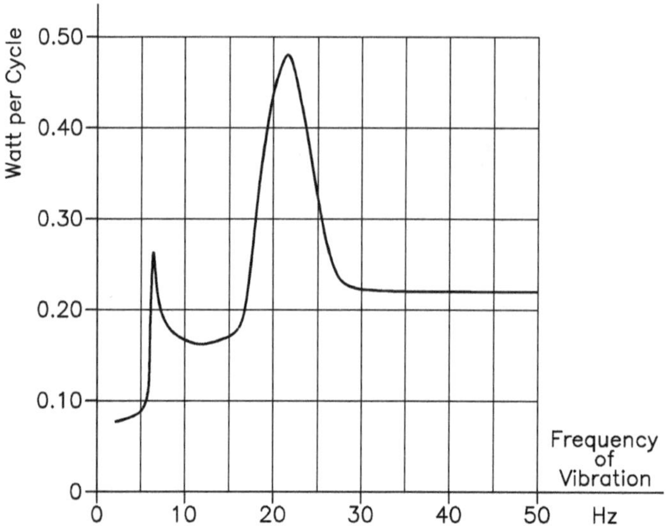

Fig. 6.14. Characteristics of a stockbridge type damper

Within the boundaries where the somewhat erratic looking damping curve may still be approximated by a straight line through the zero point of the graph, commensurate with Eq. (6.5), this widely used device has been found to work reasonably well.

Monitoring vibration damping

Line-mounted aeolian vibration monitors like in Fig. 6.15 record a cable's displacements relative to a massive cylinder which inertia keeps it from vibrating along. They are installed with a conventional cable clamp at places where highest vibrational amplitudes are expected to occur.

A pin, reaching down from the clamp into the bore of the inertial mass, sustains the latter between a pair of prestressed, inversely wound, helical springs. Fixed farther up on the pin is the crossbow with a pair of bilaterally toothed, vertical rack and pinion drives reaching down from its ends, which maintain the inertial mass moving on a straight track and always parallel to itself. One pair of the pinions is interlocked by two engaging gear-segments (not shown in Fig. 6.15). Such rack and pinion guiding mechanisms don't gall and, operating on rolling rather than gliding friction, slide easier than their conventional counterparts. To minimize the ripple effect of gear drives in general, toothing between left and right rack is offset by half a pitch. The left-side guide pinion drives the pen that records vibrations on a strip chart.

In a different setting, a piezoelectric strain gauge is interconnected between the cable clamp and the support pin of the weight. The weight's inertia opposing the motions of the cable causes deformations of the crystal, polarizing it. True to Newton's second law, this makes signals from the strain gauge proportional to acceleration rather than displacement of the cable. This is acceptable because the

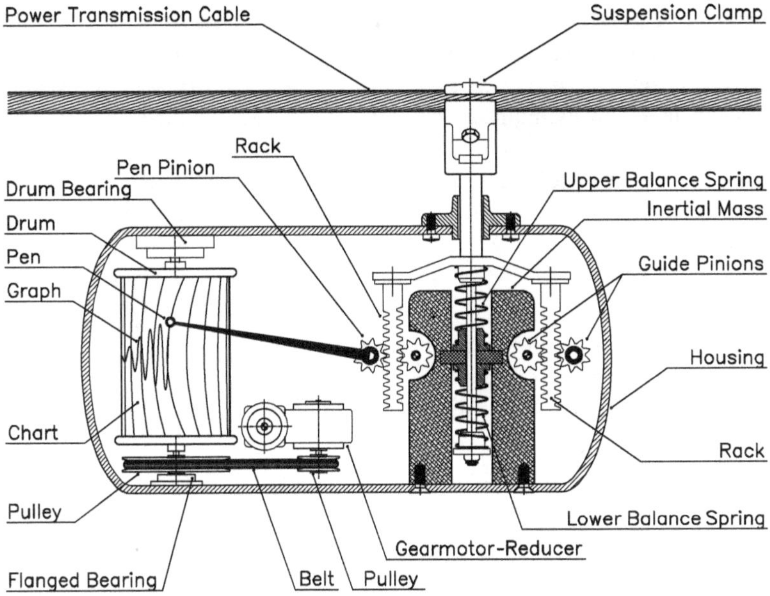

Fig. 6.15. Aeolian vibration recorder for field tests

ideal oscillation, $y = a \sin \omega t$, generates the acceleration of $\mathrm{d}^2y/\mathrm{d}x^2 = -a\omega^2 \sin \omega t$, which is likewise proportional to $\sin \omega t$.

Signals from the strain gauge can be radio transmitted to a ground station, or injected into the proper high-voltage network. If that seems a mismatch like King Kong and his glamour antagonist, never mind. As early as the 1960s, a system known as "music from wall sockets" had become popular for distributing FM programs through electric power lines. The keyword here is *frequency*: 60 Hz for power, and 88 to 108 million hertz for frequency-modulated music. A capacitor between the two sources of electricity sorts them out, because the reactance of a capacitor, $Z = 1/(2\pi fC)$, is inversely proportional to frequency. For a 0.01 μF capacitor, the 60 Hz impedance is $Z = 1/(2\pi \times 60 \times 0.01 \times 10^{-6}) = 265{,}000\,\Omega$, while the same formula, for 100 MHz of frequency, yields 0.16 Ω, a signal to noise ratio of $265{,}000/0.16 = 1{,}650{,}000 : 1$, enough to make the typical 60 Hz hum inaudible, and passing radio signals "loud and clear."

This principle is old news in interstation communication through power transmission networks, and works equally well with signals from instrumentation. And yet, don't bet on finding a million volt capacitor in your electronic treasure trough!

Terra not so firma

Oscillations in cables and strings are unique insofar as they are unidimensional, while most other wave motions, such as sound and light, propagate spherically through three dimensional space. Likewise do seismic waves, although confined to the interior of the Earth. Since most of us see the presence of *terra firma* under our feet as an unalienable birthright, it may surprise that in reality the life of us

Earthlings rather resembles a "dance on a volcano." Mind me, some 5000 earthquakes per year are strong enough to be felt, including one-hundred that cause physical damage. And seismographs register an average of more than 100,000 per year.

The eldest documented earthquake, actually a landslide, happened in 1348 in the otherwise peaceful Carinthian town of Villach, Austria, but the most devastating earthquake in known history happened on 15 August 1950 along the Chile–Peru geologic trench. With magnitude of 9.5 it became the cause of more than 2000 recorded deaths and an up to 25 meter high tsunami.

Most earthquakes result from the motions of tectonic plates, of which the specifically heavier ones lie under the oceans, while those of lighter material support the continents. Less frequent causes of earthquakes are volcanic eruptions and occasionally the collapse of subterranean voids.

Regardless that the drift of tectonic plates seldom exceeds an average of 1 to 2 centimeters (3/8 to 3/4 inch) per year, its effects build up until a major reshuffling of masses sets in as soon as the ultimate compressive strength of the base material is exceeded.

Compressive strength, or the lack of it, keeps the mountains on our planet from "growing into the sky," as the saying goes. If the land-masses that folded the Himalayan Plateau would continue their squeeze, the 8848 meter high Mt. Everest would sink – in principle – down by one inch for every inch gain on the top. By contrast, Olympus Mons, the highest elevation on Planet Mars, reaches 25,000 meter of height above the planet's median radius, but gravity on tiny Mars is only 38% that on Earth, which makes that this mountain weighs the equivalent of a $0.38 \times 25{,}000 = 9500$ meter high mountain on Earth. Which shows that ground pressure under Mt. Everest nearly equals that under Olympus Mons.

The place of maximal intensity of an earthquake is called the *epicenter*, while the point underground where the quake originates, the *hypocenter*, lies usually some 100 km below the surface. Sometimes however, it locates as deep down as 700 km (about 440 miles).

Waves emanating from the hypocenter, called P-waves (P from primary), are *longitudinal* like sound waves, but propagate at much higher velocities – typically between 5 and 15 km/s. Diffraction causes a pronounced curvature in their paths if they cross layers of different density between the hypocenter and the observer's location.

S-waves, also called secondary or transverse waves, travel at 4 to 7 kilometers per second and compare to P-waves much like aeolian vibrations compare to sound waves. And still there are the *surface waves* (Rayleigh waves) which, confined to the Earth's crust, make the ground swing up and down. Their 90° siblings, called *Love waves*, oscillate horizontally in transverse mode.

Discontinuities in the propagation pattern of P- and S-waves occur at the boundary between different layers within the Earthball, such as crust, mantle, outer core, and inner core. The latter is thought to consist of highly compressed, molten 85/15% iron–nickel alloy, which is the reason for the high specific gravity of planet Earth, 5.5 gram per cubic centimeter, highest among all the bodies in the solar system.

Physical exploration of the earth's crust is limited by the rise of temperature with depth, the *thermal gradient*. The deepest borehole so far was drilled in a bid for natural gas at the Kola peninsula near Murmansk, Russia, and goes 12,260 meter deep down. Heat at the bottom reaches 300 °C (572 °F). The deepest research drill-hole locates in the German Oberpfalz and hit in 1994 the depth of 9101 meter (29860 feet).

Earthquake detection

In most measurements, your point of observation is steady and the object of your measurements is in motion. The inverse situation happens in the measurement of earthquakes. Who shakes here are the observers, along with everything they may have hoped to hold on to.

For that reason, a seismograph is designed around a balanced heavy mass, marked "Bob" in Fig. 6.16, to stand like a dead center within this shaking and shaken world we live in. The keyword here is *inertia*. Once again, Newton's second law, force $F = am$, rewritten into $a = F/m$, shows the acceleration a of an object inversely proportional to its mass, m. Simply put, the heavier an object, the harder it is to move. If that sounds like explaining the obvious, consider that the metric system is built around Newton's second law, and so is a great deal of the equations of mechanics and, for that matter, physics in general. Applied to the detection of earthquakes, this theorem invokes a heavy mass as point of reference, which inertia doesn't allow for sudden changes of position. The heavier that mass, the less it gets unseated by seismic forces, and the better an instrument results. For that reason, older, entirely mechanical seismometers employed bobs of up to 20 tons of weight. But the event of electric recording mechanisms, which function with a fraction of the power it takes to actuate mechanical recorders, made the use of weights of only a few kilograms possible; miniaturized versions often make do with as many grams.

An earthquake detection device from AD 136 is credited to a Chinese, named Choko. It was succeeded in AD 132 by Chang Heng's *Dragon Jar*, a copper urn with eight dragonheads around its brim. Each held a stone ball between its teeth which would roll out at the slightest pitch. Though that device accused merely the *occurrence* rather than the *magnitude* of a tremor, we must give it credit for being directional within $360/(2 \times 8) = \pm 22.5°$ of arc.

Remarkable is Luigi Palmieri's (1855) use of mercury filled U-hoses in ways we still honor in civil engineering for leveling two or more distant points (though with water rather than mercury filling). When a ground movement made the mercury level lower in one of the legs and rise in the other of Palmieri's U-tubes, it made contact with an electrode which stopped a clock and put a recording drum in motion. To top it all, a float on the mercury's surface actuated a recording pen. One could argue that mercury's specific gravity of 13.6, or nearly twice that of iron, would introduce a certain degree of inertia to the system, which might have limited its capability to register the entire band of seismic frequencies. On the other hand, direct measurements like this are inherently immune to the many sources of errors that sometimes plague sophisticated instrumentation.

The seismograph in Fig. 6.16 senses *vertical* ground fluctuations. Detection of *horizontal* shifts affords two more seismographs, one operating in the direction

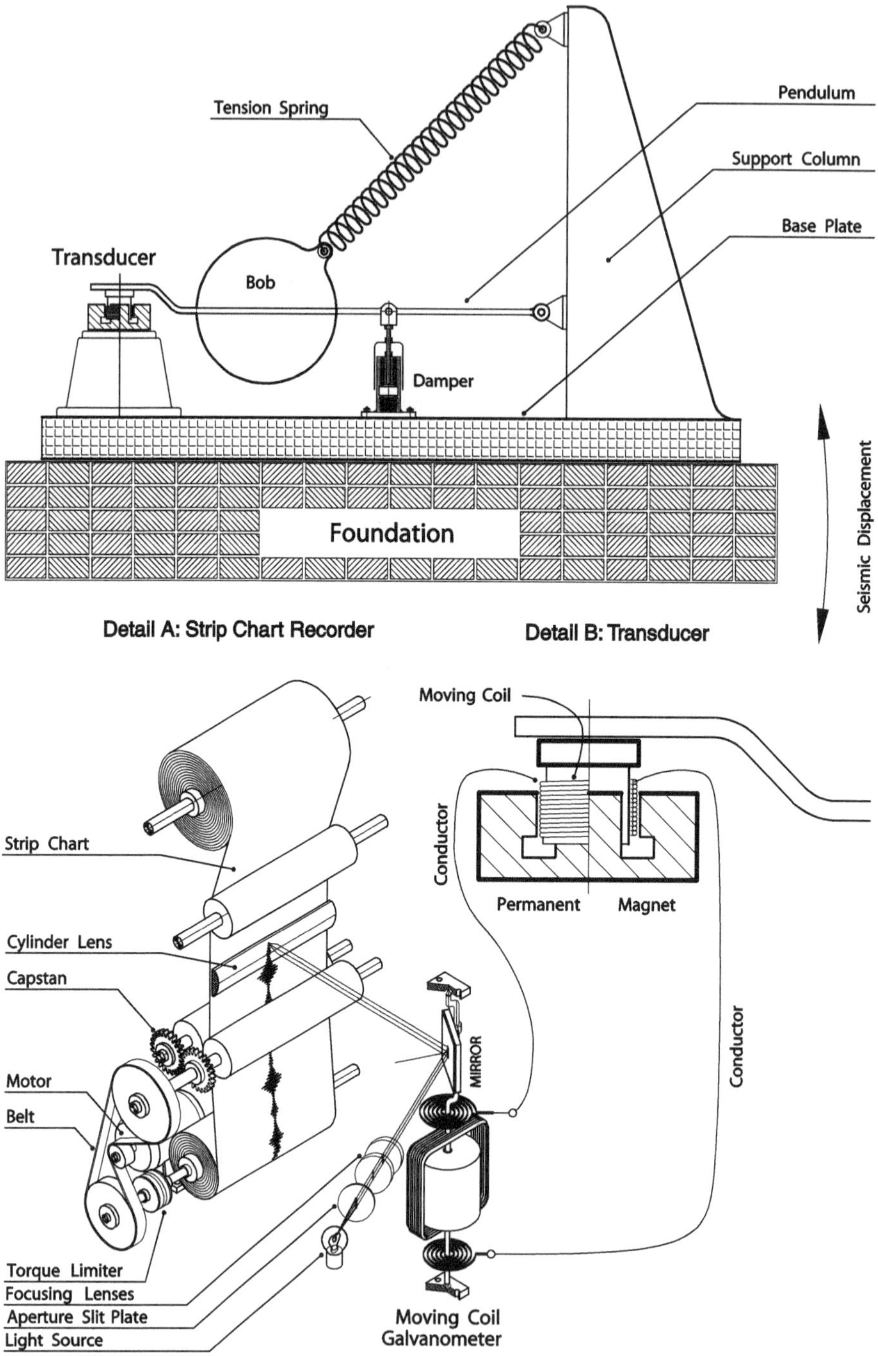

Fig. 6.16. Seismograph with strip chart recording system

north-south, the other east-west. The principle remains the same in all cases, save that the horizontally oriented seismographs have their inertial mass (bob) mounted on a folding grill like frame with its axis (from hinge to hinge) slightly off plumb. This makes it return to its rest position once the dance is over.

A seismometer's foundation must be massive enough to become an integral part of the ground it's sunk into. It is usually bricks, natural stone, or concrete, mounted in bedrock pertaining to a continental plate with no geological cracks in between. The structure resembles that of a jib crane insofar as the tip of the column supports the spring that counterbalances the weight of the bob.

The machine's sensor (Fig. 6.16, detail B) consists of a single-layer-coil of copper wire which dips into the void between the poles of a strong permanent magnet. It resembles an audio speaker in reverse: In a speaker, electric currents make the coil swing along with the tune; in a seismograph, the motions of the coil generate electricity.

The light-beam galvanometer in Fig. 6.16 shows the intensity of the induced electric current by the direction of a ray of light that reflects from a mirror mounted on the galvanometer shaft. That simple device combines amplifier and needle into one unit: It intrinsically doubles the galvanometer coil's deflection, because any turn of a mirror causes the angle of the reflected beam to change by twice as much. And it simulates the sensitivity of a meter whose pointer would be as long as the distance between mirror and the recording drum.

Recording happens generally on photographic film, which however makes it necessary to shield the strip chard from ambient light. The light beam from a projection lamp or laser source passes through a vertical slit aperture and is focused on the recording matter by a cylinder lens, which warrants a sharp focus over the entire width of the recording film (Fig. 6.16, detail A). The same roll of film can be recycled times again by shifting the spool's position sideways after each rewind, an important feature for a machine that operates day and night for years in a row. And regardless that most recordings will be background noise, feeding speed must be high enough to keep the traces from getting indistinguishably from each other when – ahm – the bell tolls.

In addition to the elements in Fig. 6.16, a seismograph needs a time base or clock, set to put marks on the recording strip in regular intervals. Knowledge of the exact time of a quake from station to station in a worldwide network allows for the construction of a spatial model of the propagation of seismic waves. Most of our research of the inner Earth rests on such projections.

To keep the bob from following random trepidations of the spring and its related components, damping by the counter-electromotive force (e.m.f.) in the recording coil may not suffice and the addition of a mechanical damper, as in Fig. 6.14, may become necessary.

A number of different seismometer designs have been tried, the simplest consisting of a bob hanging from several feet of steel wire. Here too, a mirror attached to the wire deflects the ray of light which traces vibrations on the readout screen.

Lately, the availability of sensitive linear converters allows for direct measurement of seismic motions from the spacing of a pair of piers, set at distances from 1 meter to 1 kilometer from each other.

The recording device of the information age is – of course – the computer, fed with signals from an analog to digital converter, something found already installed on most recording sound-cards.

The magnitude M of an earthquake is expressed by the Richter Scale (Charles Francis Richter, 1900–1985), which classifies the intensity of quakes by numbers from 0 to 10, and like the decibel scale for sound, is logarithmic. A number 5 quake is 10 times stronger than one of number 4, and 100 times stronger than a number 3 quake, etc. The field of the less significant earth movements stretches from 1 to 5, while the destructive ones lie between 6 and 8.7. The magnitude of the strongest earthquake in American history, which occurred in 1964 at Prince William Sound, near Anchorage, Alaska, was 8.4. A magnitude 10 quake would open a San Andreas like fault-line circling the Earthball, and a hypothetical magnitude 12 quake would (Heavens beware!) open a rift halfway down toward the Earth's center. Energywise, the magnitudes 4 to 4.7 resemble the explosion of nuclear fission bombs, while fusion devices, vulgo H-bombs, range from 7 to 7.5.

The intrinsic wave energy E of an earthquake is related to its magnitude M on the Richter Scale by the formula $E = 12 + 1.8M$ and is measured in units of erg. The erg is the unit for energy in the centimeter-gram-second system, the first ever absolute system of measures, and relates to the joule (watt-second) by $1\,\text{erg} = 10^{-7}$ J.

Vibration sensors

Back on solid ground, most everything that turns also oscillates to a certain degree. This includes the machinery, industrial and domestic, which once triggered the industrial revolution and with it, the spread of civilization. Drivers of older cars will remember the "critical speed," usually in the 70 to 80 miles range, where the entire vehicle tended to shake. The ruthless and the daring may also recall that at still higher velocities the vibrations miraculously calmed down. The explanation is – once again –, no, not the dreaded speeding ticket, but resonance. Normally, each of the elements of a vehicle in motion vibrates at its own pace. But at the critical speed, the majority swings in unison, some parts at their eigenfrequency, others at harmonics thereof.

The need for vibration-free operation made that the bodies of machine tools are several times heavier than the wherewithal for controlling static forces alone. Mind me, it is the onset of vibrations that limits the cutting depth in chip removing machinery, and the precision of an engine lathe or a miller would be lost if the machines' bed vibrated to the slightest degree.

Humans can detect vibrations from 10 to 800 Hz thanks to Meissner and Pacinian corpuscles in our skin. The former locate at the borderline between epidermis and dermis, and respond to vibrations in the 10 to 60 Hz band. The epidermis, the outer layer of skin, is from 0.05 mm (on the eyelids) to 1.5 mm (at palms and soles) thick. The dermis lies underneath and is composed of living cells. While the prime function of the dermis is respiration as tiny blood vessels end here and feed the outer skin layer, the Pacinian corpuscles housed deep inside the dermis sense the band of the higher vibrational frequencies of up to 800 Hz.

The range of sensory vibration detection is dwarfed by electrical and electronic devices, from the record player pick-up and the microphone to a great variety of industrial and scientific sensor heads. Some types react to displacement, others to the speed of displacement, and still others to the acceleration of the oscillating elements.

The venerable crystal pick-up heads act on *displacement*, because the piezoelectric voltage they generate depends on the degree of pressure exerted on the crystal. Inductive pick-ups, like that on the seismograph in Fig. 6.16, generate voltage according to the *speed* of the coil's displacement. And finally, sensors, such as the ones which release automotive air bags in case of abrupt deceleration, are inertia operated.

Displaying signals

The traditional oscilloscope you find on the workbench of most every repairperson and electronic experimenter is the ideal instrument for visualizing electrical signals by waveform, frequency, and amplitude. The instrument's built in ± 0.5 volt square-wave reference allows for voltage checks, while frequency can be found by counting the number of peaks relative to the horizontal scanning frequency.

Interesting waveforms can be generated with a non-linear time-base, such as the 60 Hz alternating current from a wall socket, transformed down to 6 or 12 VAC. Utilities maintain their networks frequency very accurately at 60 Hz AC and keep the sinusoidal shape of their voltage free from harmonics which presence would make electric motors hum and occasionally heat up.

Connected to the x-input terminal of an oscilloscope, network AC makes for a time base of the type $x = a \sin(120\pi t + \alpha)$, where a stands for peak voltage. Harmonic oscillations such as $y = b \sin(\omega t + \beta)$ show up on the screen as the so-called *Lissajou's Figures*, which range from a straight line to a circle and a plurality of eye-catching loops (Fig. 6.17), revealing the frequency of the input as 60 times the figure's count of positive peaks.

Not surprisingly, a 60 Hz signal in phase with the 60 Hz scan traces a slanted straight line (Fig. 6.17 a). The loops with 2, 3, and 4 peaks (Fig. 6.17 d–f) indicate respectively frequencies of 120, 180, and 240 Hz.

A phase difference of 90° (Fig. 6.17 b) converts the straight line into a circle, and other phase differences draw ellipses (Fig. 6.17 c), which we can slenderize by varying the phase angle. Oh my, if thinning the human body were *that easy*!

Frequency measurement instrumentation

Figure 6.18 shows a handy mechanical frequency meter, predominantly used for checking network frequencies. The core of the instrument is an array of vibrating tongues (reeds), which react to the alternating field of a strong electromagnet much like a tuning fork acts upon mechanical excitation. With the center blade's resonance frequency at 60 Hz, the blades to the left are tuned to 59, 58, 57 Hz respectively, and those on the right to 61, 62, and 63 Hz, etc. The tongue that swings with the highest amplitude marks the frequency of excitation, if the adjacent blades left and right oscillate at mutually equal amplitudes. If two adjacent blades vibrate at identical amplitudes, take the average, such as 59.5 Hz if the 60 Hz and 59 Hz reeds swing in unison. With some experience, even $1/4$ Hz frequency shifts can be estimated.

Digital frequency counters, which excel by their precision and range, use an electronic counter controlled by a sequencer chip, comparable with a stop-

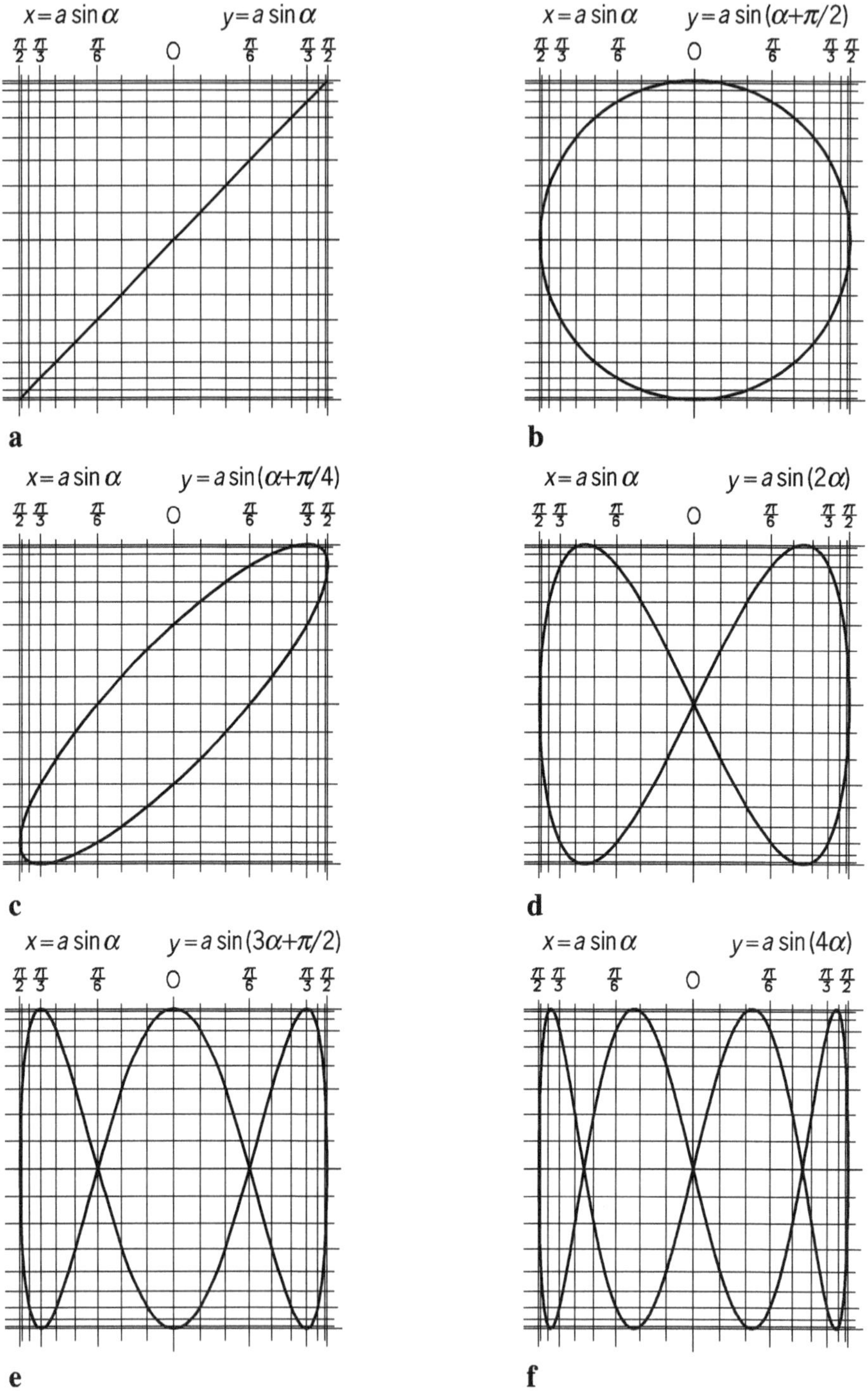

Fig. 6.17. Lissajou's figures

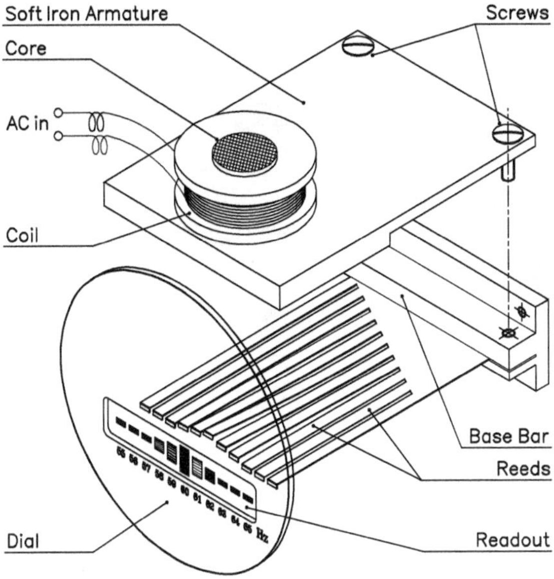

Fig. 6.18. Tongue frequency meter

watch and a pulse counter for finding the frequency of a signal. You start with resetting the counter to zero, then trigger the stopwatch and the counter simultaneously, and subsequently turn the counter off as soon as the watch reaches a predetermined time, say, 10 seconds. Connected to a wall socket, the counter would show 600 counts in 10 seconds, which, not surprisingly, gives $600/10 = 60$ Hz for the utility frequency. Abroad, where utility standards are 50 Hz, your count would end at 500.

In electronic counters, the stopwatch is replaced by a digital clock, usually designed around the classical 555 chip or its numerous modern clones along with a few external components, which values determine the clock-frequency. Next comes the sequencer chip, which resets the counter to zero at the start and stops it when the period of counting is over. With a one second time reference, the readout will show frequency directly. Higher signal frequencies allow for time bases of 0.1 or 0.01 seconds and, if necessary, can be inputted through one or several *divide by 10* chips.

Noise-contaminated inputs and stray signals must be fed through a comparator, which blocks all inputs – noise and signal alike – below a certain threshold, not unlike the familiar Dolby circuit. That leaves for the count only the peaks that stick out of the noise zone. The comparator also sets the instrument's input impedance, which up to 35 MHz should be high enough as not to become a significant load on the signal source. Above that, the particularities of very high and ultrahigh frequency transmission must be taken into account, in particular the cable termination impedances, such as 50, 70, or 100 ohm, in order to prevent reflections. For weak signals, a 741 type operational amplifier can be interceded as long as noise amplification doesn't cause erratic outputs.

Accuracy is highest with frequency meters with a long time base, but response time of the instrument should be kept within reasonable limits. On a 1 second time base, a 1 kHz signal will be counted with $1/1000 = 0.1\%$ accuracy, but a 10 Hz signal, counted on the same time base, would come out with an unacceptable 10% error margin.

Forced oscillations

Fundamentally different from oscillations with "a life of their own" are mechanically induced periodic motions. Harmonic (sinusoidal) oscillations can be generat-

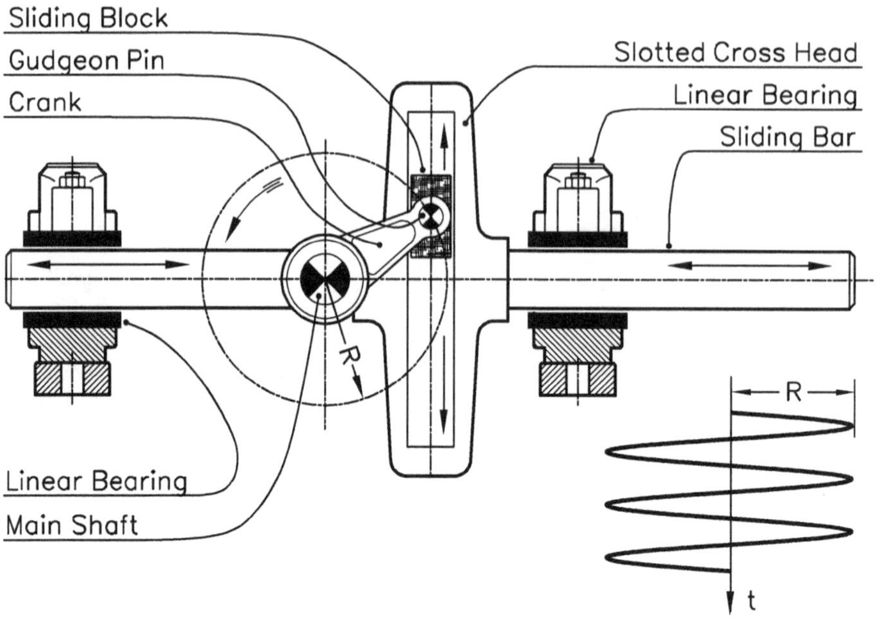

Fig. 6.19. Harmonic motion generator

ed by a mechanism called *Scotch Yoke* or, more scientifically, *Crank and slotted Cross-head*, shown in Fig. 6.19. With the crank in uniform rotation, the sliding block moves up and down the slot in the cross head, guiding it to the left and to the right in ways that resemble the plot of a sine line on paper (Fig. 6.19, lower right).

More complex than the crank mechanism in gas and diesel engines, the Scotch Yoke is left for special applications, such as a steam powered pump with pistons on both ends of the same plunger. The steam piston mounted on one end of the reciprocating road powers the device, while the water piston on the other end does the pumping.

A flywheel smoothens the operation of the machine and provides energy storage and recovery insofar, as the energy drawn from the flywheel for accelerating pistons and plunger during the first half of the stroke is recovered in the following half-stroke where the linear momentum of the mass of the latters helps in accelerating the flywheel. Without the flywheel, the energy for the repeated speed up and slow down of the mass of pistons and plunger would be lost.

With $\omega = 2\pi f$ for angular velocity, a for amplitude, and f for frequency, the equation of harmonic oscillation reads $y = a \sin \omega t$. The velocity of displacement follows as

$$a \frac{\mathrm{d}y}{\mathrm{d}(\omega t)} = a\omega \cos \omega t \,,$$

and acceleration as

$$a \frac{\mathrm{d}^2 y}{\mathrm{d}(\omega t)^2} = -a\omega^2 \sin \omega t \,.$$

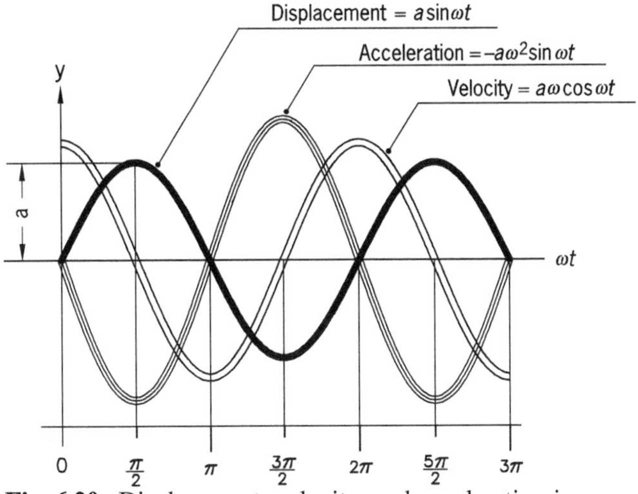

Fig. 6.20. Displacement, velocity, and acceleration in harmonic oscillations

The trigonometric identities $\cos\alpha = \sin(\pi/2 - \alpha)$ and $-\sin\alpha = \sin(\pi + \alpha)$ prove that the functions $\sin x$, $\cos x$, and $-\sin x$ differ from each other merely by phase shifts, while their sinusoidal shape remains unchanged. Therefore, displacement, velocity, and acceleration of harmonic oscillations follow *sinusoidal curves*, yet differ in phase. In the Fig. 6.20, the curve drawn in bold shows displacement; the twin-line curve, velocity; and the triple-line curve, acceleration. The shape of the three curves is the same, but the similarity ends here, because the curves' dimensions are meter (m), meter per second (m/s), and meter per second squared (m/s^2), respectively.

Introduction of the term *radian frequency* $\omega = 2\pi f$ simplifies the equations for harmonic oscillations into $y = a\sin\omega t$, $v = a\omega\cos\omega t$, and $A = -a\omega^2\sin\omega t$, where A stands for acceleration.

Crank mechanics

Due to its simplicity, the crank mechanism in Fig. 6.21 has become the heart of the machine that takes care of our daily ride to the office and back. James Watt would have used the same type of mechanism in his steam engines hadn't it been that some competitor of his held a patent on a crank drive from an earlier attempt at building a heat engine. But the quick-witted inventor and engineer he was, Watt used a twin rack and ratchet drive to convert the linear displacement of the pistons of his steam engines into the rotation of the flywheel. Regardless of such legal obstacles, Watt's machine and its successor, the internal combustion engine, opened up the world's resources of thermal energy to an increasingly energy hungry human race. In 2001 alone, 226 million automobiles crowded our country's highways and keep proliferating to the tune of about 800,000 per month.

And yet, early projects of an internal combustion engine must have looked like the typical case of an in principle correct design, yet one that would never work under real-world conditions – a verdict backed by sound reasoning – mind me. Just take the cylinders, internally exposed to the heat of exploding gases and some 1000 psi of pressure, while water chills its outside! And the piston, immersing itself into that brazing heat and still getting enough lubrication to slide and seal without wearing down in no time?

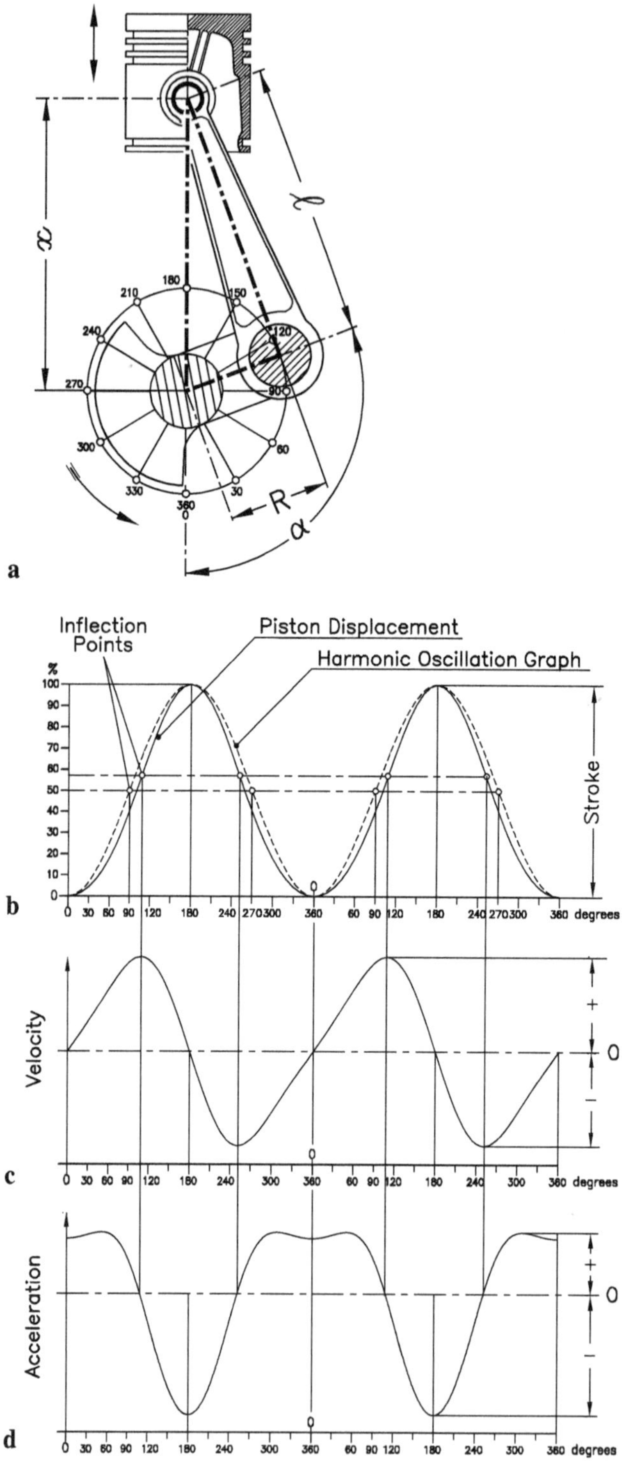

Fig. 6.21. Crank mechanism

And last but not least, the peripherals! Spark plugs supposed to operate in compressed air–fuel mixtures and still maintain their electrodes clean through hundreds of millions of discharges? The ignition switch triggered at the same rate without wearing down its contacts? And still that device they call carburetor, supposed to vaporize in every cycle just the right amount of fuel for maintaining the 1 : 16 fuel–air ratio in cold and hot weather, with the engine idling or racing at full power?

And to top it all, the proposed engine couldn't even be turned on like its steam-powered siblings or electric motors, but had to be cranked at the risk of breaking your arm in the not so unlikely event of the engine's backfiring! "Guilty on all Counts" would be the verdict for any kind of machinery of similar complexity. Except that this wasn't just "any kind of machinery." It was the *dream machine* of most every human being on this Earth: The offer of *individualized transportation.*

Only such massive demand could trigger the investment of a century of the labor of thousands of top engineers and machinists into an engineering "mission impossible". Let's send a mental "thank you note" to this army of unknown tinkerers, when our car's engine runs so quiet that it becomes hard to tell whether it's on or off.

Unlike the steam engine, where a crosshead guides the piston in a straight line, the typical crank drive of internal combustion engines leaves that to the piston itself by making it long enough to avoid galling. In Fig. 6.21, the effective radius (eccentricity) of the *crankshaft* is marked with an R, the length of the *connecting road* with ℓ, and x stands for the *displacement of the piston pin* relative to the center-axis of the crank shaft. With f for *frequency* or rotations per second in the present case, the angular position of the crank relative to the lower dead point is the angle $\alpha = 2\pi f$. In the scalene triangle composed of R, ℓ, and x, traced in bold centerlines, the corresponding internal angle is $180 - \alpha$. With the identity $\cos(180 - \alpha) = -\cos\alpha$, and the law of the cosines, we get for this triangle $\ell^2 = R^2 + x^2 + 2Rx\cos\omega t$, which with the identity $1 - \cos^2\alpha = \sin^2\alpha$ is reshuffled into a mixed quadratic equation with the solutions

$$x/R = -\cos\omega t \pm \sqrt{(\ell/R)^2 - \sin^2\omega t}$$

The substitution $f(\omega t) = \sqrt{(\ell/R)^2 - \sin^2\omega t}$ simplifies this into

$$x/R = -\cos\omega t \pm f(\omega t)\,. \tag{6.6}$$

This equation breaks down the piston displacement (relative to crank radius R) into a harmonic function, $-\cos\omega t$, and the inharmonic term $f(\omega t)$.

The velocity v of the piston's displacement is the differential $\mathrm{d}x/\mathrm{d}(\omega t)$. The differential of the first term of Eq. (6.6), $-\cos\omega t$, is $\omega\sin\omega t$, and differentiation of the second term yields

$$f'(\omega t) = \frac{-\omega\sin 2\omega t}{2f(\omega t)}\,,$$

so that we get

$$\frac{v}{R} = \omega\sin\omega t - \frac{\omega\sin 2\omega t}{2f(\omega t)}\,. \tag{6.7}$$

Here again, we have the harmonic term $\sin \omega t$, followed by a correction term dictated by the design of this particular kind of crank mechanism.

Since the cause of vibrations is not velocity, but acceleration A, we need the differential $\mathrm{d}v/\mathrm{d}(\omega t) = A$, which we derive from Eq. (6.7):

$$\frac{A}{R} = \frac{1}{R} \times \frac{\mathrm{d}v}{\mathrm{d}(\omega t)} = \omega^2 \cos \omega t - \frac{2\omega^2 \cos 2\omega t \times 2f(\omega t) - 2f'(\omega t)\omega \sin 2\omega t}{4f^2(\omega t)},$$

or simplified

$$\frac{A}{R} = \omega^2 \cos \omega t - \frac{\omega^2 \cos 2\omega t}{f(\omega t)} - \frac{f'(\omega t)\omega \sin 2\omega t}{2f^2(\omega t)},$$

or still

$$\frac{A}{R} = \omega^2 \cos \omega t - \frac{\omega}{f(\omega t)} \left(\omega \cos 2\omega t - \frac{1}{2} \sin 2\omega t \times \frac{f'(\omega t)}{f(\omega t)} \right). \qquad (6.8)$$

The first term in this set of crank equations (6.6)–(6.8) is in all cases an expression for harmonic oscillation. Only the second term introduces the deviation between an ideally harmonic drive and the real-world crank drive, regarding displacement, velocity, and acceleration. If those deviations seem small in the graph for piston displacement (Fig. 6.21 b), they build up as we get into velocity (Fig. 6.21 c) and more so for acceleration (Fig. 6.21 d). Most significantly, the acceleration graph has four peaks and four valleys of different amplitudes rather than the two of the other curves.

Since acceleration and inertial forces are proportional to each other, all that math boils down to vibrations. The sway of the pistons is a source of linear vibrations at the ratio of the weight of the piston to the weight of the engine. If a 100 kg heavy one-cylinder engine with a 1 kg piston were free to move, it would sway contrary to the piston's displacement at the rate of one millimeter per ten centimeters of stroke. That's the raison d'être for the familiar engine mounts of energy-absorptive material, such as polyurethane, which we recall from its use in electric power line hardware.

A different matter are torsional vibrations. According to the law of preservation of the angular momentum, ωR^2, a freely suspended vehicle would spin in response to the turn of the crank shaft and its associated elements. With R for the radius of gyration and ω for angular velocity, let $\omega_1 R_1^2$ represent the angular momentum of the crank and transmission shafts, and $\omega_2 R_2^2$ the opposing momentum of the car's body. Because both were zero before the start, the condition $\omega_2 R_2^2 - \omega_1 R_1^2 = 0$ is maintained while the engine is running. That means that, suspended in thin air, the entire car would rotate contrary to the engine's rotation at the ratio of $(R_1/R_2)^2$. That's why helicopters must use their tail-mounted propeller to save them from spinning horizontally.

In a one-cylinder, four-stroke engine, only one half turn of the crankshaft in every two actually drives the engine, which feeds during the three remaining cycles on energy from the flywheel. Engine speed fluctuates accordingly and so does the countertorque of the entire vehicle in an erratic vibration pattern.

The source of such vibrations is voided in the four-cylinder engine, where one of the cylinders is active at any time. But the pressure generated by combustion of the air–fuel mixture in the cylinders varies over the length of the stroke of the piston. These residual oscillations must be kept from resonating with the crankshaft at the latter's eigenfrequency. Crankshafts are thus designed to resonate at far higher frequencies than the sequence of ignitions.

While vertical mass-forces mutually annul in two- and multicylinder engines as the pistons move in opposite directions to each other, longitudinal tilt on the motor block remains because of the different positions of the pistons along the length of the crankshaft. The so-called boxer motor of the venerable air-cooled "beetle" reduced vibrations by accommodating the engine's four cylinders in pairs opposite to each other. For in-line engines, six cylinders are considered of lowest vibration. Still more stable engines have greater numbers of cylinders usually in V-configuration.

Some fascinating crank-driven machinery

Beyond the automotive environment, forced harmonic motion is present in material handling machinery, such as vibrating feeders and sieves, and not to forget such oddballs as the *Walking Beam Conveyor*, which hauls goods step by step on a pair of swinging sidebars. Walking beams take over in environments which would destroy other types of feeding devices. An example is the continuous furnace in Fig. 6.22, used for annealing heat-treated tool-steel bars, a process to reduce brittleness while sacrificing hardness to some measured degree.

Annealing temperatures are in the range of 180 to 300 °C, depending on the desired toughness, and can be visually checked by the surface-color of the blank steel parts: Light yellow for 200 °C, where a slight loss of hardness sets in, to bluish-gray at 350 °C as a warning flag that crimson glow and a substantial loss of hardness are next.

Figure 6.22 is the semischematic drawing of such a walking beam (stepwise feeding) conveyor, feeding steel bars through an annealing furnace. The machine-base doubles as a fixed conveyor table, while the swinging side beams (one in

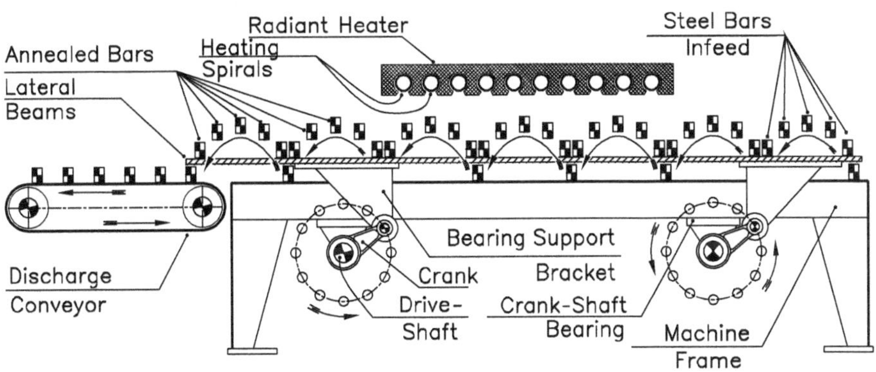

Fig. 6.22. Walking beam conveyor

front and the second on the back of the table) are actuated by a twin crank-drive which makes that any point on a beam emulates the circular path of the crank pins and remains parallel to itself throughout the full cycle. The load, consisting in the case in point of steel bars (drawn checker faced) placed square on the side beams, does the same: During the upper semicycle, the beams lift the bars off the table, carry them along, and deposit them on the holding table when the beams dive below the level of the tabletop. The next forward cycle starts when the beams reemerge from underneath.

With this system, the load remains at rest half of the time, which might be desirable in processes where a certain time of exposure is mandatory, such as in heating and cooling equipment, but would be counterproductive in applications where swift material handling is desired. Here, the principle in Fig. 6.22 must be modified for uninterrupted motion by introducing a second pair of side beams, timed to oscillate in counterphase. That second pair of beams picks up the load when the first ones dive below the fixed side beams, and vice versa.

To this end, the cranks in Fig. 6.22 make way for the pair of crankshafts in Fig. 6.23. This figure's exploded view may give the impression of a highly sophisticated design, but the complexity of the crankshafts isn't any worse than that of the crankshaft of a four-cylinder car engine, save for one extra crank that keeps the two shafts turning in unison. Positioned at 90° relative to the other cranks, its driving force is greatest at the sidebars' dead points and thus minimizes torque fluctuations.

While one crankshaft is motor driven over worm reducer and timing belt, the

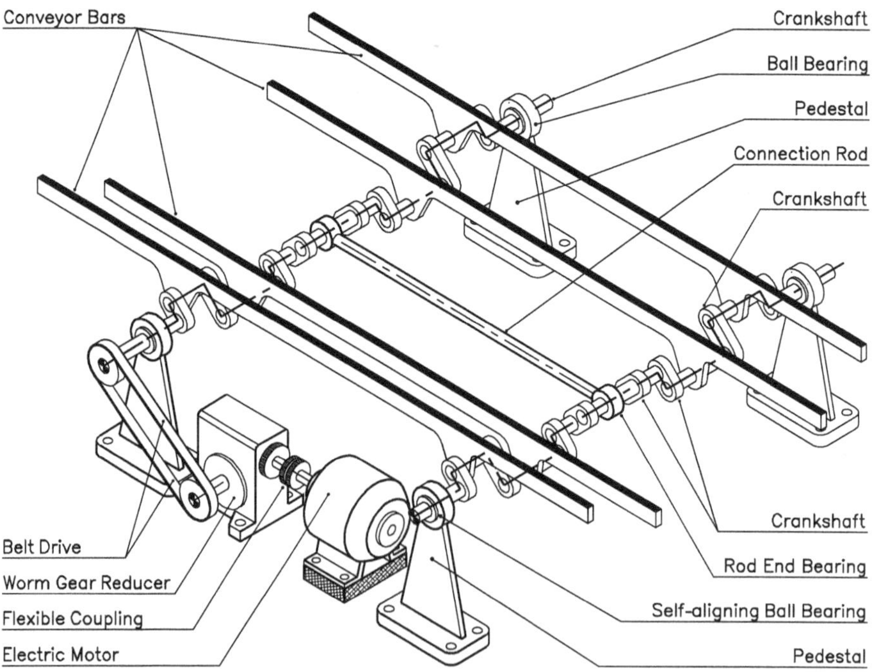

Fig. 6.23. Walking beam mechanism (exploded view)

other is carried along by rigid coupling through the oscillating side-beams and the connecting road between an extra crank on each shaft at 90°.

Analog frequency meters

While the frequency of mechanically generated periodic motions can be derived from the rpm of the drive mechanism, independent vibrations need to be measured by appropriate instrumentation, such as the aeolian vibration recorder or the tongue frequency meter. Though the presence of individually tuned reeds suggests otherwise, the latter is still an analog instrument because it seamlessly follows frequency variations.

A truly analog counter is the electricity meter in your backyard. The pointers on its five dials (Fig. 6.24) indicate the kilowatt hours used up since the latest reset or the instrument's installation[4]. They are directly geared to each other at the ratio of 1 : 10, resembling the hands of a clock, if it weren't for the latter's ratio of 1 : 12. Since mating gears turn opposite to each other, each pointer of the electric meter turns in reverse to its neighbors, one clockwise, the next counterclockwise, and so on. On the first dial (left in Fig. 6.24), the number 1 stands for 10,000 kW h, the number 4 on the second dial for 4000, the 7 on the third dial for 700, etc. Accordingly, you read the integer on one dial and find the value of the decimal remainder as the integer on the next dial to the right. Use "guessometry" for fractions on the outermost right dial.

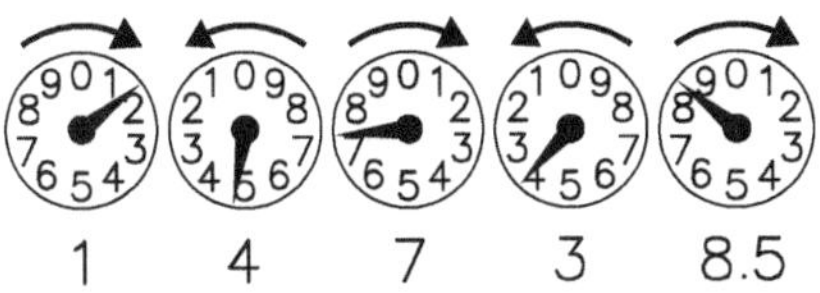

Fig. 6.24. Readout of electric dial meter

Digital metering

Unlike analog instruments, which uninterruptedly follow the changes of the measurand, digital frequency counters refresh their readouts in steps which can be quite conspicuous at low frequencies. Only if the numbers change at least as fast as the frames of a movie projector do we get the illusion of a continuously auto-correcting display.

The mileage counters (odometers) embedded in our cars' speedometer panels are digital instruments, insofar as each of their cylindrical dials remains put until the one to the right completes a full turn and actions a latch which shifts the former forward by one pitch. That goes on throughout the row of scaled dial cylinders which, combined, display the vehicle's total mileage since it rolled off the production line; apropos a legally protected number.

Similar mechanical counters were once the building blocks of a hierarchy of calculators, from cash registers to hand-cranked machines that performed the four basic operations but, if handled by an algorithm-savvy operator, could be used to figure logarithms and even trigonometric functions.

4 1 kW h = 3600 ws = 3600 joules

In comes the computer!

A basic 1 : 10 ratio building block like in mechanical counters is missing in electronics and electrical engineering. The logical elements the designer can pick from are on/off switches, and that's about that.

Connected to a power source of, say, 5 volts, a switch can generate two different signals: one of 0 volt in the *open* position, and another, of +5 V, when *closed.*

Sets of two switches, properly connected, operate as *gates* (Fig. 6.25). In series, they combine into an *and-gate*, because both have to be closed to conduct a signal, such as powering a panel light. If wired in parallel, either switch does that job, which earned this setup the name *or-gate*. Substitution of the *normally open* switches by the *normally closed* type inverts their functions: they become *nand-* and *nor-gates.*

Relays, i.e., switches actuated by electromagnets, made it possible to build rudimentary computers, known for their sewing-machine-like beat. When multichannel electronic tubes, such as the 6AU7 twin triode[5], took the place of relays, the noise subsided but the limited lifetime of the tubes' filaments cut into the half-life

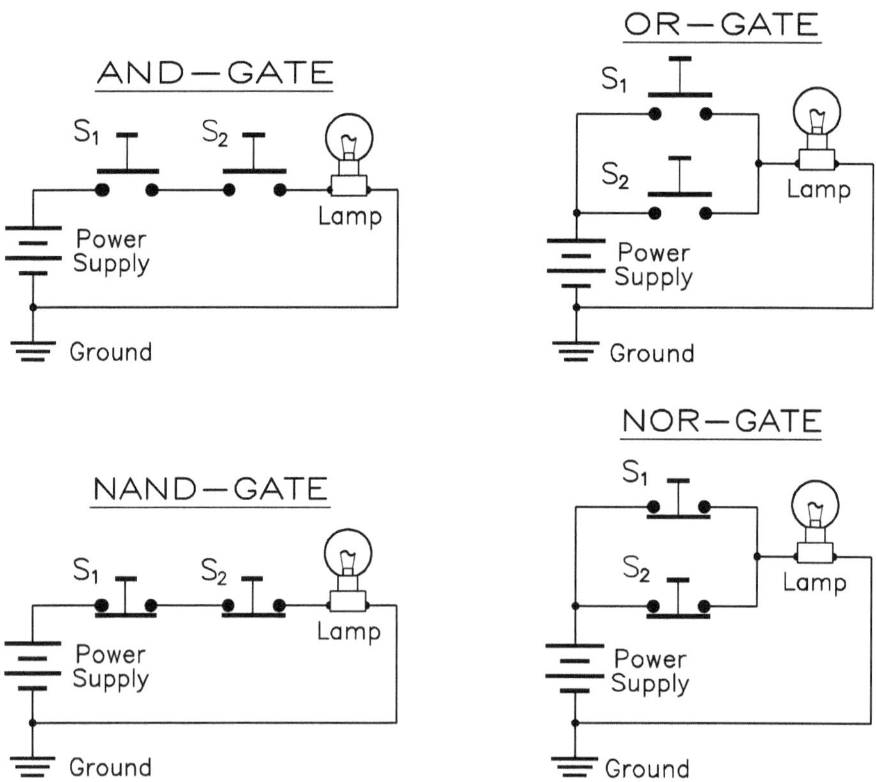

Fig. 6.25. Gates

5 twin vacuum tube with 3 active electrodes, e.g., cathode, grid, and anode.

of the devices that employed thousands of them. Contemporary cartoons showed an engineer laboriously operating a wall-sized computer until the panel lights display $2 \times 2 = 5$, while another shows a salesman trying to make a grocery shop owner cram a wall size computer into his already overcrowded premises with the argument: "The amount of initial investment shouldn't keep you from enjoying the amenities of electronic bookkeeping."

Well, that was then. Semiconductors made it possible to cram thousands and sometimes millions of elements onto a single chip, among them *flip-flops*, the two-state devices that reverse when pulsed. A counter assembled from two-state circuits operates at the 1 : 2 ratio from element to element, rather than the 1 : 10 ratio of mechanical counters, and its outputs consist of strings of the numbers 0 and 1. Used in ways of the numbers 0 to 9 in our base ten system, they combine into *binary numbers*. The numbers 0 and 1 are identical in both systems, but 2 in binary is spelled out 10, 3 becomes 11, and 4 turns into 100.

Binary numbers, such as 1, 2, 4, 8, 16, 32, etc., add up to any desired value without repetition. Once upon a time that led to a legislator's proposal to coin the country's currency in a binary pattern, which – he hoped – would minimize the government's expenses with the National Mint. Lazy cashiers who defeated the idea wouldn't have guessed that binary was set to take over their work and yet, rather than drying out the job market, generate new jobs in far greater numbers than those lost.

To display binary counts in decimal, a chip called *1-of-10 decoder* (such as the TTL 7442) converts the state of its four inputs, weighed 1, 2, 4, and 8, into the placement of ten outputs, numbered from 0 to 9. Counts from 10 to 15 (1010 to 1111) are ignored in this system, fittingly called *binary coded decimal* (BCD). In Nixie display tubes, the ten outputs of the 1 of 10 decoder get connected to the tube's ten corresponding pins that connect inside the glass dome to miniature gas discharge tubes, shaped respectively after the numbers 0 to 9.

For use with seven-segment LED (light-emitting diode) displays (Fig. 6.26), each of the ten outputs of the 1 of 10 decoder must light a selected set of the display's seven sections in a pattern resembling the intended numeral. The number 1 is formed by two vertical sectors, 2 by two verticals and three horizontals, arranged like a question mark, and so on. The matrix converter, a passive chip operating exclusively with diodes, performs this trick. A quick witted, yet timid electronic experimenter allegedly built a diode matrix that, at the turn of a multistep rotary switch, successively displayed the letters: I L O V E Y O U. But let's leave such details for a later chapter.

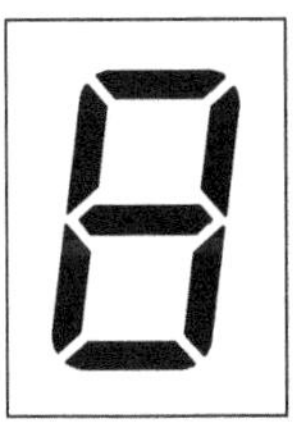

Fig. 6.26. 7-Segment readout chip

Noncontact measurements

We usually take it for granted that the energy of the phenomenon under investigation greatly exceeds the energy required for its detection and measuring. We neglect a priori the electric energy spent on deflecting the needle of a galvanometer, or on feeding the oscillation of the reeds in a tongue frequency meter, just like we hardly worry about the gas we spend on keeping the car's odometer spinning. But not always are things that simple. For instance, vibration tests on a variety of scaled-down prototypes from airframes to automotive chassis could have their outcome perceivably altered by the inertia of attached sensors.

Noncontact signal sensing eliminates such risks of distorted results. Simple and rugged devices, they are all around us: As counters and open/closed triggers at the gates of movie theaters, stadiums, train stations, baggage claim areas, the majority of buildings with public access, and so on. More sophisticated applications include numerically controlled machine tools and automated conveyor systems and, last but not least, safety systems on presses and garage doors.

The most common contactless sensors include a light source and a light sensor, the latter built around a photoresistor, photodiode, or phototransistor. Where a separate light source would not be practical, a reflecting surface can do.

Sonar and radar systems use the reflection of radiated sound or radio wave beams to detect and locate the object of interest. Here, the frequency of the radiated signals counts. Early police radars for instance could be confused by a tensioned steel wire ahead of the front bumper of the car: As the wire vibrated in response to eddy formation, its frequency became comparable with that of the police radar, causing bungled readings. The problem was easily solved with the use of higher radar frequencies, but yet, response time remained limited.

Frequencies of lightwaves are still the highest and provide unsurpassed resolution when used for length measurement. Their characteristics became clear when Albert A. Michelson (1852–1931) and Dr. Edward W. Morley (1838–1923) designed a device – the Michelson interferometer – intended for checking whether the Earth's velocity through sidereal space, of 30 km/s, would add to the speed of a parallel directed beam of light from a source on Earth shifting it from 300,000 to 300,030 km/s.

Michelson's light source had been an Argand burner, a quantum leap – at the times – from those then customary oil lamps with their reddish, smouldering flames. The Argand lamp's tubular wick, guided between concentric sheet-metal sleeves, provided the flame with atmospheric oxygen from in- and outside alike, while an external glass cylinder kept the burning air–fuel-gas mixture from dissipating. Rather than the green and crimson flames characteristic for oversaturated mixtures, the burner produced the bright "blue flames" typical for oxygen-rich blends of combustible gas mixtures. Argand burners even became the beacons of lighthouses, and without them, Michelson and Morley might have missed out on renewing some basic concepts of physics.

Though Michelson's experiment seemed to be set for "proving the obvious," it ended in a resounding "no"; which led to Einstein's special theory of relativity, a set of equations derived on the principle of the invariability of the speed of light, regardless whether light source and/or light receiver are in motion or at rest.

One could expect that such an independence precluded the use of light for the detection of motions or oscillations of physical objects, if it weren't that Michelson's results were related to the propagation of *photons*, the energy packets that carry light quanta through space. True to the duality principle (Louis Victor de Broglie, 1924), photon energy interacting with matter morphs into wave energy –

a principle knowingly or unknowingly used through the ages by designers of optical instruments, the Doppler vibrometer among them.

Doppler vibrometer

The Doppler vibrometer in Fig. 6.27 resembles the Michelson interferometer except for the reflection of the measuring beam by a vibrating object rather than a mirror, and the use of a helium-neon (He-Ne) laser, operating at 474×10^{12} Hz of frequency and 0.6328 μm (6328 A) of wavelength. The instrument's centerpiece is the beam splitter, a semitransmissive (thin silvered) mirror, positioned at 45° relative to the principal laser beam. The mirror's reflective coating is thin enough to let 50% of the lightbeam pass straight through, while the other 50% are bent by 90°. Thus the beam's horizontal component becomes the object beam, and the vertical component the reference beam. The latter hits a coplanar mirror and gets reflected into the photodetector. Meanwhile, the object reflects the straight-through beam toward the underside of the coating of the beam splitter, where it gets bent by 90° into the detector. Here, constructive and destructive interference alternate for every change of the object's position by $\lambda/2$. With a He-Ne laser of 6328 A of wavelength, that happens in steps of 0.3164 μm. The target's total displacement is then the product of 0.3164×10^{-6} times the number of voltage pulses from the photodetector.

This principle of length measurement is used for the comparison of gauge blocks, but it gets ambiguous with periodic motions, where the same train of pulses may stem from forward or backward motions of the object, such as the swinging membrane on the right of Fig. 6.27.

Directional sensitivity can be introduced with the use of 45° polarized laser light and placement of a $\lambda/8$ thick retardation plate in the path of the reference beam. The beam passes the plate on its way up and then again on its way down, doubling

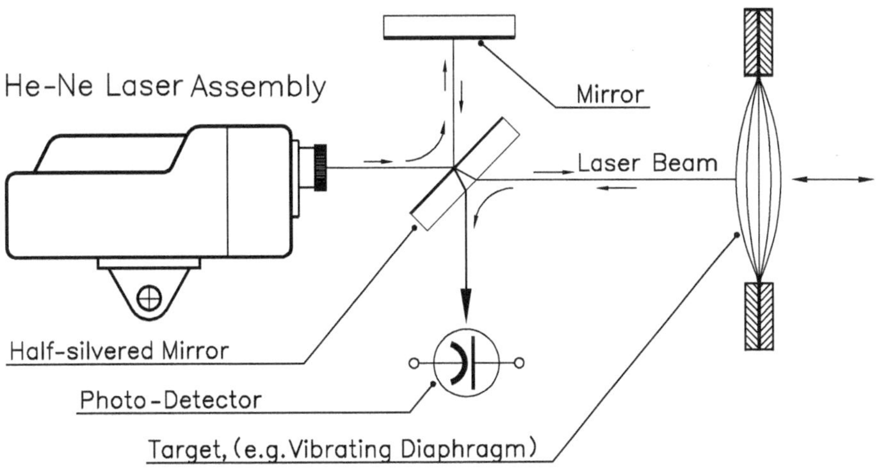

Fig. 6.27. Doppler vibrometer

the phase shift to $\lambda/4$. A polarizing beam splitter then generates two orthogonal wave components, which, as sine and cosine waves, go to individual detectors.

Next in sophistication are *scanning laser beam vibrometers*, which measure vibrational displacement and velocity at up to 250,000 measurement points in a process resembling the use of a scanning light beam in TV cameras. And finally, the combined operation of three laser vibrometers makes it possible to generate 3-D images of objects, such as automotive prototypes, which seem to breath to the rhythm of elastic deformations of their surface.

7 Thermodynamics

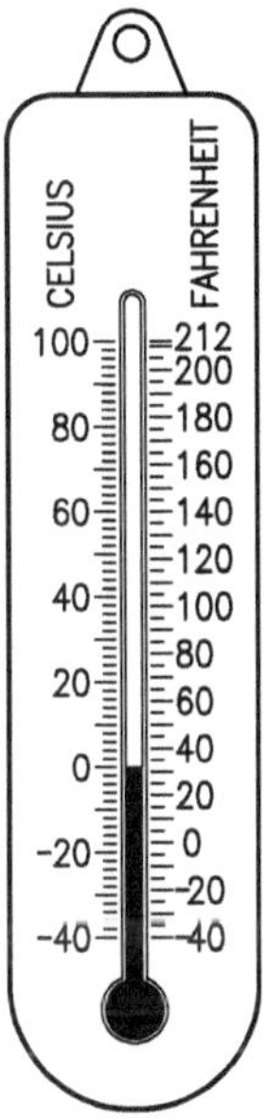

Fig. 7.1. Thermometer

My high-school physics teacher used to introduce the follow-up on the extremes of daily temperatures with the proposal for a new career path: *Thermometer Watchdog*. Those willing to apply had to possess enough willpower to read and write down the showings of an outside thermometer, such as in Fig. 7.1, for 24 hours a day (nights included).

It wasn't a shortage of willing candidates, but rather the costs of Colombian coffee the selected few would consume that led to the invention of the High/Low thermometer, also known as Six's thermometer, shown in Fig. 7.2 in its three characteristic conditions: lowest temperature, in-between temperature, and the highest temperature of the day.

Six's thermometer

The thermometer constructed by James Six (1731–1793) consists of a U-shaped capillary tube, filled with mercury around the bend, while a thermometric fluid, such as alcohol, fills the upper parts of the legs. Above the mercury mirror, the left bulb is filled with thermometric fluid to the brim, so to say, while the bulb on the right is kept at the state that pessimists define as "half empty".

The actual measurement is done by the fluid in the left bulb, which operates on the principle of the standard thermometer, mounted upside-down. With a bulb sized to hold 100 times the volume of a 10 mm (ca. 3/8 inch) stretch of the capillary, the volumetric expansion of alcohol, of 11.1×10^{-4} per degree Celsius makes the liquid column lengthen by $100 \times 11.1 \times 10^{-4} \times 10 = 1.11$ mm/°C or about 4 1/2 inches per 100 °C; provided that we neglect the much smaller effect of thermal expansion of the glass the capillary tube is drawn from. An increase in temperature makes that the expanding fluid forces the mercury column down in the left leg and up in the right leg (Fig. 7.2, center), where the respective temperature can be read from the position of the meniscus (vulgo: upper end) of the mercury column. The fluid on top of the mercury is hereby forced into the partially empty expansion bulb above. The *index*, detailed on the right of Fig. 7.2, hereby ascends as if floating on top of the mercury up to the highest point, where it retains its position, even when falling temperatures make the mercury column retreat while the liquid sneaks by through the spaces between the coils of the helical spring, which are prestressed to press lightly against the capillary's inner walls.

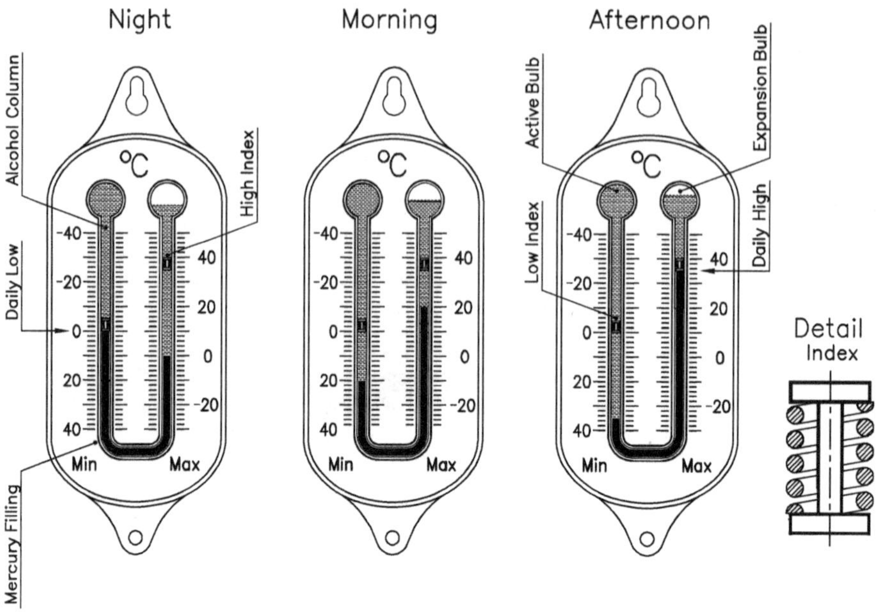

Fig. 7.2. Six's maximum and minimum thermometer

Falling temperatures – by contrast – make that the left-hand fluid column shrinks and vapor pressure in the right-side expansion bulb forces the mercury to follow. Thus, the index in the left leg is pushed up toward the position of lowest temperature, which means that the left-side scale is marked and read head down; the higher the mercury, the lower the temperature.

If simplicity can be misleading, Six's U-tube with bulbs on both ends is an example. In reality, this is a sophisticated instrument that works only if built to a set of conditions, including the size and clearance of the sliding indexes (Fig. 7.2, right), actually dumbbell-shaped metal spools encircled by a helical spring that exerts just the right pressure against the walls of the capillary as to stay in place when the mercury column retreats and the thermometric fluids (alcohol or mineral spirits) ease by, yet still allowing to be reset from outside with a strong permanent magnet.

Gas thermometer

Without special mention, we assumed so far that mercury and thermometric fluids expand linearly (regularly) with rising temperatures. Celsius' definition of the degree of his namesake temperature scale as $1/100$ of the temperature span between freezing and boiling water made the *thermometric constant*, actually the coefficient of cubic expansion of mercury, 0.54×10^{-3} per degree, not the other way round.

If that "constant" is what the term implies, or if it should be called a "standard," remains a matter of minor concern, as long as all measurements are made in accordance to this one reference, much as the *meter* standard, thought of $1/10{,}000{,}000$

of the meridian quadrant, remained unchanged even when the quadrant turned out to be 1954.5 meters longer than expected. An ideal temperature scale would show equal energy input for equal temperature differentials. For instance, the energy for heating a given quantity of water from 10 to 11 degrees should equal precisely the energy for the same quantity from 80 to 81 degrees or for whatever other span.

Research for such "ideal" thermometers led to the use of "ideal gas" as thermometric medium. By definition, an ideal gas follows exactly Gay-Lussac's and Boyle Mariotte's laws, which estimate the volume V of ideal gas as inversely proportional to its pressure, p, and directly proportional to its absolute temperature, T: $pV \propto T$, or for a gas under constant pressure: $V \propto T$. Herein, T is counted from absolute zero, the lowest possible temperature, namely, −273.15 °C below the freezing point of water.

The gas thermometer consists of a glass or platinum thermometer bulb, connected to a sensitive manometer. Use of the once customary mercury filled U-tube pressure meter adds the volume of the displaced mercury in the U-tube to the volume of the gas, and thus complicates the results' evaluation. Modern gas thermometers use membrane-actuated manometers which function with negligible volume increase; or better still, piezoelectric sensors.

Thermoelectric temperature sensors

The direct conversion of heat into electricity goes back to Thomas Alva Edison's 1883 discovery of the emission of electrons from a heated filament in vacuum, followed by the development of the first vacuum tube rectifiers, and later voltage and current amplifier tubes. This release of electrons from heated metals is indicative for the presence of free electrons within their lattice. When energy input, which we perceive as heat, boosts the vibration of the atomic nuclei in the lattice, free electrons are forced outward and migrate from the heated cathode to the positive electrode, the anode.

In a junction of different metals, the cathode–anode potential difference is created naturally by their difference in the density of the cloud of free electrodes within. They jump from the boundaries of one nucleus to the next, leaving "holes" for the following electrons to land. *Holes* simulate positive charges, but must not be confused with positrons, the positive counterpart of the negative elementary particle, the electron. The concept of electron transfer explains the slow path of electric charges in a conductor, though the electric field that drives them propagates with the velocity of light.

Each metal has its particular "electron density," given by the number of free electrons per unit of volume, which difference accounts for the *contact potential* in the junction of two metals. Like in a vacuum tube, the receiver of electrons is positive, and the emitter negative. Those who find this contradicting our scholarly wisdom of the flow of electricity from positive to negative, are correct. But the designation of polarities happened long before we learned of the existence of negative elementary charges, and was too entrenched for change when we found out.

A thermocouple converts heat energy directly into electric energy. Unlike in

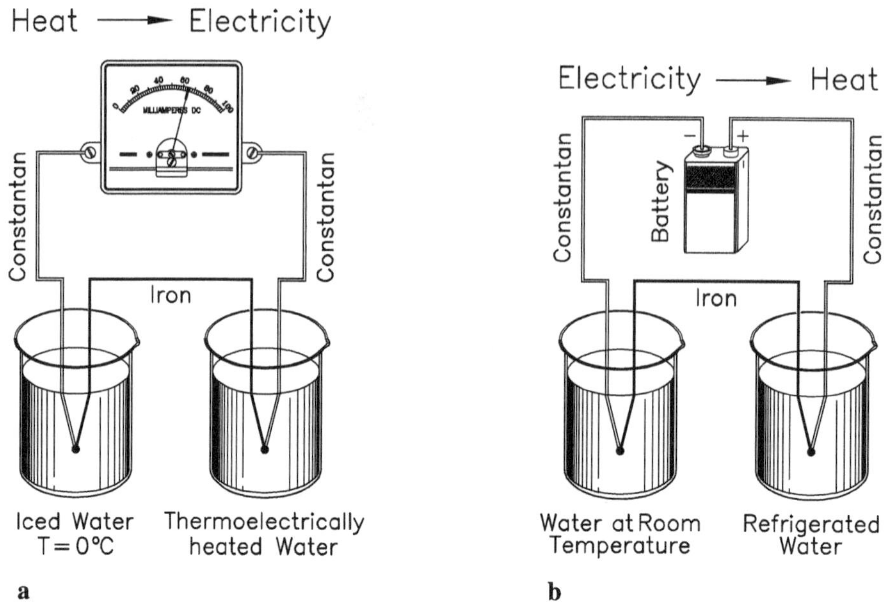

Fig. 7.3. Reversibility of thermoelectricity

heat engines, the process is reversible. Figure 7.3 illustrates the consumption of heat in one junction and the generation of heat in the other of a ferro-constantan thermocouple. In Fig. 7.3 a, hot water in the beaker on the right, which temperature is to be measured, causes the flow of electric current from the iron wire to the constantan wire through the galvanometer and back to the second junction, which is maintained at the freezing point of water. The intensity of that current rises with the temperature of the hot water, yet is far from directly proportional. Each type of thermocouple has its own characteristic curve, which must be considered in the layout of scales calibrated in degrees Celsius or Fahrenheit. Digital instruments do that with built-in circuits for the conversion to a variety of different types of thermocouples.

Figure 7.3 b illustrates the inverse process, namely, the shifting of heat energy from one reservoir to the other by the flow of electric current, just like a window-mounted air conditioner takes heat from inside a room and expels it into the environment. The capacity of such *thermoelectric refrigerators* is intrinsically small but can be enhanced by connecting some 10 to 50 thermocouples in series. Only the voltage drops in the interconnecting wiring limit their number. Nevertheless, such devices found their place in space technology, where the absence of moving parts is valued higher than efficiency.

While the use of an iced reference junction warrants the most accurate temperature measurements, it is mostly confined to laboratory work. In industrial applications, the reference junction is the connection of the thermoelectric wires with the wires to the galvanometer coil. Since such junctions are hidden inside the connection head of the thermocouple's protective tubing, the readout then becomes the temperature difference between the thermocouple itself and the connection

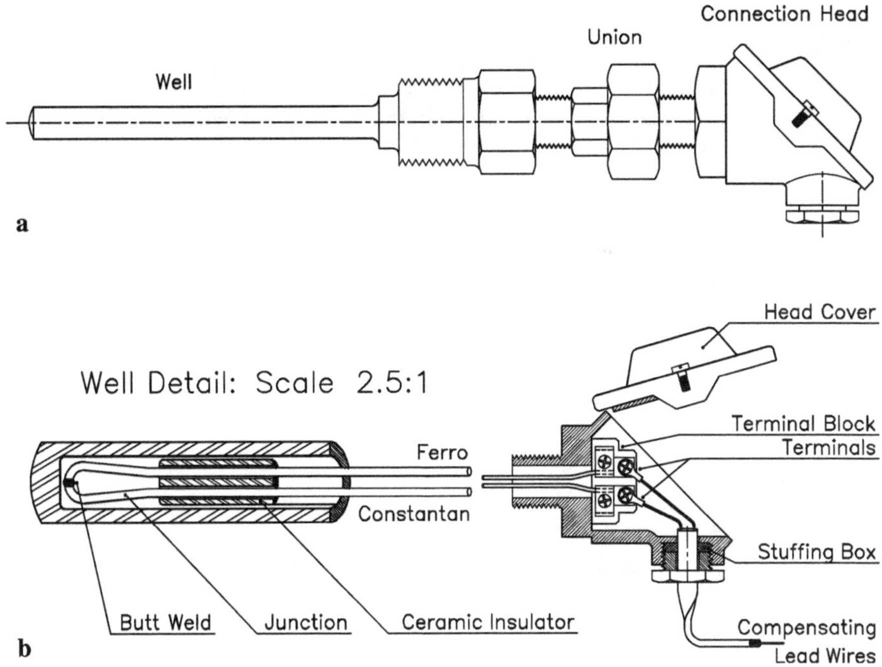

Fig. 7.4. Thermoelectric pyrometer

head. As long as head temperature remains reasonably constant, this difference is often (and sometimes unwittingly) taken as the true process temperature.

Thermoelectricity, the creation of a potential difference (vulgo: voltage) in a junction of different metals, is the discovery of Thomas Seebeck in 1821, which led to the development of the industrially most popular temperature sensor, the *thermocouple*.

The butt-welded junction of an iron wire with a wire of a 45/55% nickel-copper alloy (constantan), as shown in Fig. 7.4b generates a voltage difference of 5.269 mV between the open ends of the wires for every 100 °C it heats up.

Thermocouples maintain their characteristic voltage even when used with inexpensive low impedance galvanometers that excel by their immunity to parasitic leakage voltages which could show up as spurious readouts on top of the line high impedance instruments.

This relatively high amperage of thermocouple currents makes that the length and thus the resistance of the lead wires matter. The electrical resistivity of iron wire at 20 °C is 0.12 ohm per 1 mm^2 cross-sectional area and 1 meter of length; for constantan it is 0.496. The resistance of a pair of 1 m long lead wires of those materials with 1 mm^2 cross section (0.785 mm diameter) amounts to $0.12 + 0.496 = 0.616\,\Omega$. With Ohm's law, and 5.369 mV of voltage differential, that gets us for the electric current: $J = E/R = 5.369/0.616 = 8.716$ mA.

If the thermoelement in this example is installed at *two meters* rather than *one meter* from the indicating instrument, output amperage is cut in half. Therefore, commercial thermocouples come with standard lead wires attached, which must neither be cut nor extended. Where the price of thermocouple wires, such as pairs of platinum–rhodium, makes their use as leads impractical, wires of less valuable material are available. Their characteristics are tailored to resemble those of the thermocouple wires but cannot match the heat resistiveness of the originals.

Electronic meter circuits overcome the problems inherent to DC amplification by converting the thermocouple's voltage output into pulsed alternating current (AC) that can be used with operational amplifiers to any desired degree of gain. Originally, mechanical interrupters, built to the principle known from the electric doorbell, became standard. Elsewhere, they were used to boost those times car-batteries' 12 volt DC to 115 VAC for feeding automotive tube radios and became the most frequently replaced and painfully expensive electronic spares.

As part of instrumental amplifiers, they remained unreliable, regardless of the industry's best efforts toward improvement. But replacement by transistorized chopper circuits proved elusive, since the forward voltage-drop of typical germanium transistors, around 200 mV, and worse the 600 mV of silicon transistors, used to block the low voltages to be measured. Only the invention of the Unijunction Transistor made transistorized DC/AC converters feasible.

Table 7.1 lists the composition of the wires of the commonly available types of thermocouples, their temperature range, and voltage per 100 °C of temperature differential.

Table 7.1. Industrial thermocouple data

Type	Alloy of positive leg	Alloy of negative leg	Lowest and highest temperatures (°C)	Thermoelectric voltage per 100 °C temp. diff. (mV)
J (iron/constantan)	iron	45% nickel 55% copper	−40 750	5.269
K (chromel/alumel)	90% nickel 10% chromium	95% nickel 2% aluminum 2% manganese	−270 1350	4.096
N (nicrosil/nisil)	14% chromium 1.4% silicon 84.6% nickel	0.4% silicon 95.6% nickel 4% magnesium	−270 1300	2.774
T (Cu/Cu-Ni)	pure copper (copper/constantan)	45% nickel 55% copper	−270 400	4.279
E (Ni-Cr/Cu-Ni)	90% nickel 10% chromium	45% nickel 55% copper	−270 1000	6.319
S (platinum/rhodium)	90% platinum 10% rhodium	100% platinum	−50 1700	0.646
R (platinum/rhodium)	87% platinum 13% rhodium	100% platinum	−50 1700	0.647
B (platinum/rhodium)	70% platinum 30% rhodium	94% platinum 6% rhodium	+50 1800	0.033

Resistance thermometry

The change of the electric resistance of most metals with temperature is close to regular, which makes resistance thermometers often preferable to thermoelectric devices. Accordingly, the resistance R_T of a given length of wire at the temperature of T (°C) is given by the equation $R_T = R_o(1 + \alpha_1 T)$, wherein R_o stands for the wire's resistance at 0 °C, and α_1 for the increase in resistance per degree of temperature rise. This resembles Boyle's law if you make $\alpha_1 = 1/273$ and replace R_T and R_o by the respective gas volumes. However, while Boyle's volume curve crosses the zero line close to the absolute zero, −273 °C, the R_T curve is not so lucky. With the average value of $\alpha_1 = 0.004$ for most metals, the zero crossing would happen at the much higher point of $-1/0.004 = -250$ °C or 23 K. This suggests a certain degree of curvature in the temperature vs. resistance curve at cryogenic levels. Only copper with $\alpha_1 = 4.274 \times 10^{-3}$ remains quasi linear in the range of −50 to +150 °C.

The most popular resistance temperature sensors, better known by the acronym RTD, are bulbs with platinum filaments, thanks to this metal's wide temperature range and high stability. Platinum resistance bulbs are available with 100 Ω and 1000 Ω at 0 °C base resistance. Copper bulbs with 10 Ω (at 25 °C), and nickel bulbs with 120 Ω (at 0 °C). Typically, a 100 Ω platinum bulb changes its resistance from 100 Ω at 0 °C to 138.4 Ω at 100 °C.

For higher temperatures, the resistance–temperature relation is best expressed by introducing a second-degree term in the basic equation, which brings it into the form $R = R_o(1 + \alpha_1 T + \alpha_2 T^2)$ for the range from 0 to 850 °C. The constants vary slightly in propriety curves from manufacturer to manufacturer around the averages of $\alpha_1 = 3.95 \times 10^{-3}$ and $\alpha_2 = -0.583 \times 10^{-6}$. To extend the reach of this formula to −200 °C, a third constant, $\alpha_3 = -4.14 \times 10^{-12}$, is added, which leads to $R_T = R_o[1 + \alpha_1 T + \alpha_2 T^2 + \alpha_3 T^3(T - 100)]$.

Nickel wire, next in use to platinum, excels in sensitivity and can be used from −100 to 260 °C. Its temperature coefficient α_1 is 0.00672 Ω/°C, but there is no proven formula for the relation between temperature and resistance of nickel wires. Such values must be taken from manufacturers' tables.

Temperature-induced resistance changes can be derived directly from the voltage drop over the sensor resistor if power is taken from a constant current source. But the resulting temperature scale wouldn't start at zero. Take for instance a platinum resistor with 100 Ω at 0 °C and 139 Ω at 100 °C, connected to a 100 mA constant current source. According to Ohm's law, $E = JR$, the respective voltage drops become $E_1 = 0.1 \times 100 = 10$ volt, and $E_2 = 0.1 \times 139 = 13.9$ volt. The difference, of 3.9 volt, comprises only 20% of the range of a 20 V voltmeter, and readout precision would therefore be low. Bridge circuits are free of such shortcomings. They exist in two versions: (a) zero crossing and (b) indicating.

Figure 7.5 a shows the conventional Wheatstone bridge with the resistance bulb, R_B in the leg at the lower right, and the adjustable balancing resistor, R_V, at the lower left leg. R_V is a wire-wound precision potentiometer, or a dial mounted rheostat. The scale can be laid out in ohm or directly in degrees Celsius, but remains in both cases far from linear.

If the sensor is located at considerable distance from the bridge circuit, like in the

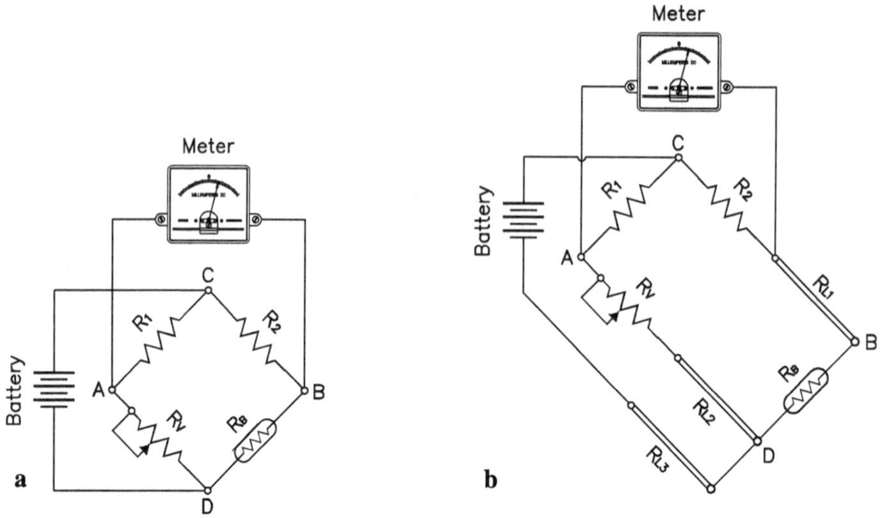

Fig. 7.5. Wheatstone bridge temperature meter

case of a furnace on the shop floor controlled from a panel in the control room, the lead wires become long enough to distort temperature readings as their proper resistance adds itself to the sensor resistance. The three-wire connection between sensor and instrument shown in Fig. 7.5 b provides lead wire resistance compensation. Here, the point D on the bridge is brought out to the sensor location, so that the lead resistances R_{L1} and R_{L2}, shown in the schematic as fine double lines, add their values equally to R_2 and R_V. The resistance R_{L3} of the wire connecting the negative of the battery to the now remote point D doesn't matter, but it equals R_{L1} and R_{L2} anyway if standard three-conductor cable is being used.

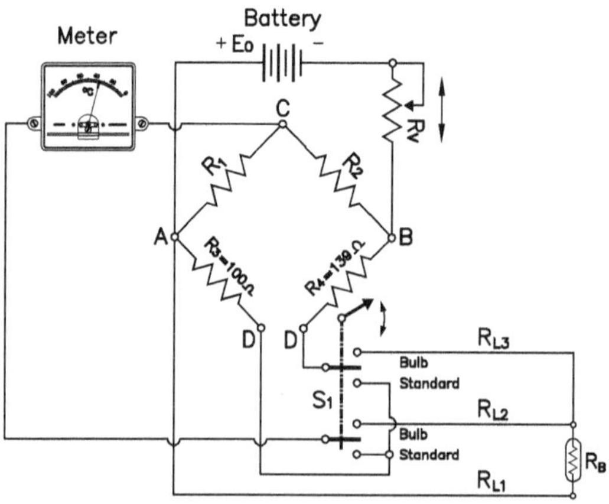

Fig. 7.6. Direct-readout temperature-measuring circuit

Where manual setting of the rheostat for zeroing the meter is not feasible, direct reading instruments, such as in Fig. 7.6, are the solution. Here again, the principle is that of the Wheatstone bridge, yet the resistors are fixed and any unbalance shows on the meter. The ratio arms R_1 and R_2 are of identical values, and R_3 is a wire-wound precision resistor equal to the basic bulb resistance, such as 100 Ω. R_4 confers with the bulb resistance at the highest temperature expected, such as 139 Ω at 100 °C for a 100 Ω platinum resistance bulb. S_1 is a two-posi-

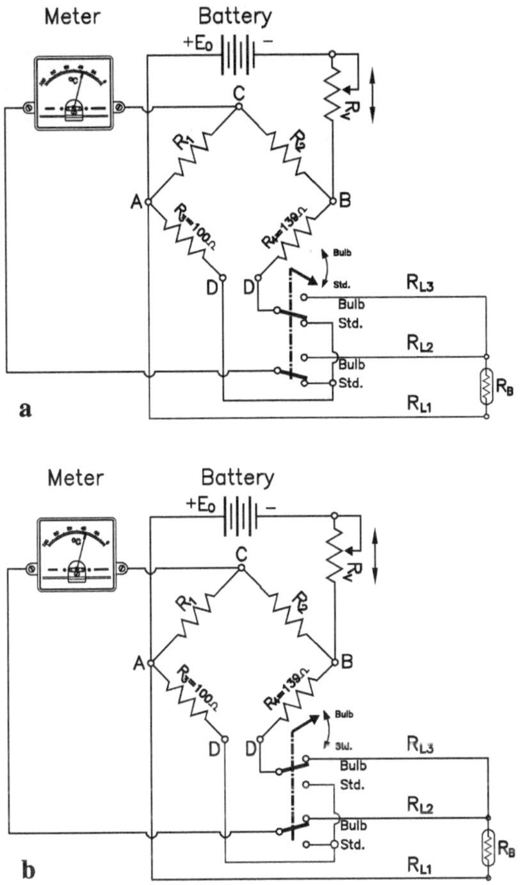

Fig. 7.7. Circuit calibration

tion double throw (TPDT) selector switch for the circuit's initial calibration. With the switch at the position marked "Std." (standard) in Fig. 7.7a, the bridge consists only of the four resistors R_1, R_2, R_3, and R_4. Setting $R_1 = R_2$ makes the voltage at point C half the supply voltage, namely, $E_o/2$. The voltage at point D then follows as $E = E_oR_4/(R_3 + R_4)$. With a 10 volt battery as the power source, and resistors as marked in Fig. 7.7, we get the voltage at point C equal to $E_o/2 = 5$ V, and the voltage at point D as $10 \times 139/(100 + 139) = 5.816$ V. The calibration potentiometer, R_V, must now be set to make the difference from D to C, of 0.816 V, deflect the meter needle full scale to the right.

Switch S_1 in the "Bulb" (upper) position (Fig. 7.7b), replaces R_3 by the bulb resistance. With the bulb at 0 °C and its resistance at 100 Ω, this again should result in full-scale deflection of the needle.

Heating the bulb up to 100 °C and 139 Ω resistance like R_4 should leave the bridge circuit balanced with zero meter current. Like in standard multimeters, the resistance or temperature scale must be read from right to left.

An interesting version of readout instrument is the "Continued Balance System," which feeds the electric current, which so far went through the meter, into a servomotor that drives the instrument's pointer. A slide-wire potentiometer in one of the arms of the Wheatstone bridge, mechanically coupled to the pointer, feeds the pointer's position back and makes the motor stop when a new state of equilibrium has been reached. A chopper-amplifier circuit boosts the error signals enough to drive the motor.

Servomotor-driven instruments are highly resistant to vibration, often omnipresent in industrial environment. In a conventional temperature controller, which turns the load on and off according to the position of its lightweight needle, vibrations may cause sporadic switching, while a motor-driven pointer holds its position at the point were the bridge circuit is balanced, no matter what.

Resistance bulbs and thermocouples in particular suffer from corrosion from the oven's inner atmosphere, which can be oxidizing, reducing, or acidic. In foundries, alloying of the batch of molten metals with the thermocouple's material leads to similar effects. Under high temperatures, iron-based thermocouples tend to scale and split open.

Whatever the cause of broken up thermocouple welds, they make the temperature controller linger on zero and fail to prevent overheating with sometimes disastrous results. That's why most instruments come with automatic zero deflection shutoff. But remember this and don't panic if the system refuses to come on in the morning as stubbornly as a frozen in car engine. Temporary shunting the controller contacts with a timer or (for the contemplatives) a push button switch gets things going again.

Protective tubing, bulbs, and wells, keep the oven's inner atmosphere away from the sensor elements but greatly increase the time it takes for the heat in the oven to reach the temperature sensor, expressed by the *thermal inertia* of the system. Belated turn-off makes the true furnace temperature overshoot the setpoint. Likewise, the turn-on comes too late. Shielding could even make an otherwise stable control system unstable and therefore must be accounted for early in the design process.

Radiation pyrometry

The problems resulting from detrimental effects of the oven's inner atmosphere on sensor elements are summarily circumvented by radiation pyrometers for remote temperature measurement. According to Stephan Boltzmann's law of radiation, heat energy Q from a body at temperature T_2 is radiated to another body, of temperature T_1, at the rate proportional to the fourth power difference of these temperatures, that is,

$$Q \propto \left[\left(\frac{T_2}{100}\right)^4 - \left(\frac{T_1}{100}\right)^4\right].$$

The quotient of 100 is meant to ease the use of the formula; without it, the fourth power of absolute temperatures, such as 273 K for 0 °C, would become awkward to handle. In compensation, the proportionality constant, of $C_s = 4.95\ \mathrm{Cal/m^2} \times \mathrm{K}^4$, is 100^4 times the original. K stands for the absolute temperature in kelvin.

The formula shows heat transfer by radiation as far more sensitive to temperature changes than heat transfer by conduction, which is proportional to the first power of temperature differences, rather than the fourth. For instance, heat exchange between the walls of an oven at, say, 1000 °C, with the charge we imagine at 900 °C, happens in proportion of the rate $1000 - 900 = 100$. α, the coefficient of heat transmission, can be estimated from $\alpha = \sqrt[4]{24\Delta T + 14}$, which gives for a 100 °C temperature gradient $\alpha = \sqrt[4]{24 \times 100 + 14} = 7.0\ \mathrm{Cal/m^2} \times \mathrm{h} \times {}^\circ\mathrm{C}^4$. That makes the hourly heat exchange by conduction equal to $7 \times 100 = 700\ \mathrm{Cal/m^2} \times \mathrm{h}$.

But radiation would, under equal condition, transfer the amount

$$C_s\left[\left(\frac{1273}{100}\right)^4 - \left(\frac{1173}{100}\right)^4\right] = 7329.3 \times 4.95 = 36280\ \mathrm{Cal/m^2{\cdot}h}\ \text{of heat}.$$

That is $36280/700 \approx 50$ times the rate of heat transfer by conduction.

The worldwide success of the clinical infrared ear thermometer is proof for the reality of that figure as it takes a bare 100 to 300 milliseconds to come up with the patient's blood temperature, while early mercury thermometers took up to 10 minutes for reliable (and repeatable) results, even as used in the most unlikely regions of the human anatomy.

Infrared clinical thermometers reliability rests on checking the thermal radiation emitted by the eardrum, located close to the hypothalamus, the body's temperature regulating organ. Since radiation is accumulative, the instrument uses a shutter, opening for a short but closely controlled amount of time to let radiation charge a thin pyroelectric crystal, which then discharges into the built-in electronics for amplification and conversion into a readout.

But such electronic sorcery was still a century in the future when the American painter and inventor Samuel F. B. Morse (1791–1872) patented his *Disappearing-Filament Pyrometer* in 1899 (Fig. 7.8). The instrument is built like a telescope, yet has a carbon filament light bulb placed in the focal plane of the eyepiece, where conventional telescopes have the crosshairs. Initially, the eyepiece is focused on the lamp's spiral filament; next, the ocular tube with its rack and pinion drive is set for a sharp image of the source of radiation, such as the inside of an annealing

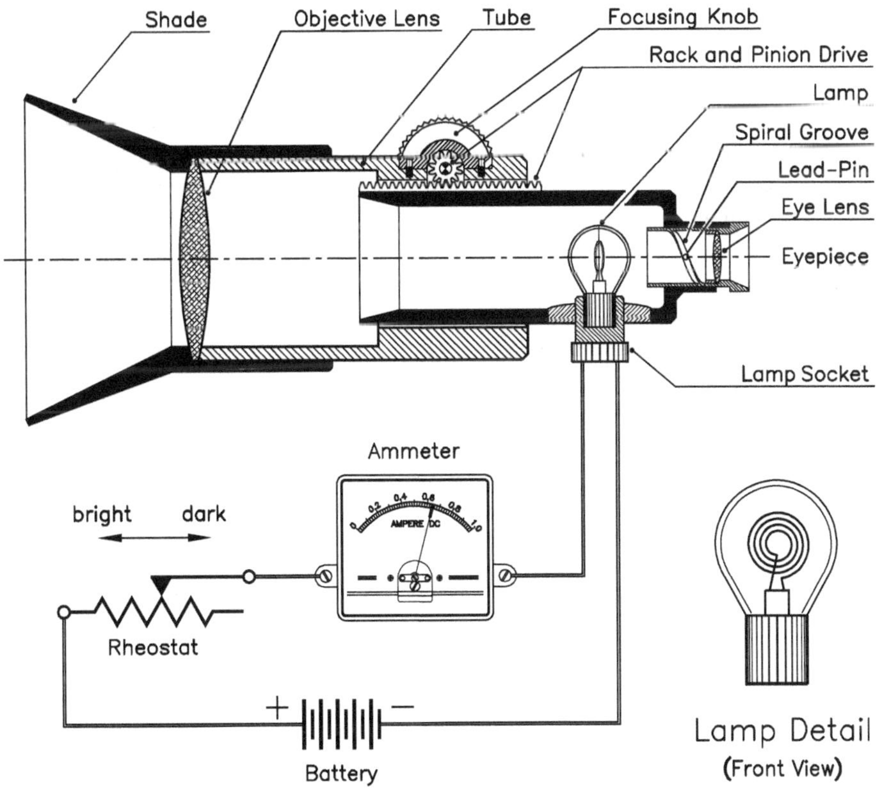

Fig. 7.8. Disappearing-filament pyrometer

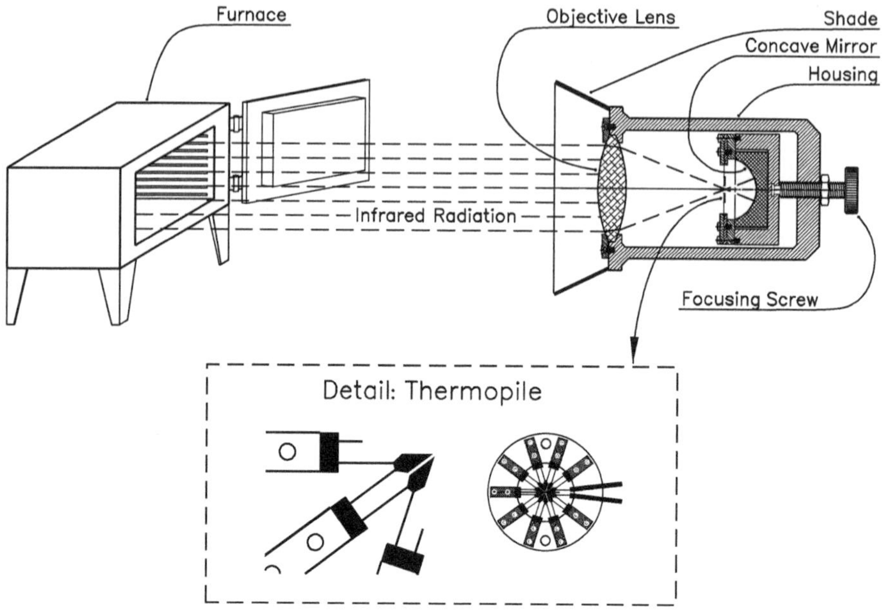

Fig. 7.9. Thermopile radiation pyrometer

oven or the stream of molten steel at the discharge of a blast furnace. For best results, a narrow band of the infrared spectrum is singled out for observation by placing a filter of red copper-oxide-dyed glass with the eyepiece lens. The objective lens is of calcium fluoride glass, transparent to long-wave radiation.

The circuit is being balanced by adjusting the rheostat until the lamp's filament fades into the glowing background, which happens when the temperatures of the source of radiation and the lamp's filaments are identical. At this point, the ammeter is read and the respective temperature looked up in conversion tables.

Direct readout pyrometers (Fig. 7.9) have the image of the source of radiation projected on a concave mirror, which reflects and further concentrates it on the center of a thermopile – an array of serial connected thermocouples, which individual thermoelectric voltages add up.

The junctions of the elements of a thermopile are flattened and blackened for highest energy absorption. The "reference junction temperature" is that of the mounting flange and is kept constant by internally heating the tube to a predetermined, closely regulated level, typically 50 °C, which must ultimately be added to the readout. An alternative is to shunt the output terminals of the thermopile with a nickel wire spool, dimensioned in ways that the variations of its resistance compensate for changes in ambient temperature.

Note that the size of the radiating body must be sufficient to fill the field of vision of the pyrometer. For most commercial models, that amounts to at least 5% of the distance between the radiating object and the objective lens of the instrument. Pyrometers mounted to the structure of a furnace must be water-cooled for heat protection and principally to maintain the reference temperature.

Principal applications of radiation pyrometry are in measurements of very high temperatures that would destroy conventional sensors; for temperature checks on moving objects, such as molten metals flowing from crucibles; and for measurements in highly corrosive and poisonous atmospheres.

While the blackened heads of the thermopile array can be expected to absorb the full amount of the radiation energy they receive, and thus act like *black bodies*, the same is not granted for the source of radiation. At room temperature, most materials reflect part of the incident radiation and therefore heat up less than expected. The proportion of energy received and energy emitted is the *emissivity* of a material. It is low for metals with reflective surfaces, such as (blank) aluminum (0.03), nickel (0.05), and steel (0.08). Oxidation brings their emissivity up to 0.11, 0.31, and 0.80, which implies that unintended oxidation to various degrees might greatly distort measurements. An object's emissivity also depends on the angle of observation and can markedly change above 30 to 40 degrees of arc.

Astronomers call the emissivity of their objects, such as the moon and the planets, *albedo*. The lowest albedo, of 0.06, has Mercury, followed by the moon (0.14), and Mars (0.16). Venus sets the record at 0.76, while Earth (0.39), Jupiter (0.34), and Saturn (0.33) hold the middle ground.

Hot bodies up to 260 °C can be reliably (if not accurately) checked by previously covering them with black masking tape, which equalizes their emissivity at about 0.95. *Platinum Black*, an oxide of the rare metal, best known from its use as chemical catalyst, is the "blackening agent for the rich." But take heart. All radiating bodies come close to resembling *black bodies* if heated high enough.

Thermistors

Much to the dismay of electronic circuit designers, semiconductors inherently increase their conductance with rising temperature. That makes transistors prone to burnout by the so-called *avalanche effect*: a slightly overloaded transistor gets hot, augments its conductance and carries higher currents than normal. That in turn makes it heat up further, which again . . . you guessed the not so happy end of the story.

This otherwise nonsensical semiconductor characteristic is brought to good use in heat detectors with semiconductor elements, called thermistors. Thermistors react far more violently to temperature changes than resistance probes and thus can be brought to show, in certain applications, changes as small as 0.001 °C.

Manufactured from oxides of the so-called transition metals, such as copper, nickel, cobalt, and manganese, and on the other hand, barium titanate, thermistors divide into two groups, depending on the sign of their temperature coefficients. If the latter is negative, we have an NTC; and if it's positive, a PTC. The former provide gradual decrease of resistance with rising temperature, the latter an increase, which can be made to abruptly alter at a predetermined temperature threshold. This makes them ideal for heat sensing switches, such as the ones that, squeezed between the windings of electric motors and transformers, shut the machine off if excessive heating occurs. In photography, "smart flash guns" use thermistors to match the time of film exposure with the intensity of the flash.

The negative temperature coefficient of NTC thermistors led to their use for compensation of the positive temperature coefficient of copper wire, principally in galvanometer coils. Instruments can be "temperature hardened" by a piece of NTC wire in series with the coil. Since the temperature curve of an NTC is much steeper than that of copper, the length of the NTC wire needs to be only a fraction of that of the coil windings.

Attached to the cargo of high-altitude balloons, NTC thermistors measure the temperatures in the upper atmosphere. Power NTCs are used as current limiters, holding down those infamous inrush currents at the turn-on of electric motors, which often exceed four- to fivefold the motor's rated amperage. Some of us still remember the dimming of lights when heavy motors were switched into the network. Further, NTCs lend themselves to the use as surge suppressors in a wide range of applications.

The relation between resistance R and temperature T of thermistors is given by the *Steinhart–Hart equation* $1/T = a + b \log R + c(\log R)^3$, where a, b, and c are device-specific constants.

Calorimetry

Some of the laws of physics can be derived intuitively by logic deduction, a. k. horse sense. For instance, the law of hydrostatics is self-explaining if we think of pressure as the weight of water in a vertical pipe of unit cross-sectional area.

Even the inverse square law of gravitation can be guessed, considering that the surface area of a sphere grows in proportion to the square of its radius; if we assume that gravitation from a central source, such as the Sun, loses intensity commensurate with the area it subtends, the inverse square law is the logical conclusion.

By contrast, the idea of convertibility of thermal and kinetic energy did not slumber inborn into the human mind, just waiting for the kiss of the fairy prince to rise and shine. The concept of energy was entirely "plagiarized" from nature's ways of doing thinks, by – look at the irony of fate – a physician! His name was Julius Mayer (1814–1878), and the breakthrough occurred on a trip to Batavia, then capital of the Island of Java, where (therapeutic) bleeding of Mayer's co-travelers showed less than the usual contrast in color from arterial to venereal blood.

Compared to the conspicuously red blood from arteries, venereal blood taken in the tropics had darkened to a lesser degree than what the doctor recalled from his homeland. He concluded that the human body works harder on maintaining its blood temperature in a frigid environment than in the heat of the tropics; a concept that, then for the first time (1842), contained the equivalence of thermal and mechanical energy. Modest as such beginnings might seem, they led Mayer already into a prediction of the mechanical equivalent of heat: the amount of thermal energy to raise the temperature of a given mass of water by one degree on the Celsius scale was to equal the kinetic energy developed by an identical mass dropped from 365 meters of height.

Make this mass 1 kg (or equally 1 liter) of water, and you have the calorie, which Mayer thus made the equivalence of 365 kgf$\times$m of work. For a first estimate, this comes surprisingly close to the true value of 427 kgf$\times$m/Cal. Subsequently, the Calorie was specified as the heat capacity per degree Celsius of 1 kg of pure (de-aired) water at 15 °C, with the symbol Cal. By contrast, cal stands for 1/1000 Cal, and refers to 1 gram of water. We find it mostly in older physics textbooks, along with the cm/g/s system of measures. With relation to electrical and mechanical units, the calorie equals 4.186 joules (or watt-seconds), or still 427 kgf×m. And 860 Cal are the thermal equivalent of 1 kWh of electricity.

Does that mean (for Heavens sake!), that every lousy calorie in our meals should enable us to lift 427 kg by 1 meter, or nearly 1000 pounds by more than 3 feet? Take heart. Most of our body's energy input goes into maintaining blood temperature (much like one third of the expensive gas in the tanks of our cars is inevitably wasted on keeping the glycol in the cooling system hot). Only some 5% are biologically converted into mechanical energy. Still, even counting that in, the remaining 50 pounds weight-lift per Cal should suffice to damp our appetite for a typical 1000-calorie meal.

The calorie's dietary applications aside, it's a unit that opens ways to simple solutions of heat energy related problems. For instance, you can figure the heat for raising the temperature of, say, 80 kg of aluminum (of 0.214 Cal/kg specific heat capacity) in an electric crucible furnace by 100 °C with the multiplication $0.214 \times 80 \times 100 \approx 1712$ Cal. The conversion factor of 860 Cal/kWh makes that $1712/860 = 2.0$ kWh of electricity, which at the industrial rate of, say, 5 cents per kilowatt-hour, will cost you a scant 10 cents.

But in all that, it took the work of James Watt for the plunge from understanding thermal energy and physically converting it into usable amounts of kinetic energy. You may or may not recognize Watt as the originator of the idea of steam engines, but undoubtedly it was him who developed a viable design in all its uncounted details for the primary power source that was to propel civilization. If it weren't for the utilization of the thermal energy stored in our vast resources of coal and oil, human lifestyle would still resemble idyllic Dutch paintings of a farmer who carries a sack of grain on his shoulders to a windmill. Albert Einstein himself once envisioned the potential for development of any country in the world as dictated by the size of its deposits of coal.

But not all sources of heat energy are born equal. Hydrogen is heading the list with 33,900 Cal/kg, but there is no such thing as deposits of hydrogen gas on Earth, unless you count on the tenuous uppermost layers of the atmosphere. The hydrogen available to us has either been produced by electrolysis of water – in a process resembling the one that supplies the International Space Station with oxygen – or in a rather complex series of chemical reactions, starting with the gasification of coal or coke, which proper thermal energy is hereby being used to obtain – in ultimate analysis – a lesser amount of hydrogen-bound thermal energy; here again, you get what you pay for – or less.

By contrast, anthracite, the purest form of coal, herds 8400 Cal/kg, while petroleum derivates, such as gasoline and diesel oil, range between 11,900 and 10,600 Cal/kg, respectively. Wood comes in as a distant third with 2800 Cal/kg, and yet was once the combustible of choice for wheeled steam engines in farming and logging operations.

But before you multiply such figures by 427 and start marveling at the enormous energies hidden in every kilogram of combustible, remember that the thermal efficiency of heat engines in earthbound environment is inherently limited to 40% at best. And further that the air that fans our fires consists to 80% of the inert gas nitrogen that, like it or not, is being heated along with the active reaction components to the highest process temperature.

To bring all this down to Earth, imagine the layout of a coal-fired power plant of 1000 kW capacity. Combustion of 1 kg of coal renders 6600 Cal, which with the conversion factor of 860 kWh per calorie is the equivalent of $6600/860 = 7.67$ kWh/kg. Generation of 1000 kWh per hour, or simply 1000 kW, would thus take (at 100% efficiency) $1000/7.67 = 130$ kg of coal per hour. Considering the true efficiency of such systems, typically 35%, consumption becomes $130/0.35 = 371$ kg of coal per hour for generating the desired 1000 kilowatt of electricity. That's why electric room heating costs so much more than the good, old, coal or wood cast iron stove.

Berthelot's bomb calorimeter

Measurements of the "true" heat energy of combustibles presuppose the use of pure oxygen, supplied for good measure at about 280 psi (20 kgf/cm^2). The need for "force-feeding" of oxygen lies in the reversibility of chemical reactions. Even

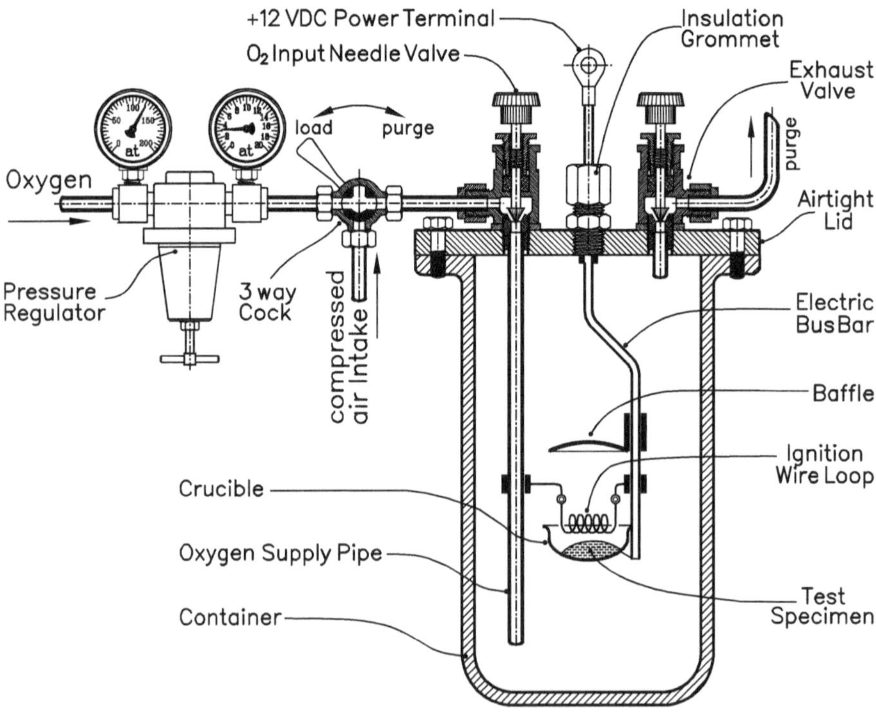

Fig. 7.10. Berthelot's calorimeter

such straight forward reactions, such as $C + 2O = CO_2$, occur in parallel with the dissociation of $CO_2 \rightarrow C + 2O$. At room temperature, the first version is dominant, but as temperatures rise, the second gains strength. The outcome is incomplete combustion, leaving our car engines' exhaust gases with all those widely discussed impurities, carbon monoxide among them.

The "*bomb calorimeter*" (Fig. 7.10) introduced in 1877 by Pierre-Eugène Marcelin Berthelot (1827–1907), is based on the combustion of the test specimen in a pressure vessel charged with compressed oxygen. The vessel holds a crucible with the probe, and an electric fuse to ignite it. The heat energy created in the process is measured by placing the vessel in a water bath (not shown in the illustration), which temperature rise is commensurate with the heat of combustion of the specimen. But unfortunately, the water bath is not alone in heating up; actually everything, from the pressure vessel itself and its accessories to the thermometer, picks up certain amounts of heat as each warms up at its own pace. Use of a water container built to the principles of the thermos-bottle stems heat losses into the environment, and a motorized circulation pump equalizes the temperature throughout the water mantle, yet the mechanism of heat losses remains too complex for theoretical prediction. Their magnitude must be found experimentally for each instrument. To that end, substances of well-known heat of combustion, such as naphthalene ($C_{10}H_8$) or sugar, are initially burned and the real temperature rise in the water mantle is compared with the theoretically expected to get the instrumental constant.

The end-products of the combustion of hydrocarbons include carbon dioxide and vaporized water, as in $C_{10}H_8 + 12O_2 = 10CO_2 + 4H_2O$ for naphthalene, and $C_nH_m + (m/4 + n)O_2 = nCO_2 + (m/2)H_2O$ for the general case of hydrocarbons, C_nH_m, such as coal and oil. Water created in this process still holds its heat of vaporization, of 539.2 Cal/kg, usually lost with the exhaust gases. But in a calorimeter, the end-products are allowed to cool down all the way past the 100 °C threshold, where the condensation of steam liberates this particular portion of heat energy. Therefore, the values from the calorimetric bomb are higher than what we get in real life, much like the interest rates you pay for loans are higher than what you cash in for deposits.

For that reason, we assign two different values to the heat of combustion of flammable materials: the *gross calorific value*, which we get from the calorimeter, and the *net calorific value* under real world conditions. The difference is the heat of vaporization of the amounts of water generated by the reaction. For instance, gross and net calorific values of naphthalene are 9600 and 9260 Cal/kg, respectively.

Oxygen used in calorimetric tests is available compressed to 2200 psi in steel cylinders of usually 1.625 cubic feet capacity. With 14.696 psi for standard atmospheric pressure, 2200 psi are equivalent to $2200/14.696 \approx 150$ times air pressure (atm), so that we can estimate the volume of oxygen in a cylinder filled to capacity as $150 \times 1.625 \approx 244$ cft. The pressure regulator valve (upper left in Fig. 7.10) brings the cylinder pressure down to manageable levels, usually $20\ \text{kg/cm}^2$ or about 280 psi. The three-way cock allows to selectively fill the calorimeter tank with oxygen or to purge its contents with compressed air when the test is over.

A needle valve at the intake controls the flow of oxygen through the filler tube that leads down to the bottom of the tank. With the purge valve open, the level of the heavier-than-air oxygen rises gradually, pushing out whatever the tank held before.

The oxygen filler tube doubles as the electric "ground-conductor" in the process of lighting the fuse and igniting the test specimen. For the positive pole, an isolated copper or stainless steel bar is used that also supports the crucible and the baffle plate. The latter is meant to limit spatter during combustion, which happens much more violently in an atmosphere of pure oxygen then what we are used to in air. Which brings to mind painful memories of the astronauts Virgil Grissom, Edward White, and Roger Chaffee, who died in an experimental capsule right here on Earth due to a short-circuit in the wiring hidden in the pilot's seat, because the capsule's oxygen atmosphere made the resulting fire spread so fast that the helpers outside couldn't open the access door soon enough to save the lives of those trapped inside.

The ignition coil can be thin steel wire, which overheats due to its proper electrical resistance in response to a current surge from the power supply, such as a 12 V car battery. An alternative is a platinum heating-coil with wrapped around steel wire. The latter melts and vaporizes at temperatures that don't yet affect the platinum filament. The ignition energy, drawn from the battery, must be figured into the final results.

Let's keep in mind that bomb measurements are inherently made at constant volume. Combustion under constant pressure, like in flames, yields somewhat less heat. The difference is the energy used up in the expansion of the exhaust gases.

The *adiabatic calorimeter* circumvents the need for quantification of heat losses by preheating the water bath to the point where the flow of heat in the course of the reaction is kept at zero. This calls for very accurate temperature checks, but allows to work with flammable materials of extremely low heat of combustion.

Specific heat capacity

The watched cattle never boils, says the adage, but as every Wagner opera buff recalls, teenage hero *Siegfried* gets his magic sword hot enough to forge its blade in seconds, regardless that both – the watched cattle and the sword – draw their thermal energy from open fires.

Whether or not you believe in legend, the reason for this discrepancy is – beyond a multitude of other factors – that heating up one kilogram of water takes nearly ten times the energy afforded for an equal mass of iron or steel. Important for our well-being is the relation between water and soil, of circa 5 : 1, responsible for the mild oceanic vs. the harsh continental climates: The waters of the sea retain the summerly heat into the winter; and inversely, stay cool during the sweltering summer months.

The specific heat capacity of a given material can be found by heating up a weighted piece of it, drop it into a polystyrene cup of water (or for better thermal isolation, two cups inserted into each other), and measure the resulting increase in the water's temperature. For instance, take a 200 gram slab copper with the specific heat capacity of $c_{Cu} = 0.094$ Cal/kg °C, heated to 100 °C. It then holds $Q_{Cu} = 0.200 \times 0.094 \times 100 = 1.88$ Cal. Put into a bath of 100 cm^3 of 20 °C water, which heat energy is $0.100 \times 20 = 2.00$ Cal, the total heat energy stored in the system becomes $1.88 + 2.00 = 3.88$ Cal, while the summed up heat capacity of the components amounts to $0.200 \times 0.094 + 0.100 \times 1.00 = 0.1188$ Cal per degree. Thus, the heated copper bar will get the water temperature up to $3.88/0.1188 = 32.66$ °C. In the process, the copper slab lost $Q_{Cu} = 0.200 \times 0.094 \times (100 - 32.66) = 1.266$ Cal, and the water bath gained $Q_W = 0.1 \times 1 \times (32.66 - 20) = 1.266$ Cal. Thus, the law of conservation of energy, called here *the first law of thermodynamics*, is fulfilled by $Q_{Cu} = Q_W$.

By the same rationale, the specific heat of copper can be deduced as

$$c_{Cu} = \frac{(T_W - T_0) \times c_W \times m_W}{(T_{Cu} - T_W) \times m_{Cu}},$$

where T_0 and T_{Cu} are the initial temperatures of the water and the object under test, and T_W stands for the water temperature after the preheated test specimen had been submerged. Not surprisingly, this formula yields in the present case:

$$c_{Cu} = \frac{(32.66 - 20) \times 1 \times 0.100}{(100 - 32.66) \times 0.200} = 0.094 \text{ Cal/kg} \times^\circ \text{C}.$$

Best precision can be achieved with the use of kinetic energy (read: electricity) to heat the probe, because the transformation of electricity into heat in a resistor happens at virtually 100% efficiency, as does the heat transfer from a submerged resistor to the surrounding water-bath. And finally, laboratory grade digital instrumentation allows for measuring electrical input with typically 0.025% of accuracy or better.

Nernst calorimeter

The Nernst calorimeter in Fig. 7.11, a design of Walther Hermann Nernst, has the heating wire wound around a cylindrical piece of the material under test, which fits into the longitudinal center bore of a bigger cylinder of that same material. Inner and outer electrical insulation of the filament is granted by wrapping it in thin glassine paper or, for higher temperatures, Teflon sheet. For best heat transfer, the remaining gap is replenished with paraffin wax. For close to perfect thermal isolation from the environment, the specimen hangs in an evacuated container or bomb.

If the heater spiral is made of thin platinum wire, it can double as resistance thermometer, which principles are described earlier in this chapter. This avoids the need to raise the temperature of a mercury thermometer along with that of the probe. Likewise, the intimate contact between the test specimen and the filament guarantees a virtually zero temperature gradient between the two.

$c \times m \times \Delta T$, the calorific value of the test specimen of mass m, can be figured

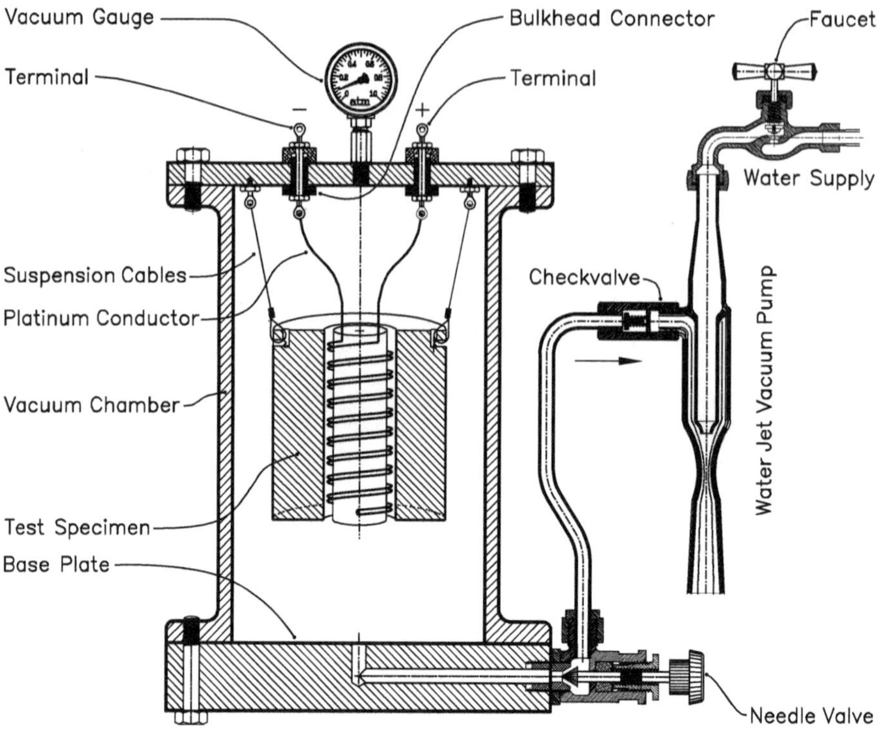

Fig. 7.11. Nernst calorimeter

from the amount of electrical energy used up for raising the temperature of the probe by ΔT.

If voltage and amperage could be held constant for the duration of the experiment, the electric energy used up in heating the probe would simply be $JEt = J^2Rt$, with J for amperage, E for voltage, and R for heater resistance. However, the platinum wire's resistance R varies along with its temperature, so that readings of J too will vary during the time-span t. Therefore, electrical energy must be totaled by the integral $E\int_0^t J \times \mathrm{d}t$. One way of doing that is by taking a reading of the amperage J in, say, 30 to 30 seconds intervals throughout the course of the experiment. The average of J is then the *root mean square* of the n readings, expressed by $J_{\text{eff}} = \sqrt{J_1^2 + J_2^2 + J_3^2 + \ldots}/n$.

A programmable instrument executes such operations continuously. Note that the results for work come in watt-seconds (joule), with the conversion factor as 4.186 joule to the calorie.

Precision of specific heat capacity measurements is particularly critical at cryogenic levels, where c becomes highly temperature dependent, as exemplified in the Table 7.2 for copper and aluminum.

Accordingly, the standard formula for heat energy, $Q = c \times m \times \Delta T$, is reasonably accurate only within a relatively narrow window of temperatures. And

Table 7.2. Specific heat of metals

Metal	Specific heat (Cal/kg) at temperature (°C) of:							
	−200	−100	0	20	100	200	300	500
Copper	0.040	0.082	0.0906	0.0915	0.0947	0.0969	0.0994	0.1049
Aluminum	0.075	0.175	0.210	0.214	0.224	0.235	0.241	0.26

since the term "heating by 1 °C" in the definition of the calorie does not take that into account, it has been changed into "from 14.5 to 15.5 °C", because measurements at other temperature levels would yield somewhat different results. Here is where the accuracy of the Nernst calorimeter is essential.

Circuits for measuring the input of electrical energy and the related temperature gains in the Nernst calorimeter are shown in Fig. 7.12. Digital high-impedance instruments are used for voltage measurements, and a low impedance ammeter to measure electric current. Any kind of battery can be the power source as long as its capacity is high enough for keeping its voltage stable when the filament current comes on.

In Fig. 7.12a, the ammeter is connected in series with the heater, and the voltmeter in parallel. As the filament heats up, amperage will drop in proportion to the heater's raise in resistance. Though this drop ΔJ is indicative for the temperature of the heater, it would be too little for an accurate deduction of heater temperature. However, the bridge circuit in Fig. 7.12b singles out the difference ΔJ and herewith obtains the desired precision.

Heater and bridge circuits would interfere with each other if connected simultaneously. However, we can alternately turn them on and off with a two-position selector switch or, for continuous readings, use alternating current (Fig. 7.13) in ways that the negative half-wave goes into the power circuit, which includes the heater spiral, while the positive half-wave supplies the measuring (bridge) circuit.

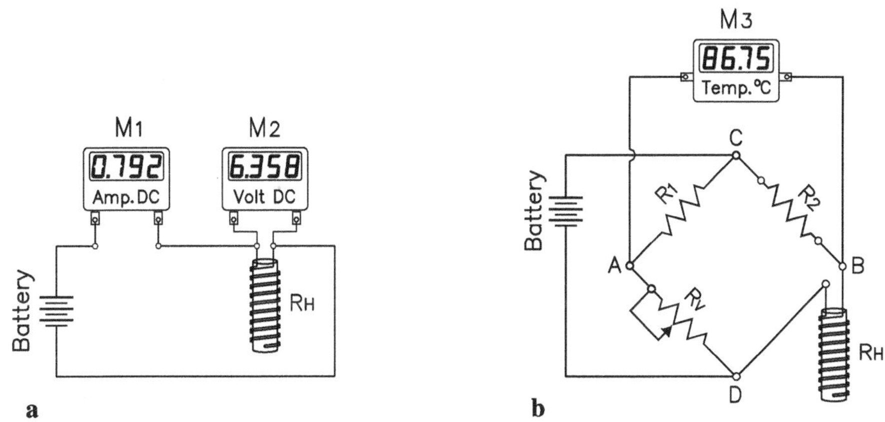

Fig. 7.12. Circuit diagrams for Nernst calorimeter

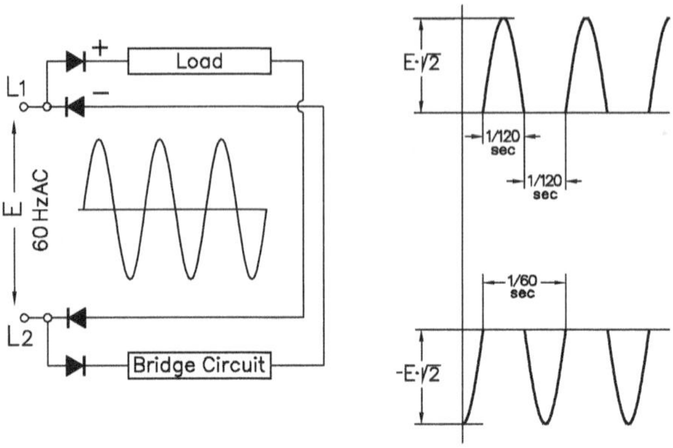

Fig. 7.13. Separation of power and measuring circuits

With 60 Hz network frequency, each circuit branch is on for 1/120 s and off for an equal time-span.

In the circuit in Fig. 13, a pair of silicon switching diodes, connected forward and backward, does this separation. The second pair of diodes shown is there for balancing the silicon diodes inherent 0.6 V forward voltage drop between cathode and anode throughout the circuitry.

The circuit in Fig. 7.14 functions on these principles. Heavy lines indicate power conductors, while measuring circuits are shown with light lines. Meters M_1, M_2, and M_3 respectively show amperage, voltage, and temperature, but certain precautions must be taken with alternating-current-powered measuring circuits.

The scales (or readouts) of AC instruments normally show the equivalent DC value for identical energy transfer. In case of sinusoidal AC, this is the root mean square of voltage and amperage, respectively, but this relation does not stand for other waveforms, such as the AC from commercial radio transformers, distorted by the nonlinear magnetization curves of their iron cores. Autotransformers with air core are free of such shortcomings, but carry far bulkier windings. And as long as no load has been connected, the low voltage tap of an "economy transformer," like the one in Fig. 7.14, carries the full network voltage of mostly 115 V in the Americas, and 220 V in Europe.

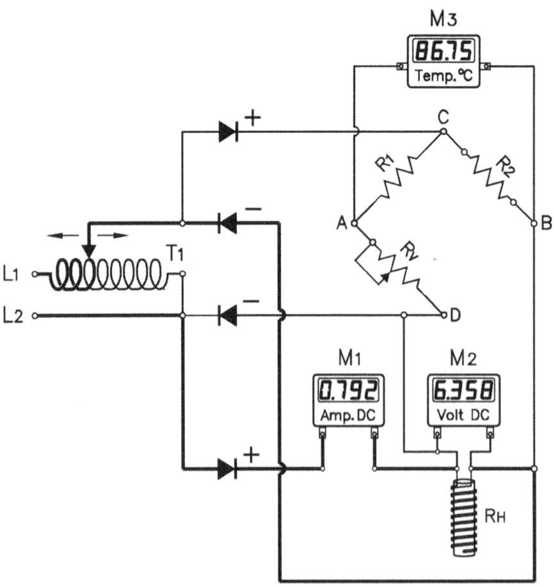

Fig. 7.14. Combined power and measuring circuits for Nernst calorimeter

A heater coil wound as double helix avoids errors caused by the vector addition of the coil's resistive and inductive impedance and, as shown on the lower right of Fig. 7.14 bringing the beginning and the end of the windings to the upper side of the coil, simplifying their connection to the terminals.

The principles of the Nernst calorimeter can likewise be applied to the measurement of the specific heat energy of liquids if we submerge the heater coil in a metered quantity of the liquid under test and store it all in a heat-retaining container, such as the double-walled *Dewar* flask. The temperature increase of the liquid from a given measure of electric energy is then a measure for the heat capacity of the quantity under test.

With an electrically well-insulated heater, the outfit can run on DC, but if the reliability of the insulation between heater and liquid is in doubt, an AC power source is preferable as to avoid the effects of electrolysis. If a thermometer is used to get the temperature of the bath, its calorific value must be taken into account.

Thermophore

He who has a mercury barometer rather than an electronic weather station hanging in his vestibule in the belief that instruments with the least number of components operate best, will love the *thermophore* (Fig. 7.15).

Imagine a thermometer with a 5.76 cm inner diameter bulb, holding as much as 100 cm^3 (1.36 kg) of mercury. Such a monster bulb boosts the instrument's reading accuracy to 0.01 °C but would make the capillary uncomfortably long if it weren't for a widening the shape of a bubble somewhere at midrange. Our thermometer has no scale save for the two marks, one for the lower and one for the upper temperature of the range wherein specific heat is to be tested. The first is situated below the bulge, the latter above. And that's about that.

Initially, the bulb is slowly heated until the mercury column reaches past the upper mark, where the heating is stopped. As the mercury retreats, the thermophore is dipped into the liquid which specific heat capacity is to be measured at the very moment when the mercury column passes by the upper mark, and quickly removed when the column gets to the lower mark. At that instance, the temperature of the liquid is read.

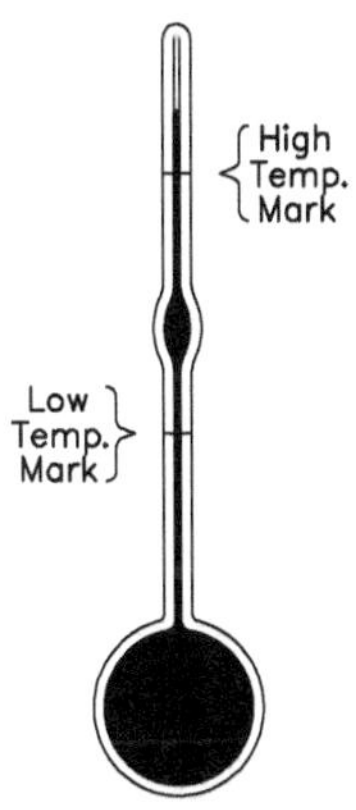

Fig. 7.15. Thermophore

For calibration, the thermophore is pretested with pure, deaerated water. If the water's temperature raise is measured as ΔT_w, and an identical weight of the liquid under test gives ΔT_x, the specific heat energy of the liquid is $c_x = \Delta T_w / \Delta T_x$.

Based on the specific heat energy of 1.000 Cal/kg × °C for water, it is 0.550 for ethanol, 0.531 for gasoline, 0.598 for propylene glycol, and 0.033 for mercury. The Olympic record for the specific heat energy of liquids is held by liquefied hydrogen with $c_H = 6\,\text{Cal/kg} \times {}^\circ\text{C}$.

Specific heat energy of gases

Unlike solids and liquids, which specific heat we consider pressure independent, gases show a difference between measurements at constant (unchanging) volume vs. constant pressure. This because the transfer of a given amount of heat energy, dQ, into a certain volume, V, of gas, results in a raise in temperature *and* in expansion. Herein, the energy spent on expansion is $p \times dV$. In a hermetically closed container, we have $dV = 0$, and therefore $dQ = dU$, where dU means the energy used for heating the gas by dT (°C).

However, if pressure is maintained while the gas volume is allowed to change, the first law of thermodynamics $dQ = dU + p \times dV$ applies. Shifting the terms of this equation into $dU = dQ - p \times dV$ shows that the temperature gain from a given supply of heat energy is smaller under constant pressure than at constant volume: a certain gain in temperature costs more energy at unchanging pressure than at unchanging volume. In other words, $c_p > c_V$.

The ratio of c_p/c_V, symbol κ, lies between 1.40 and 1.41 for two-atomic gases, such as air, hydrogen, nitrogen, etc., except for the group of halogens, including chlorine, bromine, and iodine, which already fall into the realm of three-atomic gases, with κ between 1.26 and 1.32.

The specific heat energy of gases is measured by conducting a metered quantity of the preheated gas through a helically spiraled copper or silver tube, meandering through a water sump. The resulting rise of the water temperature and the related drop in the gas temperature allow to figure the specific heat energy of gases as follows.

From T_{in} and T_{out} for the input and output temperatures of the gas under test, the energy transferred by m_G kg of gas flowing through the spiral tube is $Q_G = m_G \times c_p \times (T_{in} - T_{out})$. This should equal the heat needed for raising the sump temperature from T_1 to T_2. With m_W for the mass of the water and $c_W = 1$ Cal/kg, this amounts to $Q_W = m_W \times (T_2 - T_1)$. Equaling the two terms gives the specific heat of the gas as

$$c_p = \frac{m_W}{m_G} \times \frac{T_2 - T_1}{T_{in} - T_{out}}$$

The flow of gas through the system must be slow enough to allow a measurable amount of heat getting transferred from the gas into the sump. To this end, the spiral pipe is stuffed with fine metal chips, leaving just enough space in between for the gas to squeeze through. This adds to the length of the passages through the piping, while friction between gas molecules and the chips slackens the flow.

Apparatus of Clément and Desormes

The equation for adiabatic changes of state in an "ideal" gas:

$$p_0 V_0^{\kappa} = p_1 V_1^{\kappa}$$

is used in the apparatus of Clément and Desormes for finding the ratio $\kappa = c_p/c_v$

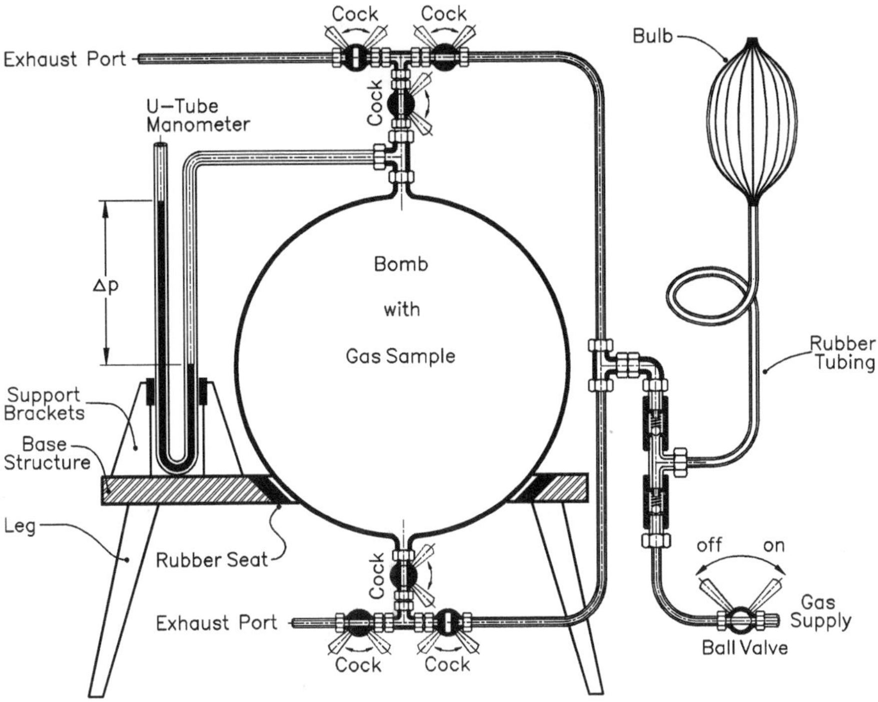

Fig. 7.16. Apparatus from Clément and Desormes for c_p/c_V measurements

of the specific heat energy for gases at constant pressure vs. at constant volume.

The system (Fig. 7.16) includes a gas container or bomb, an U-tube manometer, and a device for gas compression, which may resemble the rubber balls we remember from perfume spray bottles and blood pressure checks.

Depending on whether the gas under test is lighter or heavier than air, the container is filled from above or, respectively, through the bottom port. In the first case, the lighter gas, such as hydrogen, accumulates in the upper regions of the bomb and gradually forces the remaining air out through the lower spud. For the case of gases heavier than air, such as carbon dioxide, the situation is reversed; the incoming gas settles at the bottom and air is let to escape on top. The three cocks on upper and lower port must be set in each case to accommodate these conditions. Filling should take place with a generous overdose of the gas under test in order to expel any residues of air or gas from previous experiments.

Measurements of c_p/c_V then start with the shut-off of the exhaust port and the pumping of an extra amount of gas into the container, until the U-tube manometer reads some 40 mm of mercury column (Hg), the equivalent of $40 \times 13.6 = 544$ mm of water head. If sulfuric acid, 1.831 heavier than water, is used as barometric fluid, the same pressure will show as a $544/1.831 = 297$ mm H_2SO_4 column.

Next is a momentary opening and closing of the upper exhaust port, which makes the internal pressure drop to zero in a process we consider as adiabatic, because the time-span between opening and closing the valve is too short as to allow for any significant heat exchange between the gas in the bomb and the environment. Thus, the kinetic energy afforded for the gas to escape is drawn from the gases internal energy, which shows by a drop in its temperature from T_1 to T_2.

In the following, the gas is allowed to warm up to T_1, the temperature of the surroundings, which makes the pressure in the bomb raise from p_1 to p_3. This happens without change to the amount of gas in the container, which makes $V_3 = V_2$. Since the allover process ends at the same temperature as it started, it is isotherm and can be written as $p_1 V_1 = p_3 V_3 = p_3 V_2$.

For the deduction of κ, we elevate this equation to the power of κ, getting $p_1^\kappa V_1^\kappa = p_3^\kappa V_2^\kappa$, and divide it by the adiabatic equation $p_1 V_1^\kappa = p_2 V_2^\kappa$, from which we get with some reshuffling

$$\frac{p_1}{p_2} = \left(\frac{p_1}{p_3}\right)^\kappa \quad \text{and} \quad \kappa = \frac{\log p_1 - \log p_2}{\log p_1 - \log p_3}\,.$$

Upgraded versions of this system have the U-tube manometer replaced by digital pressure gauges, actuated by a bulging membrane or the voltage from a piezoelectric crystal. They avoid the slight increase in gas volume caused otherwise by the shift of the liquid column in an U-tube.

The coefficient κ for air at 18 °C has been found as 1.4053. An interesting connection between this constant and the speed of sound exists in the form of the equation $v_{\text{sound}} = \sqrt{\kappa p/\rho}$, where p stands for atmospheric pressure of 1 atm, or 101320 newton per square meter, and ρ for the specific gravity of air, 1.293 kg/m^3, both at 0 °C. With these values, we get $v_{\text{sound}} = \sqrt{1.4053 \times 101320/1.293} = 331.8$ m/s, in striking conformance with the values obtained by direct measurements of the speed of sound.

Entropy and the heat-death

Notwithstanding the dire predictions of doomsday-sayers, our Earth's stock of unused energy is enormous. As an example, let's take the temperature gradient between surface and bottom water in the oceans. Water at 4 °C, the temperature of its highest density, tends to sink to the ocean floor and there to accumulate, while somewhat hotter water amasses at the surface where sunlight heats it up. If a thermal machine could be built to convert the water's internal energy into kinetic energy, our ships wouldn't have to rely on burning oil to propel them. All they would need is a long hose reaching down into the 4 °C regions and a pump to circulate the liquid up and down.

In principle, such an outlandish offspring of James Watt's venerable steam engine could be built by replacing the boiler water with a suitable liquid, such as dichlorofluoromethane, $CHC\ell_2F$, which boils at 8.9 °C. The exhaust of the engine would be pumped down into the 4 °C zone and condense into liquid, pumped up into a boiler floating barely submerged in surface water of, say, 24 °C, that is, $24 - 8.9 = 15.1$ °C above the boiling point of the working fluid. This temperature gradient converts into an absolute $CHC\ell_2F$ steam pressure of 1.72 atm, or $(1.72 - 1) \times 1.033 = 0.74\ \mathrm{kg/cm^2}$ relative to the surroundings. Enough to drive a reciprocating engine engineered for these conditions.

But the thermal efficiency, η, of an engine operating between the absolute temperatures of $273 + 8.9 = 281.9$ K and $273 + 24 = 297$ K, would be $(297 - 281.9)/297 = 0.051$ or about 5%, barely enough for pumping the water up and down between the cold and warm regions. 95% of the available energy cannot be converted, and how much – if anything – would be left for driving the ship's propellers is everybody's guess.

The fact that η gets better the more the working medium is heated, lead to the assignment of *value* to thermal energy. Unless you make part of some Polar Bear Club, you will find a bathtub filled with water of 40 °C (100 °F) far more useful than two tubes with 20 °C (68 °F) water, though the amount of heat energy is in both cases the same.

The *value* of heat energy was named *entropy*, *s*, and defined by $\mathrm{d}s = \mathrm{d}Q/T$. Accordingly, entropy gets lower the higher the temperature of the working medium. A heat reservoir of lowest entropy is the best choice for energy generation.

Pursuant to the second law of thermodynamics and everybody's personal experience, heat flow is unidirectional from hotter to cooler regions or objects. The resulting drop in average temperature amounts to a steady, unstoppable raise of entropy, unless energy from external sources is injected. For the world as a whole, entropy must be expected to rise with the passing of time, until a uniform[1] overall level is being reached. This irreversible state of world affairs is called the *heat death.*

Discovery of nuclear energy may explain why the 15 billion years of age of the universe weren't long enough for the heat death to occur. But even nuclear transformation of mass into energy with its mind bogging mc^2 conversion factor will not go on forever. Somewhere in the future, stars will have exhausted their inner nuclear engines, and those with supernova potential will have exploded and cooled down, and the highest possible entropy will reign over the world.

But continuing this rationale, we would expect that the world at the instant of the "big bang" was in the lowest possible state of entropy, which leaves the question how the jump from highest level of entropy in a previous, energywise extinguished world, to the lowest was performed previous to the birth of the world "as we know it". If we lack an answer to this kind of question, it is that the very concept of the term "question" has been generated in that "world, as we know it" and thus has no meaning outside its realm.

Let's thus keep our positive attitude and go on living in hopes that somewhere, sometime, entropy might be unobtrusively shrinking.

1 but extremely low.

8 Pressure

In 1654, in the city of Regensburg, Otto von Guericke performed one of the most famous experiments in the history of physics. He took two copper hemispheres whose diameters were approximately 14 inches and sucked the air out from between them (Fig. 8.1). Two teams of eight horses apiece were yoked to the hemispheres and tried to pull them apart, but to no avail. This became known as the Magdeburg-hemisphere experiment, named after the town in Saxony from which Guericke hailed, and it was an immediate success. In the audience that day was King Ferdinand III, who would later become the patron saint of engineers. Guericke, though, would take this show on the road: In 1657 he would repeat the experiment in the Emperor's court in Vienna, Austria, and later on to the German Elector.

The secret to Guericke's success was the extreme power of air pressure. The earth is swaddled in a thick blanket of air, about 10 miles deep. Air is a gas made up for the most part of nitrogen and oxygen, and so has weight. This weight generates a pressure at the Earth's surface. At the top of Mount Everest, about 29,000 feet above sea level, air pressure is a lot less than at sea level. At the summit, at an altitude of roughly 6 miles, there's only a thin layer of air above a climber that can exert a force, so air pressure is less. That leads to a problem for mountaineers: the temperature at which something boils depends on the local atmospheric pressure. Water, as any tea drinker knows, should be boiled at 100 °C. At high altitudes, boiling might well occur at temperatures as low as 70 °C, yielding a disappointing brew to a mountaineer after a day of braving ice needle-loaded hurricane strength winds.

In Guericke's experiment, the two sets of eight carthorses each were not sufficient to pull apart the hemispheres. The sphere has only a vacuum inside it. Outside, the air is at pressure p. The radius of the spheres was 7 inches. Barometric pressure in Regensburg, 342 meter above sea level, averages 14 pounds per square inch. So, the force exerted by air pressure on the spheres, the product of pressure p and surface area, was[1] $7^2\pi \times 14 = 2155$ pounds. Assuming that Guericke's pumps reached 90% evacuation, the difference between external and internal pressure on the semispheres must have been $0.90 \times 14 = 12.6$ psi, and the holding power of the device $2155 \times 0.90 = 1940$ pounds.

Otto von Guericke was not the first scientist, though, to be captivated by pressure. A few decades earlier, an Italian physicist, Evangelista Torricelli, came up with a device that would lead to an indispensable tool for modern meteorologists, the barometer. In his own words, written to scientist friend Ricci in 1644, "We have

1 in units of our fps system of measure.

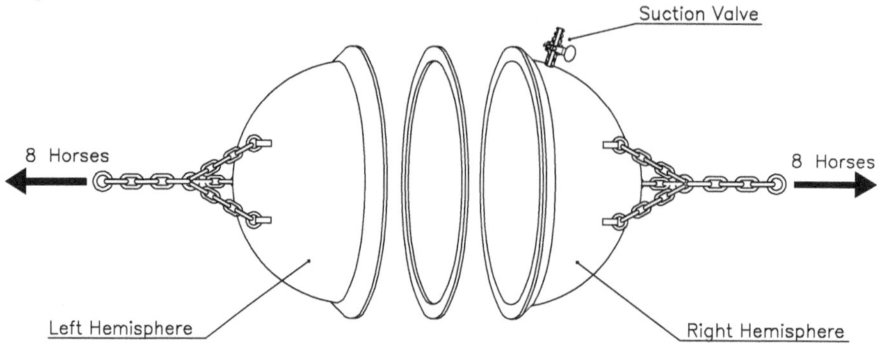

Fig. 8.1. Guericke's hemispheres

made many glass vessels ... with tubes two cubits long. These were filled with mercury, the open end was closed with the finger, and the tubes were then inverted in a vessel where there was mercury ... We saw that an empty space was formed and that nothing happened in the vessel where this space was formed ... ".

It is a simple device (Fig. 8.2). Air pressure pushes down on the mercury in the bowl, which causes some of the mercury to rise up the inverted tube. The lower the air pressure, the less liquid rises up the tube. On a typical day, the mercury would rise some 760 mm above the level of mercury in the bowl. As the chemical symbol for mercury is Hg, standard air pressure is called 760 mmHg. This is also often called one atmosphere, or 1 atm of pressure.

Blaise Pascal, sometime philosopher, sometime scientist, put Torricelli to the test. He took the nascent barometer up to the top of the Puy de Dome, a volcano in the Massif Central mountain range in France, a peak that is popular in the Tour de France. At the top, about 407 meters, the mercury level sank quite considerably.

The key, though, is to make the barometer sensitive to tiny changes in air pressure. After all, when the barometer is falling, we expect bad storms or hurricanes. If it rises, it's time to make sure the garden plants are watered, for we might be in for a dry spell. One simple trick to increase sensitivity is to put the tube at an angle. A slight change in pressure then generates a greater change in the position of the mercury meniscus. Placing the tube at an angle of 30°, as $\sin 30 = 0.5$, doubles that change of position.

Mercury barometer

Atmospheric pressure is still being measured with the classical mercury-filled U-tube barometer (Fig. 8.3), though dial and digital instruments have become popular. Previous to reading air pressure from the mercury column in the barometers top-sealed left leg, the mercury mirror in the U-tube's open-ended right leg must be brought to coincide with the zero point of the scale, shown by a mark on the instruments baseplate. A micrometer screw on the U-tube's support-bracket allows for this adjustment.

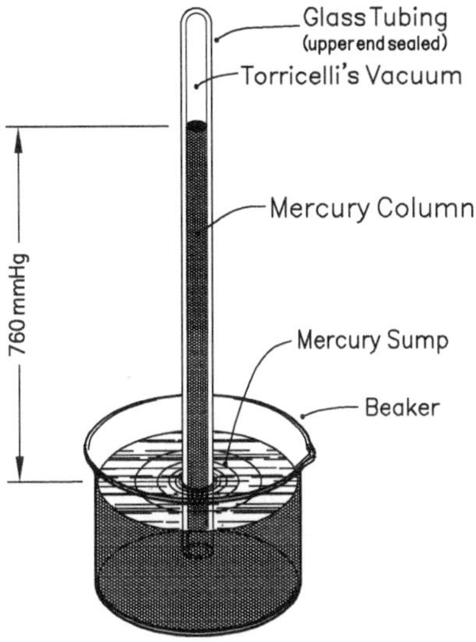

Fig. 8.2. Torricelli's barometer

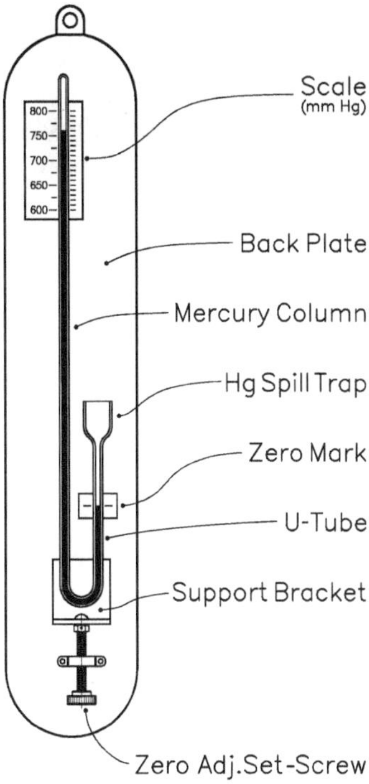

Fig. 8.3. U-tube barometer

Elementary measuring devices, such as the U-tube barometer, avoid the errors and inaccuracies of the mechanisms that convert sensor signals into readout in instruments of higher sophistication. The mercury barometer is sensor and readout in one piece, a guarantee of reliability and low maintenance.

But mercury filling has drawbacks. It's heavy, for one thing, and poisonous for another. A different gauge fluid might be ideal. Water, for example, is cheap and safe. But it is much less dense than mercury, so that 760 mmHg of pressure would cause water to rise by about 33 feet. For the typical U-tube barometer, water may be safer, but it certainly would be more cumbersome.

There is a role, though, for water barometers. Trying to eyeball a change between 0.1 mmHg and 0.2 mmHg would be extremely difficult. So, at low pressures, changing the working fluid of the barometer is a good idea. To measure pressures of 1 mmHg (called 1 Torr, in honor of the Italian scientist), a water barometer would rise by 13.6 mm H_2O.

You can't use a water barometer at low temperatures, though, because water will freeze. Benzene, whose chemical structure was discovered by Friedrich August von Kekule (1829–1896) after he dreamed about a snake swallowing its own tail, will also work as a low-pressure, high-vacuum gauge fluid. Density aside, the liquid used in any barometer, therefore, depends on the temperature range it will be used over, as well as the corrosiveness, toxicity, and flashpoint of the fluid.

By choosing the appropriate working fluid, Signor Torricelli's barometers can be used for ordinary atmospheric air pressure down to pressures of about 1 mmHg. This extraordinarily broad range makes them ideal for use in fluid mechanics. In one of the high points of classical physics, Bernoulli recognized that the pressure p of a fluid, its density ρ, and its velocity v, were linked by the simple expression: $p + \rho v^2/2 = \text{constant}$. So, if you can measure the pressure exerted by the moving fluid, then you can infer the fluid velocity. Thanks to Torricelli, there was a straightforward way to measure the flow speed of laminar or turbulent fluids. Bernoulli's principle also predicts something. As the speed increases, there can be a point at which the pressure can become negative. This marks the onset of cavitation, where bubbles are generated quite noisily. One of the great challenges for those who manufacture warships and submarines is to design the propellers so that cavitation doesn't occur, for otherwise the vessel would operate at lowered efficiency and send out an extremely strong sonar reading.

Torricelli managed to obtain a high vacuum through the honorable scientific process of serendipity, when he filled a one-side-sealed glass tube with mercury and then slowly propped it up into the vertical. As the mercury filling dropped to 760 mm, it left the space above an almost perfect vacuum.

Far off from the ivory towers of high science, a similar effect keeps the odors from underground sewage lines from leaking up through our kitchen sink. A low-tech device, known as the siphon (Fig. 8.4, upper right), does this better than the most sophisticated valve could perform. After the sink has emptied, water plugs the U-shaped section of the drain piping, sealing off the sewage line. An idea that made the siphon a basic design feature of toilet bowls (Fig. 8.4).

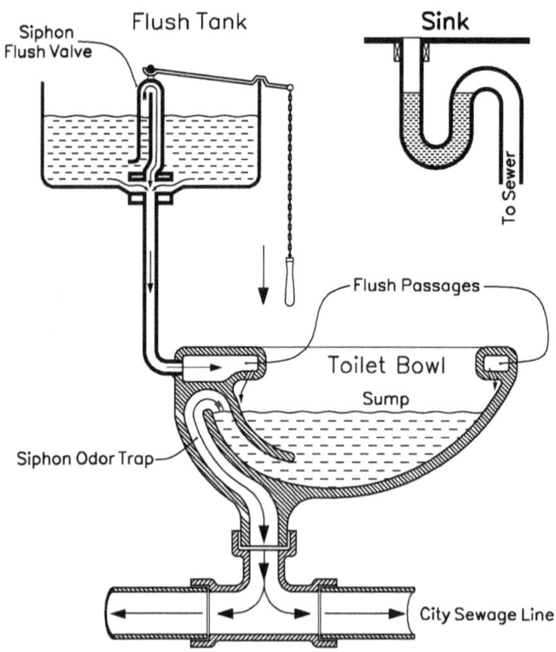

Fig. 8.4. The siphon and its applications

During the first half of the 20th century, the siphon resurfaced in water toilets as a U-shaped passage in drain valve plugs of overhead purging tanks (Fig. 8.4, upper left). When a pull on the chain lifts the valve plug, water starts flowing freely down the pipe into the flush passages and the toilet bowl. But the discharge continues even after the chain's release while the valve plug seals the flush tank's outlet, because water now rises up the U-shaped passage in the valve plug and spills into the discharge piping. Like in the U-tube barometer, the descending liquid leaves behind a partial vacuum, which ultimately soaks all the water out of the tank.

"Soaks", so to speak, because what really happens is that atmospheric pressure on the liquid surface in the tank forces the water up the siphon and down into the region of reduced air pressure in the drain-tube. So let's not get carried away by the "suction power of vacuum." It's all a manner of speech, because nature doesn't allow for *negative pressure*, and the term "suction" stands in reality for the effects of a zone of lower pressure within an environment of higher pressure. Pressure in an absolute vacuum is zero, and from there on can only get higher. If the scales of some manometers tell you differently, it is that their dials are plotted from 1 atm downward, and its divisions marked inversely. Nothing is wrong with that, as long as we keep in mind that the allocation of a scales' zero point at standard barometric pressure of 760 mmHg is a convenience and not the expression of a law of physics. If it were, said instrument should show 0 atm on the atmospherically challenged moon – but would it?

It is possible, however, to create high vacuums on purpose. Suppose you have a container filled with 10 liters of air, connected to a pump that has a capacity of one liter. With the piston's first backstroke, those 10 liters of air expand to fill the 11-liter pump-cylinder/container combo. Since a checkvalve keeps air from flowing back from the pump cylinder into the container, the now following forward stroke squeezes the remaining air through another checkvalve out into the environment.

The air that remains in the container is now 10/11 as dense as the air that was originally in it. After two cycles, we will have 10 liters of air that is only 100/121 as dense as the original. This may seem like a slow process, but after only 50 cycles, the air is down to 0.0085 of its original density. This means that if the air originally was at a pressure of 760 mmHg, after 50 strokes it has plummeted to only 6.5 mmHg. After 100 cycles, the pressure would be 72.5×10^{-6} of the starting value, or 0.055 mmHg.

In principle, the vacuum pump can be seen as an inverted version of the common foot- or hand-driven pressure pump such as the one we use for pressurizing the children's bicycle tires and (Heavens beware) our car tires in roadside emergencies. But brace yourself for a shock: Only part of your sweat from pushing and pulling the pumps handle goes into pressure gain. Some of that energy converts into thermal energy, vulgo 'heat,' rather than compression. If the heated air would be pumped "as is" into the air tank, pressure would subsequently drop while the air cools down, because pressure in a sealed container raises and falls along with absolute temperature. This is the reason for the cooling fins on the cylinders of air compressors. Industrial models can be even water-cooled and have radiators similar to the one we find under the hood of our car.

Actually, the similarities between compressors and car engines made that McGeyver type do it yourselfers sometimes assembled their personal low cost compressors from scrapped car engines. But that was previous to the times when you could buy a compressor for less than fifty bucks (electric motor and an extra length power cord included) at K-Mart and similar places.

The part to replace was the cylinder head with its domed recesses over each cylinder, known as combustion chambers. Comprising about 1/8 of the cylinders volume, those chambers assure the 1 : 8 or so compression ratio of the air–gas mixture previous to ignition and combustion. By contrast, the pistons of compressors should force the air completely out of the cylinders into the air tank, which means replacing the cylinder head by a flat plate in which the valve-seats are being bored.

Which brings us to the second big difference between motors and compressors: While the formers' valves are positive controlled by the cams on the commando shaft, the latter's are autoactuating. In a compressor, the air intake valve opens when the piston's retreat generates a partial vacuum in the cylinder. Conversely, the discharge valve is pushed open by overpressure from the piston's forward stroke.

So the compressor builder can scrap the commando shaft and its ancillary components, but at a price. Self-actuating valves must be correctly sized for handling the expected flow of air, and critically balanced to maintain correct timing and react promptly to flow changes. Moreover, too strong valve springs keep the valves from operating, while too weak springs cause them to flatter.

It might even be safer to stick with the original cam-controlled valve operating system and replace the 1 : 2 silent chain or timing belt transmission between crankshaft and commando-shaft with a 1 : 1 transfer drive. On the other hand, that could amount to trading one complexity for another.

While a compressor is meant to raise air pressure above the prevailing barometric pressure, Torricelli's barometer uses the barometric pressure of the atmosphere to lift what we could see as a fluid piston – the mercury column. But regardless its simplicity and reliability, Torricelli's basic barometer along with its modern cousin, the U-tube wall barometer, make for bulky and position-dependent instruments, inherently unfit for a host of applications, such as aeronautical speedometers, altitude indicators, etc. Neither would they readily fit the baggage of the conquerors of the world's mountain peaks.

Dial barometer

The dial barometer in Fig. 8.5, better known under the term *aneroid barometer*, is a light and portable instrument which principle goes back to the French lawyer and later engineer Lucien Vidie (1805–1866). But his design (1844) failed to gain acceptance on those times' French market, and better economic success in England was met with bitter lawsuits from competitors.

Aneroid barometers, typically operating in the range between 29 inch and 31 inch Hg, sense air pressure from the deformation of a face-to-face mounted pair of ondulated membranes – the bellows. Evacuated, these bellows compose a capsule that "breathes" in response to changes in the surrounding atmosphere.

The capsule is mounted rigidly on a spacer plate fastened on the instrument's mainframe, while a knife-edge hinge on the upper capsule actuates the displacement transfer mechanism. In Fig. 8.5, this includes the lever, the tie bar, and the

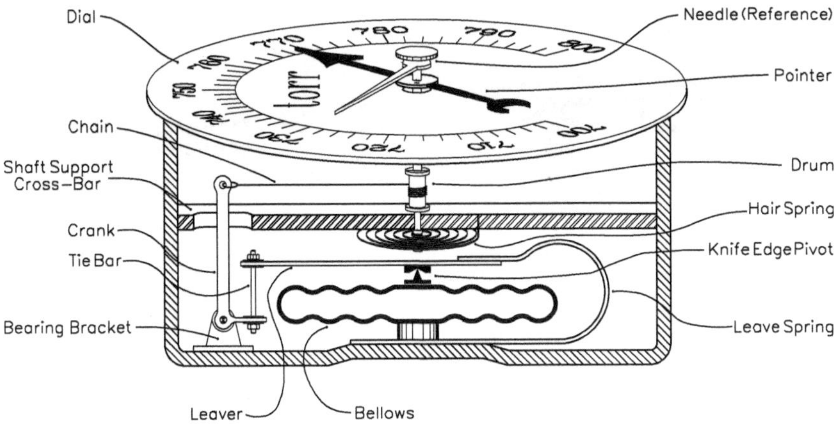

Fig. 8.5. Aneroid barometer

crank; the latter pulling a string or fine chain slung around the drum that rotates the pointer.

This multilever mechanism boosts the small swellings and contractions of the bellows and ultimately actuates the string slung around a drum on the pointer shaft. Also on the pointer shaft is a hairspring to maintain string tension and compensate for slack between drive train components. And a manually settable pointer on the barometers sight glass is meant to recall yesterday's weather for comparison with the present.

The U-shaped leaf spring arcing around the capsule preloads the bellows enough for them to respond to positive as well as to negative pressure fluctuations. Since the deflections of the membrane in response to changes in air pressure are intrinsically minute, owners of older models used to improve the sensitivity of their instruments by lightly tapping the sight glass when the needle adamantly refused to swing toward some betterment in the weather pattern on a rainy Sunday afternoon.

Low and rough grade vacuum

Low and rough grade vacuum includes the range from 760 mm down to 100 mmHg. Vacuum cleaners, which usually operate at about 2 inches (50 mmHg) below ambient pressure, are typical low vacuum devices. However, what makes a vacuum cleaner tick is dynamic pressure, $p = \rho v^2/2g$, generated in the wake of the flow of a great volume of air at high velocity (v). After all, the lowest observed barometric pressure in the eye of a hurricane (Wilma, in October 2005) was 662 mmHg or 13% below normal atmospheric pressure – not enough to explain the damage. It is dynamic pressure that makes the cows fly, if you know what I mean.

The heart of the vacuum cleaner in Fig. 8.6 is the blower, which central admission port faces the suction nozzle through a detachable filter-unit comprising a strainer and a multifold of filter paper or cloth that retains dust and impurities before they can get to the fan.

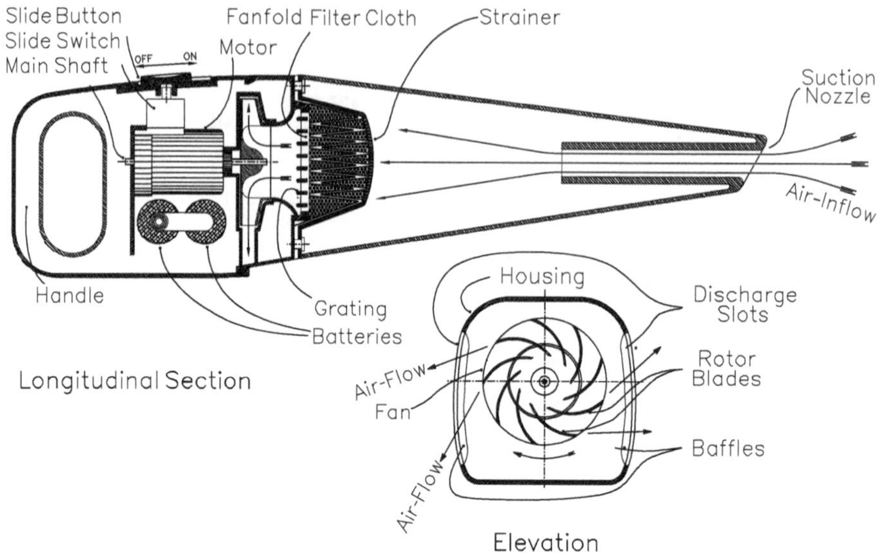

Fig. 8.6. Vacuum cleaner

Shown at the lower right of Fig. 8.6 is the face of the fan with its usually 9 vans, which curvature suits the air-flow generated by the combined action of centrifugal acceleration $R\omega^2$ and tangential velocity, $v = \omega R$. With R for the distance from the center of the rotor, and n for the rotor's rotations per minute, angular velocity becomes $\omega = \pi n/30$. Tangential air speed is thus directly proportional to center distance, which makes the vans' curvature nearly radial near the center, and close to tangential farther out (Fig. 8.6, lower right).

Unlike dynamic pressure that powers vacuum cleaners and industrial cyclonic dust precipitators, the effort of soaking soft drinks through a straw remains essentially unchanged, whether we contemplatively enjoy the drink or gulp it nervously down. In any case, we must produce some 6 to 8 inches of water-head (about half an inch of mercury column) to get the liquid up the straw. Intrepid record seekers claim that their lung-generated pressure reaches 0.5 atm (380 mmHg) to no better end than to inflate toy balloons, though he who will make it into *The Guinness Book of Records* would no doubt have to blow into a pressure gauge of the Bourdon tube or membrane type – but beware of tire gages, which block access unless a stop valve in their neck is pushed open by a stem in the tire's filling valve.

Once upon a time . . .

The need for compressed air predates by far the birth of instruments to measure its pressure. Previous to 1000 BCE, the low melting points (<1000 °C) of copper, tin, and zinc allowed to use simple ovens for melting certain alloys, such as bronze (copper-tin) and brass (copper-zinc); in part because an alloy's melting temperature is always below that of its constituents.

We may never learn when and where a fearless member of the human race found out that blowing at the bed of a campfire made the dormant flames shoot up. But the need for mechanical wind makers must be linked to the appearance of iron around 1000 BC, because the temperatures for processing iron ore cannot be reached without artificial draft. Though the reduction of iron ores, such as hematite and magnetite, occurs in the presence of carbon at about 800 °C, at least 1200 °C is needed to create raw iron through absorption of carbon from the firebed.

Long before Romulus and Remus threw their first shovels of earth and gravel on the walls of the future city of Rom, the Etruscans produced on the Apennine Peninsula iron in quantities worth exporting to most other countries bordering the Mediterranean coastline. Obviously, they had some kind of blast furnaces. But that alone wouldn't have sufficed without a good knowledge of the tricky business the production and processing of iron really is. Lo and behold, blacksmiths of the middle ages still lived and worked in the firm belief that humans would never learn the ultimate secrets of steel making, and Communist China's 20th century's action "A Blast Furnace in every Backyard" led people to build some 600,000 mini blast-furnaces which output was ultimately found unfit for industrial steel production.

Since so little is known of Etruscan culture, we are left to envision early iron makers as using a pair of hides, sewn into airtight bags, to force air into the firebeds of their ovens by the efforts of an operator who alternately lifted his bodyweight from one to the other airbag and back. But still, a checkvalve would have been indispensable to keep the air from being soaked out of the oven and back into the airbag at the return stroke, and we might never find out how the Etruscans crafted autoactuating valves some three millennia ago.

Smelters and blacksmiths in medieval times used hand-driven bellows, while waterwheel-powered air-moving devices appear in the 14th and 15th centuries. Railroad engines have jets of steam ejecting air into the firebed under the boiler.

Pressure gauges got into the picture in the course of optimizing the performance of steam engines, and principally to forestall explosions, quite common in the early days of thermal power generation. Mercury manometers were out, because something like 10 atm of steam pressure would make a mercury column shoot up by $10 \times 0.760 = 7.60$ meter or 25 feet! That's how dial manometers were born.

Bourdon tube manometer

The Bourdon tube manometer shown in Fig. 8.7 responds to pressure changes by the deformation of a flattened, crescent-shaped tube of hard drawn phosphor bronze or beryllium copper. The difference in length between outward and inward walls of the semicircular tube makes that the total of pressure-generated forces directed outwards from the center exceeds the total of inward directed forces. In short, an increase in pressure causes the curvature of the tube to "open up".

The particular type of instrument in Fig. 8.7 has the lower end of the Bourdon tube soldered to the base, leaving the upper end free to move in step with the tube's pressurization. The T-frame mounted mechanism driving the needle in the lower part of the illustration augments these intrinsically small displacements by (take a

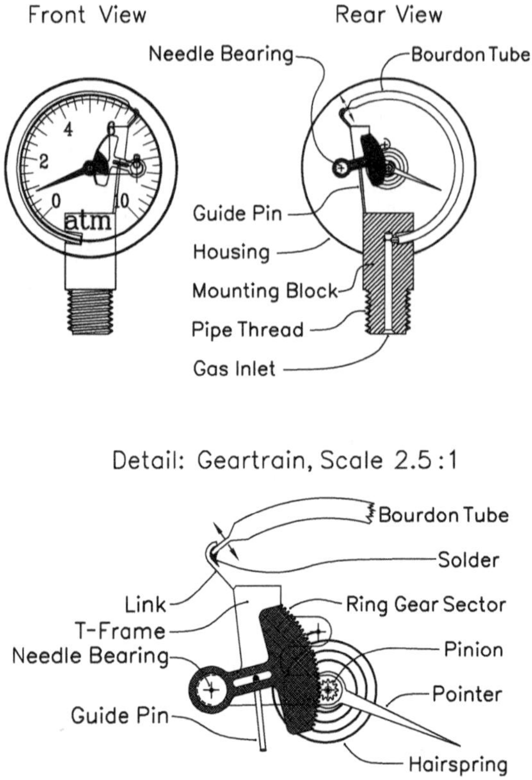

Fig. 8.7. Bourdon tube manometer

deep breath) the ratio of pitch diameter of the ring gear to pitch diameter of the pinion times the ratio of the pitch radius of the ring gear to the space between ring gear swivel point and guide pin.

The center-bearing of the ring gear sector locates on the pivoted T-frame (rather than the instrument housing) while the upper edge of that same T-frame is rigidly attached to the free end of the Bourdon tube. A stationary guide pin reaching through a slot in the ring-gear makes that it turns in step with the displacements of the T-frame. In short, it actually is the angular displacement of the *ring gear relative to the T-frame* that drives pinion and indicator needle.

Other designs hinge the centers of pinion and the ring gear sector in fixed positions on the instrument housing, which implies a hinged bar connection between ring gear and the free end of the Bourdon tube. Conversely, the floating T-frame design in Fig. 8.7 makes for a robust setup at the price of some deflection-dependent, yet minute displacements of the pointer-shaft's position.

Bourdon tube manometers are available from 0 to 0.5 kgf/cm^2 up to 0 to 1000 kgf/cm^2. For highest sensitivity, the Bourdon-tubes can be helically wound like a spring, or spiraled similar to a hairspring. In these cases, deflection of the pointer is boosted in proportion to the number of turns.

Membrane-actuated manometers

Membrane-actuated manometers offer higher sensitivity than their Bourdon tube siblings and excel by their immunity against attacks from corrosive gases, because the proper membrane seals off the instruments mechanism from the source of pressure. Therefore, the membranes are usually made of AISI 316, a high alloyed austenitic stainless steel with typically 25% Cr and 20% Ni.

The membrane manometer in Fig. 8.8 is seen from the rear with the backplate of the in strument case omitted, which explains the mirrored

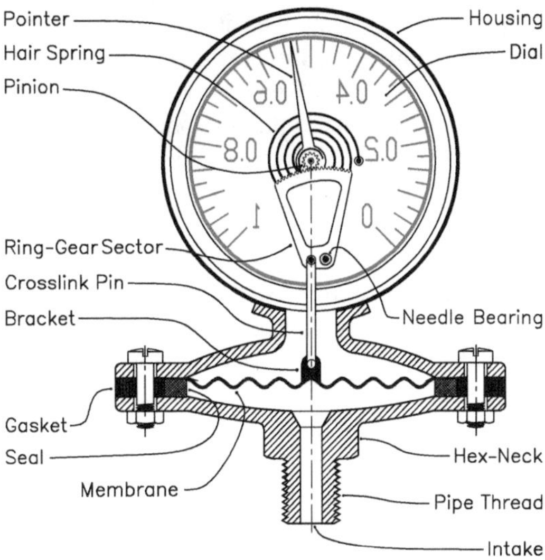

Fig. 8.8. Membrane manometer

"ghost image" of the dial. The transfer mechanism from the sensing element (the membrane) to the pointer-shaft resembles that of the Bourdon tube manometer, yet with the center of the ring-gear sector supported by a needle-bearing in affixed position. To counteract backlash, gears are preloaded by the torque of the hairspring.

Membrane as well as Bourdon tube manometers are available encapsulated with glycerin filling, which keeps out contaminants and dampens oscillations.

Energy via airmail

Iron, which has been a widely used material ever since its discovery, soared to prominence during the 19th century's Industrial Revolution. It was the era when steam power stepped in for the wind and watermills that had been sources of power for so many centuries. With steam engines, you could drill deeper, and drain mines more thoroughly, than ever before. This enabled coal mining to go beyond simple opencast mining. And with more coal, more machines could be produced to do more work, and produce more goods.

The impact of James Watt's steam engine on civilization was twofold: First, the machine ended the millennium old energy crunch, which had hampered inventors and industrialists alike. And second, save for the power of horses, the steam engine was the first location-independent prime mover ever. Windmills were restricted to areas of strong and principally steady airstreams, such as The Netherlands, some regions of northern Germany, the Baltics, and Scandinavia. Waterwheels were bound to be located along rivers in places where an abrupt slope in the riverbed lent itself for the construction of weirs.

In the mining industry, pumping groundwater out of the shafts had been left mostly to horse-actuated roundabouts, driving deep-seated water pumps. Actually, the earliest versions of steam engines were dual in line piston pumps with steam and pump pistons attached to the upper and lower ends of a long, vertically sliding piston road. They didn't have flywheels and were powered by the suction of condensing steam in the enclosed space above the piston in the steam-side section of the pump.

While the mining industry was foremost in steaming up the spread of the steam engine (pun intended), smaller enterprises, such as village forges, factories, and grain mills lacked the capital to invest in such futuristic machinery, and energy distribution became the next hurdle to progress. It began in plant with one or several under the roof mounted lineshafts carrying a number of take-off pulleys in sets of pairs of powered pulleys with free spinning ones in between – aligned with similar sets on each of the machines below. Such installations survived occasionally into the years preceding World War II, except that the power source became a huge (50–100 PS) electric motor. One setback was that those motors had to be turned on full force even when nothing worthier than one single hole in some steel part had to be drilled. Notwithstanding, individually driven machinery became the rule only when small electric motors had become affordable.

Along with the steam engine came the availability of steam-pressure for other purposes, often resembling the applications of today's air cylinders. The pivoting roof beam (Dachwippe) was common in 19th-century ironworks for lifting the front end of a brazing-hot iron slab off the anvil and rotate it into a new position under a Nasmyth's or Condie's steam-hammer. The considerable force for lifting such a slab along with the weight of the rocking beam came from a steam cylinder, mounted on top of one of the building's structural support walls.

More entertaining was some movie makers vision of farther-reaching applications of steam pressure, such as in that *James West* episode where the hero is made to face a horde of hostile steam-powered robots, mounted by the villain to no better end than to do away with the otherwise invincible West. Needless to say that even the robots lose in that unequal battle, if not by the hero's unwavering aim, then (I guess) thanks to accumulation of condensate in their extremities.

Compressed air is free of such problems and was first used as a means of power distribution around 1860 in the construction of a railway tunnel between southeastern France and Italy under the Mt. Cenis. After three years of attempts to manually break the rocky ground, wet compressors were installed to drive the familiar type of enervating pneumatic drills with teams of workmen digging in from opposite ends of the tunnel. Each worked through 4 miles of rock until they met in the middle, which was proof enough for the viability of long-range energy transmission with compressed air as the medium.

Such results encouraged the Austrian engineer Victor Popp to install in 1886 a compressed air network in Paris, which supplied craftsmen shops and – surprisingly – public clocks and even a number of elevators. A 1500 kW compressed-air-operated energy distribution system in an industrial plant followed in 1888. Its success made many of those times engineers think of compressed air as preferable to electricity. As we now know, electricity won out in the end, but compressed air continues to drive an uncountable variety of hand tools, such as drills, grinders, sanders, nailers, staplers, etc., and regained prominence in actuating machine elements that heralded the era of automation.

Less realistic uses of compressed air were envisioned in the mid eighteen hundreds by the proponents of pneumatic railway systems in the wake of air-driven railways built in Ireland and France over short distances, while England even attempted a 7.5-mile stretch from New Cross to Croydon. Driven by a 7-inch diameter piston in a pneumatic tube between the tracks, the trains reportedly achieved in 1845 the startling speed of 70 miles per hour. The weak link here was of course the coupling between the piston in the tube with the wagons above. To that end, a longitudinal slot was left open along the upper side of the tube, through which a flattened section of the piston rod reached upward and coupled with the wagons. A pair of soft leather strips bordering the length of the slot was meant to seal it hermetically, save of course for the place where the piston rod got through.

The system had to be operated on a partial vacuum because pressurizing the tube would have blown the sealing strips apart. Conversely, the vacuum helped holding them in close contact with each other. But still, the unavoidable leakage around the piston rods made the system too costly for the project to last. Equally short-lived was an identical system, connecting the cities of Exeter and Newton Abbott.

Next came Thomas Webster Rammell's 1860 patent that placed entire carriages inside a pneumatic tube with the remarkable feature of applying suction in front and pressure at the rear of the carts. The inventor demonstrated his idea on a cargo mover traveling in a 30-inch diameter generously curved tube laid out on the ground. It worked well enough to be adopted for postal parcel transport over a 1/3-mile stretch between the London's Euston station and the Eversholt Street post-office. A capsule with up to 35 bags of mail needed less than one minute for the trip, and 13 trips a day became the rule. But still, operating costs remained above those of conventional mailcarts. Other ideas, such as a collar of bristles sealing the lead car with the walls of the tunnel never made it beyond the prototype stage.

A steam-engine driven underground railway existed in London as early as 1863, but running those smoke-pouring monsters in the enclosed environment of railway tunnels must have been a life-threatening undertaking. Thus it shouldn't surprise that as late as 1867 Charles T. Harvey, inventor of an elevated-track cable-operated transport system, convinced a Senate Select Committee to authorize a half-mile trial segment from Battery Place northward. Construction started that same year and on 7 December 1867, Harvey personally drove a trial vehicle above Greenwich Street, drawn by a cable that wound in several loops around a steam engine-powered underground drum. Off that drum, the cable run over guide-rollers up to the line's elevated structure where it pulled a series of so-called travelers, four-wheel bases on a small-gauge track central to the structure. They had upward projecting engagement horns, which hooked into retractable follower brackets mounted on the underside of the wagons.

After the wagons had been rolling on their flat-faced wheels and reached the rated speed of 15 miles per hour, the line got permission to extend until close to Cortlandt Street. On 6 June 1868, the commissioners, and subsequently Governor Fenton and Mayor Hoffman, got their chance to personally test the new concept. And yet, cable systems didn't last. Only the famous San Francisco cable cars have survived into our days, along with a cable-pulled train up and down the 800 meter high Serra do Mar to the beaches of the port-city of Santos, Brazil. Slung around a big pulley at the upper station, the cable holds one wagon on each end, so that their weights cancel out, and so do equal numbers of up and down passengers. Only an excess of upward oriented riders costs energy. A rudimentary, yet for sure energy wise approach!

The first tunnels for what became the New York subway were started no sooner than 1904, preceded in 1890 by what Londoners baptized The Tube.

The "incompressible" fluids

Air pressure drops off rapidly as we leave the surface of the Earth and head for outer space. The opposite is true if we decide to plumb the depths of the ocean. Humans' urge to learn what to expect down there and early attempts at finding out reflects in Friedrich von Schiller's poem *Der Taucher* (The Diver). Set in an age when the term "oceanographer" had yet to become a household word, a certain king, eaten up by his own curiosity, throws a chalice of solid gold into the surf punishing the cliffs below the royal castle, and challenges his staff of knights to retrieve it for keeps.

When his generosity finds no taker among the anointed, a knave out of rank and file steps forward. The crowd watches him in awe vanishing amongst the furor of waves and spray while the king's daughter – an enchanting young princess, of course – silently and covertly prays for him. And sure enough, shortly before the time man can survive without breathing has passed, this muscular hulk of a youngster surfaces and – cup in hand – approaches the knowledge-thirsty king. But rather than bathing in the light of his newly acquired notoriety with sensational newscasts, he announces in (here freely translated) reverberating voice:

Don't challenge the Deities, oh man –
For down there is a cryptic den –
And never dare to invade the room –
Graciously hidden in night and gloom.

But it turns out that the contested chalice had settled on top of a submerged cliff from where the young hero had gotten no more than a fleeting glance at the Loch Ness type monsters luring in the deep. Frustrated by the meagerness of the diver's briefings, the king ups the ants, but even the fearless knave shows little enthusiasm for another dive into the boiling surf. Only when the princess' hand in marriage is made the prize, he goes after the royal chalice for the second time.

Not surprisingly, his luck has run out, and the crowd waits in vain, while the princess pledges to spend the rest of her young life in a monastery ... Oops – hold it right here – Schiller's poem says nothing of that kind. It's just that the translator's fantasy was running wild in the saddle of Pegasus.

Actually, the art of diving began quite soberly with the diving bell, a device comparable to an bucket turned upside down. We can try the principle ourselves by lowering an inverted glass cup into the water of a full sink. Water does not enter the cup, because the inside has already been taken by trapped atmospheric air.

Like so many ingenious devices, the diving bell comes up in Aristotle's writings of the 4th century BC. But a millennium had to pass until a bell developed by Guglielmo de Lorena in 1535 helped in the salvage of the Roman Emperor Caligula's pleasure Galleys from the bottom of Lake Nemi in Italy. Of small size, it merely enclosed the diver's upper body, which allowed him to stick out and use his lower arms from underneath.

At greater depth, the air in a diving bell gets compressed in step with the weight of the water above, which rises by 1 kg/cm^2 for every 10 meter, meaning that at 10 meter of depth, the air in the bell gets squeezed to half its original volume. To keep the bell empty, 1 at of compressed air must be pumped in, rising the absolute pressure to about twice the air pressure above. Keeping the bell dry takes higher and higher pressure the deeper we go.

A diving bell for laying the foundations of bridge pillars – named caisson in this application – sits on the ground and is pressurized enough to keep the workers' feet dry – so to say. Humans enjoy a surprising capability to function in such pressurized environment, derived from our body's capacity to build up internal pressure commensurate with the surroundings. Mind me, even the normal atmospheric pressure of about 1 kg/cm^2 would compress most of our body-mass (short of the bones) into useless molasses if it weren't for the equilibrium between internal and external pressure. One problem however remains. A surfacing diver's body releases its internal pressure at its own pace, and sudden decompression causes pain in joints, limbs and abdomen, a condition called the bends. Severe cases may lead to vertigo, choking, and ultimately – death. Decompression chambers for workers who finished their shift provide gradual accommodation to environmental conditions while the next shift is already busily at work in the caisson.

The further down we venture, the higher the water pressure becomes. In the Second World War, many submariners lost their lives because their vessels could not withstand the pressure of the water, especially if enhanced by the pressure wave generated by a depth-charge blast. To explore the ocean floor, which often lies several miles below the surface, requires human beings to invent robust vehicles capable of withstanding enormous pressures. To that end, submarines consist of two building blocks: the internal, generally cylindrical pressure hull, made from thick plates of high-strength steel such as the HY-80, encircled by the streamlined, unsealed and therefore pressure-unaffected outer hull.

At great depths, the "shrinking" of liquid volume with rising static pressure can no longer be ignored. For water, compressibility, the relative decrease in volume per unit of pressure, has been found equal to 46.4×10^{-6} per atmosphere, or $46.4/1.033 = 44.9 \times 10^{-6}$ per at. The volume of 1 liter (1 dm^3) of water at 20 °C, based on these figures, is shown for pressures of up to 10,000 atm in Fig. 8.9.

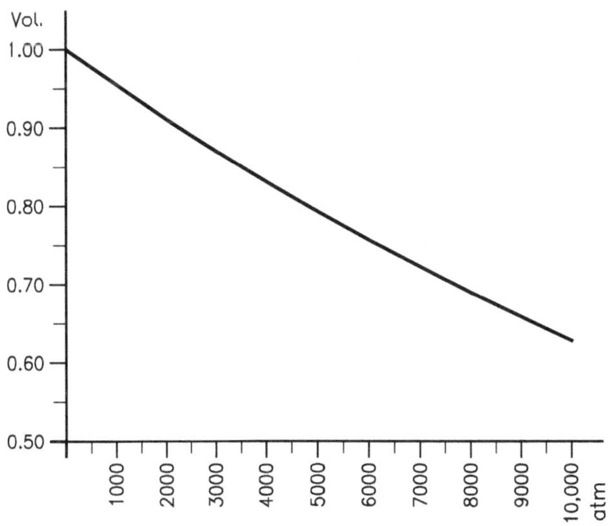

Fig. 8.9. Compressibility of water

The deepest point of the oceans, slightly over 11 km below the surface, lies within the Marianas Trench in the northwestern Pacific. With 10 meter of water-head to the technical atmosphere (kg/cm^2, or at) pressure at 11,000 m would be $11{,}000/10 = 1100\ kg/cm^2$ if water were incompressible. But the minor difference between incompressible and compressible states is enough to keep the actual level of the world's oceans 30 meters below what it were if water were incompressible. 5 million square kilometers (about 2 million square-miles) of the present landmass would then be flooded.

A device based on the water's resistance to compression is the hydraulic ram.

Hydraulic ram

The hydraulic ram is a pumplike contraption which uses the inertial force of the water column in the down-piping from a reservoir or well to push some of the flow up to a level way above where it came from.

The key element of the hydraulic ram is an autobalancing valve that drains the supply line just long enough for the water flow to reach full speed and then shuts it off while opening a bypass into the uphill tubing to the site of the consumer. Since inertia inhibits a momentary halt of the descending water column, and a checkvalve blocks backflow, upward is the only way the surge of water can take. But as always, such amenities come at a price: The quantities of water used up by the hydraulic ram exceed by far its output further up.

Conversely, the incompressibility of water was the reason for the replacement of the cock, which shuts off with a quarter of a turn, by the familiar slow-acting screw faucet. If cocks were still the rule in our widespread public water systems, simultaneous shutoff of a great number of outlets, bound to happen sooner or later, would cause pressure surges severe enough to rupture the city piping.

Down and under . . .

But we don't have to go under water to find high pressures. The same is true if we begin to dig deep, something that occurs quite often as humans try to find and exploit deeper layers of natural resources. Grecian mythology explains the world under our feet in terms of a dispute between three early power brokers – Zeus, Poseidon, and Hades – on their respective spheres of influence. In a draw of lots, Zeus comes out as the CEO, and Poseidon gains control of the seven seas. Hades, apparently new to the game, draws worst and must content himself with the underworld.

But rather than suing for a rigged election, Hades lives up to the challenge fate has dealt him. He makes Thanatos, the personification of death, his adjutant of orders, and kidnaps the adorable Persephone, Apollodoro's wife, for her charms in the widest sense of the word.

As soon as he gains a foothold on his reign, he patronizes warlords for the population growth they provide his domain, and – just in case – wears a helmet that makes him invisible as needed – such as in a peaceniks attack. He backs the wealthy, which, I guess, wished to borrow his helmet every 15th of April.

Hades balances the underworld's budget with earnings from the operation of the ferry over the river Styx, pricing the ride's one-way tickets individually with an infallible instinct for the amount of money each immigrant is carrying. But Mythology falls short of telling us how far down Hades' kingdom really was, and leaves it to us moderns to find out.

The Russian Kola peninsula near Scandinavia is home to the deepest borehole humans have drilled by 2005, which reaches 12 km down. New drillings to 15 km of depth are in the making. If that seems like peanuts in comparison to the 6372 km radius of Earth, think again. Temperatures at the bottom of those holes linger between 300 and 500 °C, and pressure from the massive earth overhead reaches 4000 kg/cm^2.

Japanese scientists from the International Drilling Programme are even planning to break through the Earth's crust at one of the deepest sites in the Pacific Ocean, where it is thinnest. So far, the seabed has been drilled to 2111 meter under the seafloor, but 7000 meter of crust must still be punched through to reach the mantle.

How much deeper one could get with the help of future technology remains an open question; but even Jules Verne's SF novel *Journey to the Center of the Earth* stops short of letting somebody set foot there. What the explorers discover early into their trip is an enormous subterranean dome over a Lake Vostok-like body of water. While sailing those waters on a hastily assembled raft, a kind of tsunami floods them through a labyrinth of caves into the extinct volcano Stromboli in the Mediterranean and from there out into the sunlight. Good for them, because getting to the Earth's solid inner iron core, which begins 1229 km outwards from the center, would have exposed them to temperatures of 7000 °C and an estimated 3,743,000 kgf/cm^2 of pressure. Conditions that – if they were survivable – would no doubt have pleased the professor by the far ranging insight into the properties of matter they would provide.

In the gray light of reality, shock waves became the means to obtain data on the state of matter exposed to ultrahigh pressure and temperature. But the inherently short duration of a shockwave-generated pressure peak stood in the way of resuming the results of such tests into an "equation of state," the equivalent of the well-known equation of state for gases, $pV \propto T$. Characteristically, it houses three variables – pressure, volume, and temperature – under one roof, so to say. Historically, this equation allowed for the first fairly accurate prediction of the absolute zero point of the temperature scale, but it still misses out on changes of phase, such as the condensation of steam into water.

Reliable equations of state must cover a wide spectrum of pressure and temperature and demand a good number of reproducible measurements for their deduction. The need for this kind of knowledge became pressing with the onset of nuclear technology. While nuclear fission devices could still be developed by the trial-and-error method, nuclear fusion proved far more difficult to achieve. Explosions of early fusion devices actually produced a mixture of nuclear fission and fusion, and lacked far behind the energy output expected from pure fusion. That triggered the long series of nuclear tests we recall within the concept of the "arms race," best characterized by a Monday morning broadcast from the official Brazilian radio station, that stated laconically: "While our people spent the weekend on the soccer fields, the Soviet Union used it to explode nuclear bombs."

Actually, some 1000 of such nuclear tests were performed beyond the Iron Curtain until the superpowers agreed on a ban on explosions in the atmosphere and under water, which President J. F. Kennedy announced on July 26, 1963, with the words:

> "...For the first time, an agreement has been reached on bringing the forces of nuclear destruction under international control – a goal first sought in 1946 when Bernard Baruch presented a comprehensive control plan to the United Nations."

This saved humanity from a future of breathing radioactive air but added urgency to the need of alternative methods for predicting the behavior of materials under conditions resembling those in the core of nuclear explosions. The device that comes closest is the diamond anvil cell.

Diamond anvil cell

The diamond anvil cell – DAC for short – generates extreme pressure by using a very small pressurized area. For instance, a pair of truncated diamonds (Fig. 8.10) with 0.6 mm top diameter has a culet area of 0.283 mm^2, or about 0.003 cm^2; loading it with, say, a modest 1000 kgf of force would already generate $1000/0.003 \approx 330{,}000\ kg/cm^2$ of pressure, achieved by nothing more complicated than tightening the 6 Allen drive screws in the upper body of the device.

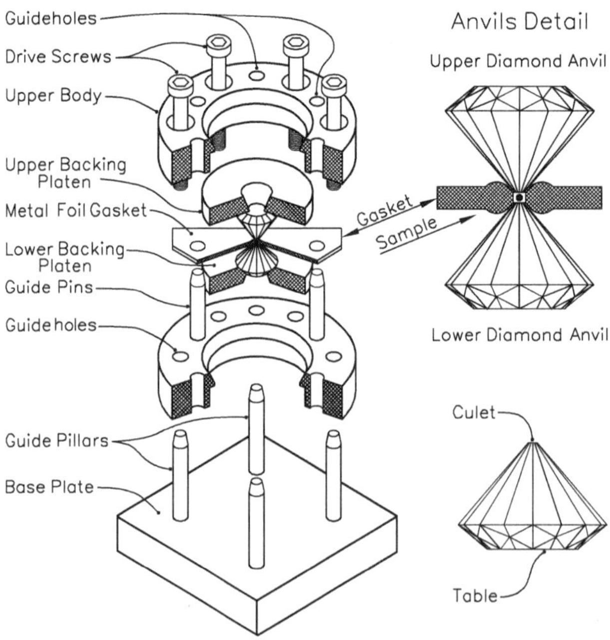

Fig. 8.10. Diamond anvil cell

The sample under test, pressurized between the culets of the opposing diamonds, is hereby contained sideways by a metal gasket with central opening, preferably fabricated of rhenium, beryllium or, for measurements at high temperature, Inconel.

Pressure fluid surrounding the sample converts the one-sided pressure exerted by the diamond tips into hydrostatic pressure, acting on the sample from all sides.

Ultrahigh pressures, such as 3,500,000 atm along with temperatures of up to 6000 °C, resembling the inner Earth and the heat of the Sun, have been reported with the use of still smaller pressure areas and, of course, higher loads. Most important, the DAC allows for the use of samples as small as 1 microgram.

Diamonds, the hardest among solids, are ideal pressure elements because they do not deform or break under pressure and are transparent to X-rays, which, in the case in point, are of the very bright and coherent type emitted by a synchrotron light source.

In pressure tests, X-rays pass freely through the diamonds, but get diffracted by the thin layer of sample material between them. Present in the pressure zone is a ruby chip laser marker which fluorescent laser light doubles as pressure indictor. The sample itself is examined with a conventional microscope, or still by spectroscopic and diffraction techniques.

Guide pillars and pins on base and lower body, gliding through honed holes in upper bodies and platens, guarantee the exact alignment of gasket and diamond.

The other face of the coin

Below 0.1 mmHg lies the region known as high and ultra-high vacuums, reaching from one-tenth to one-millionth of a torr (mmHg). If that sounds awesome, relax! Such vacuums are quite common in everyday life – even in ordinary light bulbs. If you would break the glass of a light bulb and turn it on, the filament would burn off in seconds. Then there is the Dewar flask, also known as the thermos flask or bottle. Invented in 1892 by the Scottish physicist James Dewar, it ushered in the era of high-vacuum and low-temperature physics. The double-walled dewar flask has brightly coated surfaces to cut down on the effects of radiant heating or cooling. It also has a very high vacuum between its walls, to minimize heating or cooling by convection.

The Dewar flask enabled scientists to measure the specific heat of materials, including diamonds, at low temperature. As a result, the long-held belief in the Dulong-Petit law of specific heats was shattered. In 1907, this led a theoretician by the name of Albert Einstein to propose a new theory to describe the specific heat of solids. His theory, modified soon after by Einstein himself and then by Pieter Debye, instituted the area of solid-state physics.

Industrial applications of high- and ultrahigh-vacuum techniques include the deposition of fine metal coatings on electrically nonconductive materials, such as glass or plastics, which cannot be plated by means of traditional electroplating. Known by the collective name *physical vapor deposition* (PVD), the process is based on the condensation of vaporized metals (source material) on the object being plated, here called the substrate. Just as ordinary water boils earlier in places of low barometric pressure, such as mountain peaks, metals vaporize in vacuum at lower temperatures than under atmospheric pressure. Typical gas pressure for vacuum evaporation lies between 10^{-5} and 10^{-9} mmHg.

Vaporization of the source material can be directly or indirectly. In the first case, known as cathodic arc plasma (CAP) technology, coating metals come in the form of stranded wires and are vaporized by an electric current surge. Indirect heating is done in tiny crucibles, located over a resistively heated tungsten wire filament. Crucibles for melting tin have a hole in the bottom, sized to let drops of molten metal pass through and settle on the heater spiral below, where the instantly vaporize. High-tech vaporizers use scanning high-energy electron beams to heat the probe.

Applications of vacuum vaporization, apropos the least expensive of PVD processes, include surface metallization of reflectors used in flashlights and automotive headlights, and even flat mirrors of up to 2 by 3 meters in size. Herein, condensation of aluminum vapors in a vacuum chamber renders brilliant coatings that shine like chrome plating after a protective topcoat of SiO_2 has been applied.

Selective optical filters can be made by vapor deposition of anti-glare coatings. Front-coated mirrors with up to 96% reflectivity are a must for optical instruments, where commercial mirrors cannot be used due to double reflection by their front and rear surfaces. For instance, a right angle eyepiece with a commercial 45° slanted mirror would show three images of your object, say, the planet Venus: a bright Venus in the center, bordered by fainter images above and below. Reflecting astronomical telescopes, such as Newton and Cassegrain reflectors, have face-coated primary mirrors. Even printed circuit boards are sometimes produced by vacuum deposition of the conducting metal on a plastic substrate.

Further application of the vaporized metal coating process include the shielding of electronic components from electromagnetic fields. First in line stand cathode ray tubes, as used in oscilloscopes, which precision would otherwise suffer from electromagnetic interference (EMI) by the fields from radio and TV transmitters. Shielding can involve the critical component directly, or the entire enclosure. While the sheet metal enclosures of older instruments acted as natural EMI shields, plastic enclosures, except those of polyethylene and urethane, can be shielded by PVD. Obligatory reduction of EMI intensity by shielding is 30 dB for commercial, 60 dB for military, and 90 dB for special-purpose enclosures.

Vacuum-deposited hard coatings of TiN and TiC on drill bits, milling cutters, lath turning cutters, etc., multiplicate their lifetime and allow for dramatic reductions of machining times.

And if it weren't for high-vacuum techniques, there wouldn't be electron microscopes or synchrotrons and the rest of nuclear particle smashers.

Ultrahigh vacuum starts below *one ten millionth* of a torr. An ordinary television set, a computer monitor, and the standard cathode ray tube have ultrahigh vacuums because they all must allow for unobstructed flow of electrons from cathode to anode or to the scintillating screen. In a less than nearly perfect vacuum, too many electrons would scatter off air molecules and be sent to nowhere. But even at 10^{-9} Torr, there is about one trillion of air molecules left in a 17-inch TV picture tube.

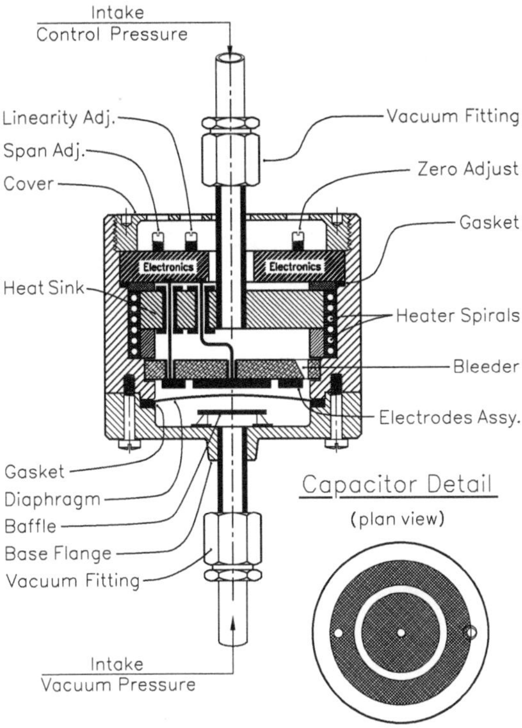

Fig. 8.11. Capacitance manometer

Capacitance manometer

The capacitance manometer in Fig. 8.11 derives gas pressure in ultrahigh vacuum from the deformation of a membrane installed under the electrodes of a flat capacitor shown in plan view on the lower right of Fig. 8.11. It consists of a circular and an annular layer of metallic foil on a ceramic base. The capacitance of this arrangement of electrodes changes with the degree of bulging of the diaphragm.

For absolute pressure measurements, the space between membrane and capacitor is evacuated and the capsule hermetically sealed. However, certain precautions are called for in order to maintain such a reference vacuum over time. Not enough that aging gasket material loses flexibility so that gas molecules from outside sneak in, "outgassing" sets free many of the gas molecules formerly adsorbed in the walls of the casing and the parts within.

Outgassing plagued Thomas Edison's production of lamp bulbs by contaminating the initially ultrahigh vacuum inside the bulbs, though the bulb's fused tip-off sealing precluded infiltration from the outside.

Here and in the manufacture of radio tubes, a process known as *gettering* was found to eliminate such residues. A so-called getter-pan, holding material prone to oxidation, such as barium or a titanium-molybdenum alloy, was placed in the head of radio tubes. After the glass enclosure had been sealed off, inductively heated vaporization and sublimation of the getter material releases metallic clouds that condense as a dark or silvery colored coating inside the upper sections of the casing as adsorbents of gaseous contaminants.

Over time, "nonfusible" getter-materials, such as certain aluminum-zirconium alloys, were found to have the same effect. Getter materials activated at liftoff of the Cassini-Huygens space mission to Saturn and its moon Titan maintained the vacuum in certain equipment at the initial value throughout the 7 years of flight.

The prevalent way of detecting small changes in capacitance is by wiring the capacitor into a radio frequency oscillator circuit (Fig. 8.12, right). Once the cir-

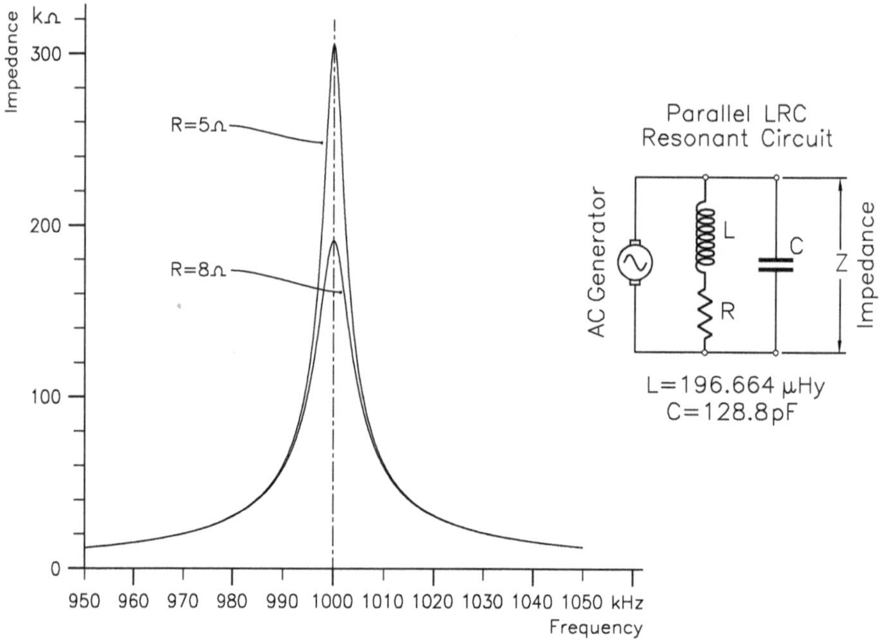

Fig. 8.12. Resonance curves

cuit has been syntonized at resonance frequency, minute variations of capacitance suffice to unbalance the state of resonance and to cause an abrupt raise of impedance and with it the induced voltage.

In radio receivers, a similar device provides station separation by means of a variable plate-capacitor. Highest output (if any) is obtained with the antenna circuit oscillating in step with the carrier of the transmitting station.

Used in capacitive vacuum meters, such *oscillating–nonoscillating* circuitry makes the instrument so sensitive that it would even show the minute variations of the membrane's shape by thermal expansion, if it weren't that the entire assembly is kept at a predetermined temperature, typically 50 °C, by a casing of heater spirals. A conventional temperature switch (not shown in Fig. 8.11) maintains the temperature at the desired level.

Less critical is the quality of vacuum in processes such as *vacuum drying* of chemicals, pharmaceuticals, food products, and granulated plastics at temperatures far below the ones needed for drying in the atmosphere.

Vacuum degassing is used in the production of epoxies, urethanes, resins, silicon rubber, etc. Water is deaerated in a degassing chamber while spread out over Raschig rings or into a thin film.

The process of *vacuum impregnation* assures the penetration of liquid compounds, such as preservatives, into the pores of solids. Food preservation aside, the process has industrial applications, among many others the soaking of the voids between windings of electrical transformers with insulating material. In paleontology, vacuum impregnation is used to stabilize specimens, principally fossils, for later research.

Vacuum preservation of wood consists in the application of chemicals, for instance, tributyltin oxide (TBTO), in a partial vacuum of periodically changing pressure. The process prevents dimensional swelling and within a couple of days dries the wood sufficiently to be painted.

Bottle and can fillers for soft drinks and carbonated beverages are equipped with filling valves that start the bottling cycle with the evacuation of the containers.

Vacuum forming includes the heating of a sheet of plastic material, such as polyethylene, until it becomes pliable, place it on a frame over the part to be molded, and evacuate the space underneath. After a short period of cooling, the form gets rigid enough to be separated from the mold. Many of those hard to open wrappings of electronic components and articles on the shelves of shopping centers and department stores are made by this process.

Applications of *vacuum distillation* include the production of lube oil and bitumen and the qualification of refinery products. The process allows for the vaporization of compounds that under atmospheric pressure would decompose or ignite at temperatures below their boiling points.

Vacuum casting can be done with the usual sand molds or shell moldings, placed in evacuated surroundings in ways that leave the hopper accessible from the outside. Since the process does not rely on gravity for filling the mold, but rather lets atmospheric pressure force the melt in, castings come out true to the finest detail.

Even plastics, such as polystyrenes (ABS), polypropylene (PP), acrylics (PMMA), nylons (PA), and several grades of rubber are often being vacuum cast rather than injection molded.

McLoid vacuum meter

Although most vacuum gauges are suitable for checking gas pressure in these applications, the McLoid vacuum meter in Fig. 8.13 stands out by the ingenious method of measurement it employs and the fact that a good laboratory glass blower can make a low-cost version of the instrument from available tubes of soft soda class glass. In principle, the instrument could be seen as a glorified offspring of the venerable U-tube manometer, safe for a flexible hose in the U-section. One end of the U-tube connects to the vacuum chamber and the other to a mercury flask. Under, say, 760 mmHg of barometric pressure, 1 mmHg in the vacuum chamber would make the liquid level difference in the legs of the U-tube 759 mm. That value is retained even if we lift or lower the glass with the mercury sump.

A tap in the U-tube connects to the sampling chamber and from there upwards into a calibrated capillary with hermetically sealed upper end. The place where the volume in the capillary, counted from the upper end down, amounts to $1/100$ of the total of the sampling chamber and the full length of the capillary, is marked with the number 100.

When we raise the bowl with the mercury sump to the point where the

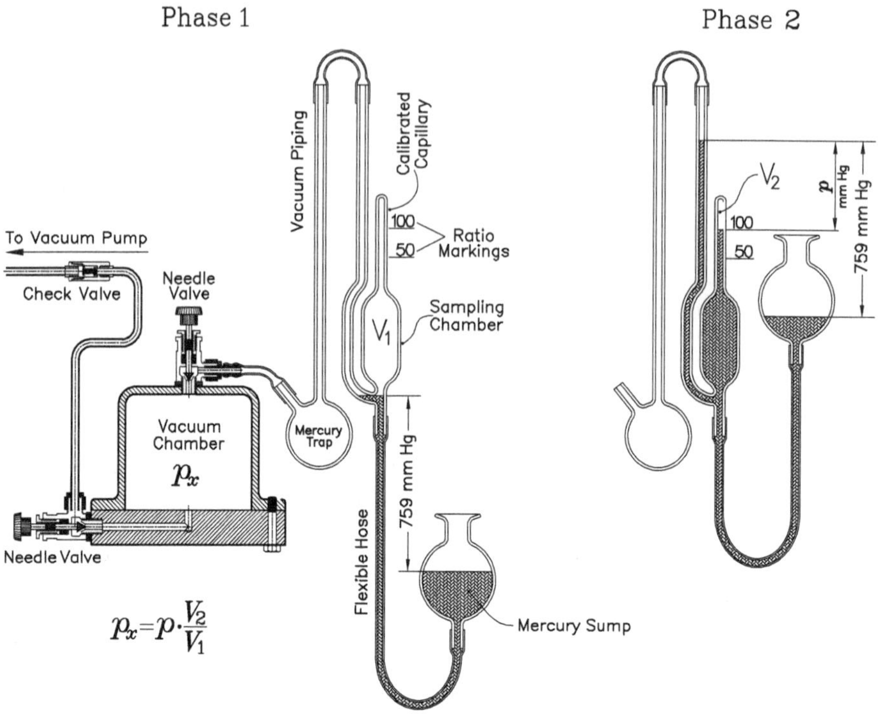

Fig. 8.13. McLoid vacuum meter

mercury column reaches the lower end of the sampling chamber, access to the vacuum chamber gets sealed off, marked "Phase 1" in Fig. 8.13. From then on, further lifting of the bowl marks the onset of compression of the gas trapped in the sampling chamber.

In phase 2, the mercury column has been leveled with the 100 mark, which makes pressure in the vacuum chamber (in mmHg) 1/100 of the level difference between the mercury in the capillary and the mercury in the tube leading to the vacuum chamber. Thus, each 1 mmHg pressure in the vacuum chamber gets us a readout of 100 mm, which, in other words, makes the accuracy of the instrument 100 times that of a U-tube manometer. Higher gains are possible with higher compression rates.

A different approach to the measurement of a wide range of gas pressures, from 750 mmHg down to 0.0001 mmHg, is derived from the degree of heat conduction in the vacuum chamber.

Pirani heat transfer manometer

The Pirani heat transfer manometer is employed in a host of processes such as freeze drying, vacuum coating: for vacuum ovens and vacuum centrifuges, and in conjunction with a host of analytical instrumentation.

As shown in Fig. 8.14, the instrument derives gas pressure from the electric energy spent to keep a filament in the vacuum chamber at 100 °C or, in

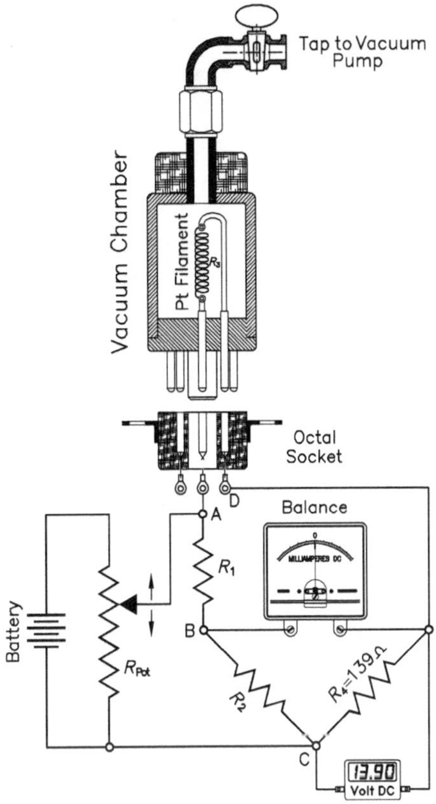

Fig. 8.14. Pirani vacuum gauge

some cases, at 150 °C. The higher the density of the gas in the chamber, the more heat is conveyed from the filament into the surroundings, and the more electricity is afforded to maintain the filament temperature constant.

Because several factors contribute to the transfer of heat – namely, radiation, convection, and conduction – the relation between gas pressure and wire temperature is quite complex. Heat energy transfer by *radiation* is proportional to the difference of the fourth power of ambient and heater temperatures, yet widely independent of the physical properties of the surrounding gas. By contrast, *heat conduction* varies with the type of gas and its density, and losses by *convection* are still less predictable.

Nevertheless, experimentally determined relations between heat flow and gas pressure in the Pirani vacuum transducer were found fairly linear within the range from 1 mmHg down to about 0.001 mmHg. This allows for deriving reliable results from comparison of the power consumption of the heater filament in the unknown vacuum vs. that in a standard vacuum probe.

Though thermoelectric controllers are widely used for keeping the filament at the rated temperature, a platinum filament can double as heat source and resistance thermometer. In that case (Fig. 8.14), the heater filament is made the leg of a Wheatstone bridge, consisting of R_1, R_2, the filament, and R_4. The independence of the balance of a Wheatstone bridge from the supply voltage allows for setting heater temperature at the rated 100 °C by adjusting the voltage of the bridge power supply without unbalancing the bridge. That means that, once the galvanometer has been zeroed in, it invariably remains in that state while the voltage to keep the platinum filament at the rated temperature is being set. Since heating to 100 °C increases the resistance of a 100 Ω platinum thermometer to 139 Ω, R_4 must likewise be made 139 Ω.

Under these conditions, the process of vacuum measurement boils down to setting the voltage splitter R_{pot} so that the ammeter (marked "Balance" in Fig. 8.14) settles in zero-position, and subsequently read the voltage-drop E over the filament on the digital voltmeter ("Volt DC" in Fig. 8.14). The power P dispersed in the filament follows from $P = E^2/R$, with R for the resistance of the heater coil at 100 °C, namely, 139 Ω. The related pressure p in the vacuum chamber is then looked up in the instrument's manufacturers p/P tables.

New age machine design

We, who once happily walked the information highway only to find it a dead end street, are left to ponder other achievements of 20th century's technology, including the transition from the halcyon days of steam- and smoke-spewing railway engines to diesel and electric locomotives; the switch from radio to television, from Morse code to E-mail, and from a sink full of dirty dishes to a dishwasher full of dirty dishes. With less fanfare came the introduction of tungsten carbide cutting tips, though their use allowed for doubling and tripling the speed of metal cutting operations – an essential factor in the process of making the means of personal transportation affordable.

And yet, the change that most mattered happened on the drawing boards of machine designers: The switch from *self-contained* to *externally controlled* machinery.

Self-contained machines, from wristwatches and vacuum cleaners to car engines and automatic lathes are locked into particular modes of operation, while externally controlled machines are built in ways that allow for one piece of equipment to cover a wide range of uses.

Each particular application is linked to a selected set of rules, known as *program,* which triggers the sequential actuation of the machine's elements in ways that make them perform the required task. Computer controlled machine tools perform operations like turning, milling, drilling, reaming, etc. according to numerical signals from computer programs. Once the machine is set up, it goes through the desired production cycle by itself, while the role of the machine operator has changed from active to passive, that is, from personally actuating the machine controls to supervision of the system's operation, checks on the state of the cutting tools, and quality checks of the final product.

The same machine that mills and bores the cylinder block of an internal combustion engine can machine the blades of a ship's propeller or whatever other parts are in demand. For instant, casepackers – typical for externally controlled machines – can be set up to pack cans, jars, bottles, etc., in an almost unlimited variety of sizes and weights, into boxes of cardboard, wood, or plastic . . . you name it.

The components of externally controlled machines are actuated sequentially in accordance with a selected program, which in every step sends electricity to a number of locations, given by relays, solenoids, electric valves, timers, or servomotors. In particular, electric valves, better known under the term of *solenoid valves,* channel compressed air into air cylinders, which in turn actuate machine elements, such as levers, slides, or cams.

The logic circuits which make such a system perform can be "hard wired" or operate under the guidance of a computer. The intricate design of logic circuitry led to a specific mode of circuit presentation, known under the term of ladder diagram.

Ladder diagram

While conventional circuit diagrams show individual components schematically, yet as such, the ladder diagram depicts each component into its elements, and places them separately at the designer's convenience. A relay, for instance, may have its coil show up in a given branch (rung) of the diagram, and each of its contacts somewhere else.

Components are identified by acronyms, such as CR for control relay, TR for transformer, T for timer, etc. Each component's contacts are numbered true to the original, for instance 2 and 7 for the leads to the coil of a typical Potter & Brumfield control relay, 1 and 6 for the relay's normally open contacts, and 1 and 3 for the normally closed ones. In this setting, *normally* stands for the de-energized state of the relay and – for that matter – the entire control system.

Timers determine the duration and spacing of individual operations. Of their three basic types, one starts timing when turned on, the other when turned off. A third type has its onset controlled through an independent signal input. Like relays, timers have normally open (NO) and normally closed (NC) contacts.

Sequencers take a machine through a number of consecutive operations, much like the clock-chip does in a computer or calculator. The typical sequencer has a number of radial cams on a common shaft, driven by a small electric motor. Each cam actuates the plunger of a limit switch, which contacts connect or disconnect one or an entire group of machine elements. The angular position of the cams is individually adjustable and determines the point within the machine's working cycle when the related element is to jump into action. The better brands of mechanical sequencers use differential gearing to allow for angular cam adjustment while the device is running.

Sequencers keep going, whether or not the operations they trigger have been completed, while timers can be wired to wait until a limit-switch signals completion.

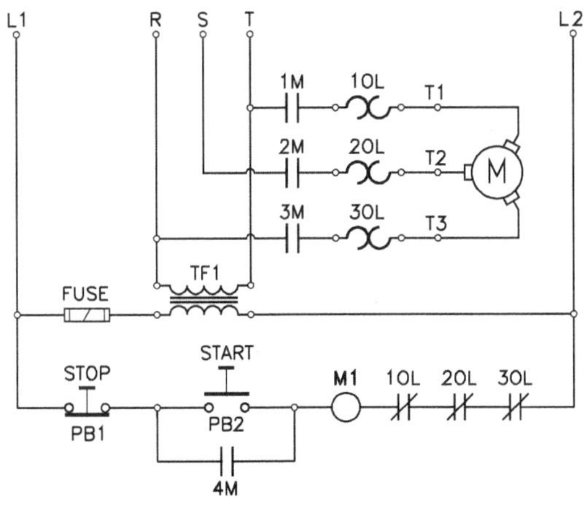

Fig. 8.15. Ladder diagram of a push button motor starter circuit

The example of the ladder diagram of a simple hard wired logical circuit shown in Fig. 8.15 actually refers to the turn-on and turn-off of an industrial three-phase electric motor by means of ON and OFF buttons. This basic circuit can be extended for gradual or star/delta turn-on of heavier motors, and for switching two-speed motors from low to high rpm and back; in a wider sense, it exemplifies the scope of logic control circuitry in general.

The upper part of the circuit diagram includes the connections between the motor and a three-phase power relay, the *motor starter*, which turns electricity to the motor on and off. Its contacts, 1M, 2M, and 3M, are shown in the OFF position. The overload protection elements, 1OL, 2OL, and 3OL, are bimetal strips that heat up and flex in case of excessive amperage. They make part of the motor starter and lie in the path of the main motor current in each phase.

This section of the circuit is generally powered by three-phase 460 volt alternating current (VAC) from the utility network, and so are the primary windings of the fused control transformer TF1, which secondary windings supply the convenient 120 VAC operating voltage (phases L1 and L2) for the control circuit, shown in the lower part of the diagram.

Since the STOP button PB1 is of the normally closed type, pressing the START button PB2 powers the coil M1 of the motor starter and the motor starts running. The normally open (NO) ancillary contacts 1M of the motor-starter shorten the contacts of the ON button. That keeps the motor running when the start button has been released.

In series with the coil of the motor-starter are the normally closed contacts of the thermal overload units, which heat up and open in case of excessive amperage that might be caused by temporary overloads and short circuits. Wiring them in series connection makes that the opening of one single contact is enough for the motor starter to disconnect entirely; something of special importance for three-phase motors, which rapidly overheat if one phase is missing while electricity from the two remaining phases still flows into the windings.

Pressing the STOP button interrupts the current supply to the coil of the motor starter. 1M, 2M, and 3M, open and turn the motor off. 4M, the ancillary contacts, also open and no longer bridge the poles of the ON button.

The not so missing link

The link between the outputs of a logic circuit – hard wired or computer controlled – and the operation of the related machine elements is in most applications the venerable air cylinder. This term includes the double acting and spring return varieties. Double acting air cylinders (Fig. 8.16) have compressed air connections on head- and endcaps, so that the piston can be pressurized either from the front or from the rear of the cylinder. Cutting off air pressure from a double-acting cylinder ends the motion of the piston, while a spring-loaded piston resets into either extended or retracted position, depending on which side of the piston the spring is installed.

At the end of the stroke, the piston's motion slows to a gradual stop thanks to the action of damper plungers on both its faces. The plungers fit snugly into matching cavities in cylinder head and end cap, so that the compression of the air trapped within cushions the piston's halt.

The sleeves of pneumatic cylinders are made from seamless, precision cold-drawn and internally lapped stainless steel (SST) tubing. The piston, of cast aluminum alloy, slides on molded-on circular bands of low-friction material, e.g., polytetra fluoroethylene, that excels by its low coefficient of *breakaway friction* – a term defined as the difference between the force needed to get the piston off its

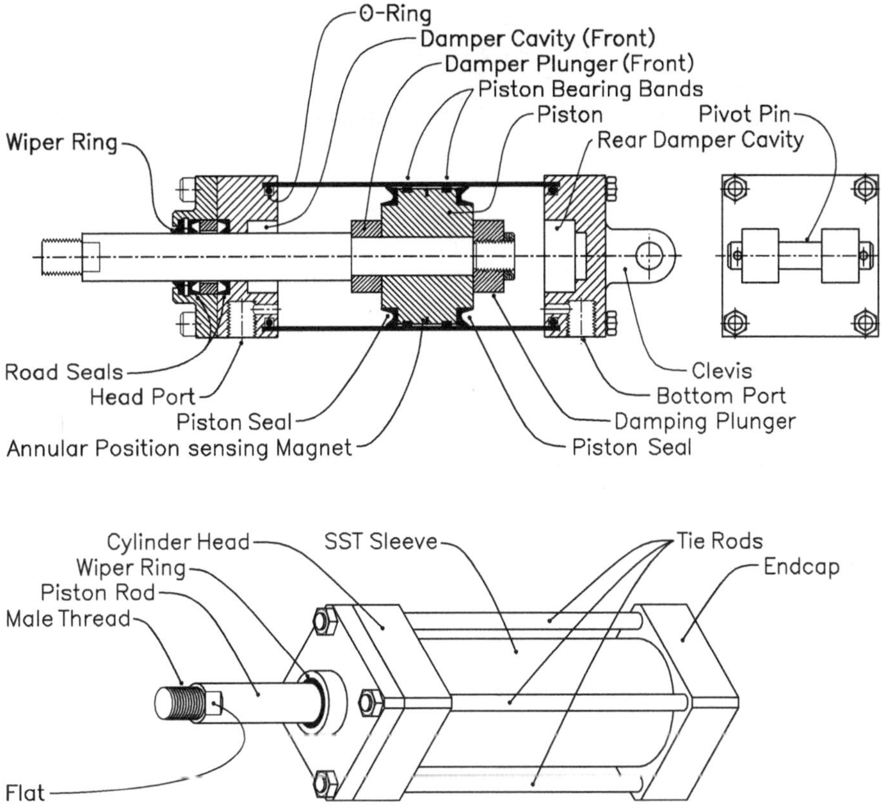

Fig. 8.16. Double-acting air cylinder

rest position and the force needed to keep it going, much like the initial push skiers use to get on their way downhill.

Air cylinders with adjustable stroke have a magnetic ring molded around the perimeter of the piston. Though the ring does not touch the cylinder sleeve, its magnetic field is strong enough to actuate an external sensor, such as a read switch, a Hall sensor, or an electronic pick up. However, when air supply is cut off, compressed air trapped in the cylinder continues to expand and keeps the piston going until the total of internal air pressure on the area of the piston equals that from the load at a hard to define position. Air cylinders that stop dead at an exact point have a pair of spring-loaded semicircular jaws spanning the piston road where it exits through the cylinder head. Controlled by a normally open magnetic valve, they click in the very moment when air supply to the cylinder is shut off.

Piston roads, guided by oil impregnated, sintered bronze bushings, often come chrome-plated for optimized wear resistance and best surface quality. Rod and pis-

2 standard nitrile resistant to petroleum-based oils and fuels.

ton V-seals are of Buna N[2]. A road wiper, equally of Buna N, located in the neck of the cylinder head, keeps dirt on the piston road from getting drawn into the cylinder during the return stroke.

Nevertheless, minute quantities of such pollutants as dust, paint chips, and in particular slush of cooling oil and water, may be drawn in and accumulate until they get into the narrow channels inside the control valves, causing erratic performance and ultimately contingency stops. With machines feeding an entire production line, consequences might be disastrous.

Frequent preventive maintenance with replacement of all plastic and rubber parts helps, and so does the installation of drip legs ending in drain traps at the low points of the piping between control valve and cylinder. But the safest way to prevent the problem from occurring is to install the cylinders "head down." With the tip of the piston road in a fixed position, the cylinder itself becomes the moving portion of the assembly, so that slush runs *off* rather than *toward* the point where the piston road slides through the front cylinder head without a chance to accumulate. Air connections to movable cylinders comprise flexible piping, such as armored PVC pressure pipes – which are easier to install and of higher immunity to vibrations than their rigid counterparts.

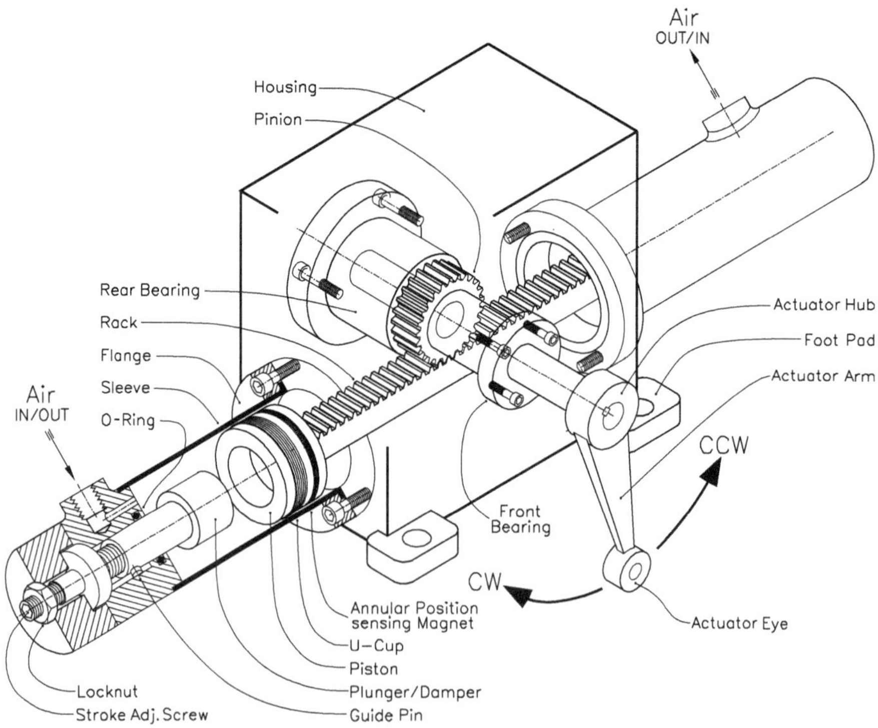

Fig. 8.17. Pneumatic rotary actuator

Typical applications of pneumatic cylinders

The rotary actuator in Fig. 8.17, also called *linear-to-rotary converter*, is built from a pair of horizontally opposed air cylinders, resembling the configuration of the cylinders in the boxer engines of original Beetle Volkswagen.

The pistons sit on the ends of a toothed rack that drives a matching pinion on the main shaft. Pressurizing the left-hand cylinder makes for counterclockwise (CCW) rotation of the main shaft, while clockwise (CW) rotation comes with pressure to the right-hand cylinder.

Both cylinders resemble standard models, except for the stroke-limiting plungers, which positioning determines the main shaft's angle of deflection. As the piston closes in on the stop, the latters thicker section seals off and then enters the recess in the face of each piston. The trapped and gradually densified air inside cushions the stop of the piston.

An interesting application of a rotary actuator is the in-line can-filler in Fig. 8.18. Herein, the actuator is turned periodically on and off for the step-by-step advance of the conveyor belt feeding the cans to be filled.

Since direct coupling between the actuator and the belt drive would result in nothing better than a seesaw motion of the belt, an overrun coupling, resembling the sprag clutch on a bicycle, is used to engage the conveyor's driveshaft counter-

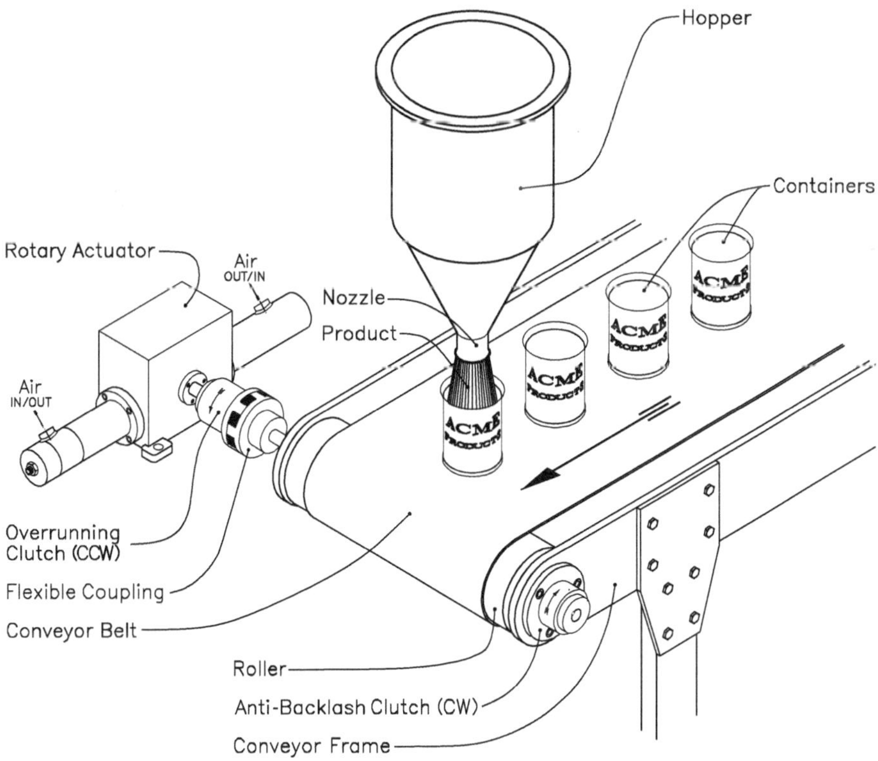

Fig. 8.18. Rotary actuator controlled in-line filling operation

clockwise and disengage it clockwise. A second overdrive, installed at the other end of the conveyor roller's driveshaft with its casing fixed on the conveyor frame, blocks backlash but lets the shaft turn unimpeded when actuated.

Not shown in Fig. 8.18 is the vibrator that keeps the product prone to move out of the hopper, and the filling valve in the neck of the hopper. A timer controls the duration of that valve's opening and with it the amount of product going into every can.

A flexible coupling between actuator and conveyor shaft compensates for unavoidable misalignments.

This is an example of the flexibility of external machine control design. Neither the designer of the rotary actuator nor the maker of the belt conveyor knew of this particular application while they developed their specific machines, which may as well become the building blocks of a variety of other kind of equipment.

Valves, the gray eminence

The manual three-way cock in Fig. 8.19 channels compressed air to one side of the piston of an air cylinder while venting the other, and vice versa. Adjustment of the height of your office chair is done by that kind of valve, controlling a hydraulic cylinder.

Of interest to programmed machine control are electrically actuated solenoid[3] valves. Among several types on the market, designers often pick spool valves (Fig. 8.20). The solenoids may act directly on the plunger or control the flow of "pilot air" to small pistons mounted at the ends of the plunger. The latter option assures sufficient force for reliable valve stem displacement with smaller solenoids. But valve testing and maintenance is complicated by the fact that pilot-operated valves do not function unless the system is under pressure.

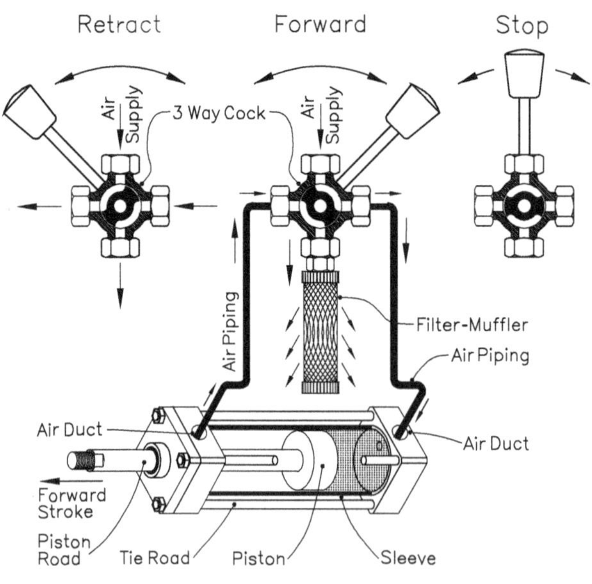

Fig. 8.19. Manual air cylinder control by three way valve

3 multilayer coil with movable ferro-electric core.

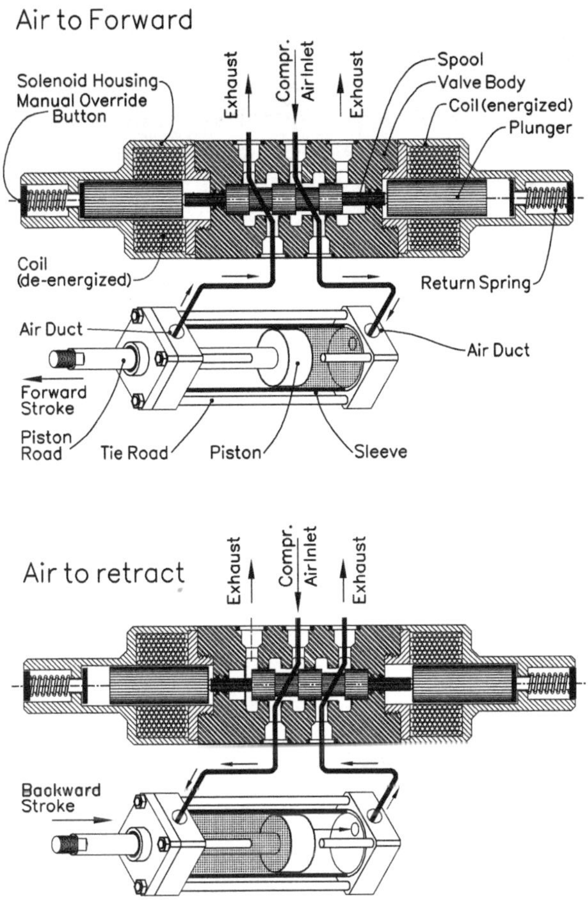

Fig. 8.20. Spool valve for air cylinder control

On the other hand, the straightforward operation of directly actuating solenoid valves calls for heavier solenoids with sufficient pull to get the plunger off its rest position, regardless of the presence of sediments, dust, and contaminants carried in with the air supply. Intrepid electricians help reluctant solenoid valves and electromagnetic linear actuators to overcome their sedentary tendencies by the use of a two-voltage power supply for triggering them.

They make a timer jumpstart the valve on a higher than rated voltage, say, 160 V for a 120 V valve, and switch it back one or two seconds later to 120 V. The related current surge is normally too short-lived to damage the solenoid's windings but provides a substantial push because the power of solenoids increases with circa the square of the applied voltage. However, this procedure is not recommended unless pre-approved by the valve's manufacturer.

Direct operated spool valves are tested by pressing one of the manual override knobs on the outer ends of the solenoid housings, regardless whether air pressure is on or off. Two-position spool valves as in Fig. 8.20 move the piston of the associated air cylinder either in the fully extended or in the fully retracted position. Three-position valves have the plunger supported between a pair of compression springs to make it return to center position and cut off air supply to the cylinder if both solenoids are de-energized.

For ease of replacement, valves are often installed on a manifold, a block of metal with internal channels that conduct compressed air to the appropriate valve ports. Defective valves can here be changed by taking them off the manifold while leaving the air piping intact.

Apart from the 2-position, single-pressure type valve in Fig. 8.20, spool valves are available as 2-position double-pressure, 3-position open-center, 3-position closed-center, 3-position pressurized-center valves, etc.

All said, compressed air lends itself best for light service, such as the actuation of machine elements, and of a host of power tools, like hand drills, hacksaws, nut and bolt drivers, air-powered bench presses with flow-controlled pneumatic cylinders, etc. He who ever took a wheel off his car with a wrench from the toolbox rather than the pneumatic nut driver professionals use, knows what the pneumatic hand-tool craze is all about.

Pneumatics and hydraulics

Compressed-air installations cost considerably less than hydraulic systems, and in many industries make part of the basic factory floor outfit. However, the economy of air cylinders suffers the bigger they become. Air compression is an energy-intensive process due to the waste heat generated in the compression of gases. And the energy content of the compressed air in a pneumatic cylinder is lost in the exhaust stroke.

For instance, the volume of an 8 inch diameter cylinder with, say, 12 inch stroke, is $8^2 \times 12 \times \pi/4 = 603$ inch^3, or 0.35 cubic feet. At the typical operating pressure of 60 psi or about 5 atmospheres, that amounts to consuming $5 \times 0.35 = 1.75$ cubic feet of compressed air in every stroke.

Boosting an air cylinder's capacity by increasing air pressure multiplies such air consumption and may cause erratic operation, in particular if the load varies along with the stroke. For these reasons, pneumatically controlled machinery usually operates within the 40 to 60 psi range. And in the (admittedly unlikely) event of broken tie-rods, the endplate of an air cylinder would likely take off like a moon rocket, while a hydraulic cylinder would merely leak oil to the tune of the pump's capacity. That's why rupture tests on pressure vessels are done with water rather than air.

And last but not least, air compressors are costly machines and inherently subjects to vibrations. As air heats up, so does the compressor itself, which calls for frequent maintenance stops and replacement of parts. Hydraulic systems, on the other hand, demand higher capital investment because they include an oil pump, oil reservoir, oil cooler, and an overpressure valve to redirect the pump's output back into the reservoir when rated system pressure has been reached and supply exceeds demand. Piping is heavier and more complex as it includes the return lines from the cylinders to the oil reservoir.

The design of hydraulic cylinders resembles that of their pneumatic counterparts, except that damping at the ends of the stroke is not needed in the inherently gradual operation of hydraulics. Centralized hydraulic systems, including accumulators for pressurized oil, are mostly carryovers from the past. Like in other fields, individually powered installations are the way to go.

Hydraulic presses have the ram descend commensurate with their pumps supply of oil, which can be regulated with valves or by varying the pump's rpm. The virtual incompressibility of liquids, such as hydraulic oils, allows for pressurizing hydraulic systems with up to 4000 psi, though 250 psi "low pressure" systems are also widely used.

Pressure sensors

While the venerable U-tube pressure meter integrates sensor and indicator in one setup, most pressure gauges have separate sensor and indicator systems, yet united under the same roof – read case. The sensor in Bourdon tube manometers is a flattened semicircular or spiraled metal tube; in membrane manometers, a deflecting membrane.

If the deflection of the sensing element is proportional to signal strength, read pressure, the dial has a regularly graded scale; otherwise, scales can be drawn to match the response curve of the instrument. For instance, pressure signals from a Pitot tube are proportional to the square of flow velocity; the scale to read flow velocity would thus be drawn to the square root of the magnitude of the input signals. More complicated would be the scale of a manometer used to show water temperature in a boiler. The relation between p, the pressure in a boiler (in hPa), and T, the temperature of the water in degrees Celsius, is given by the approximation

$$p = 6.1121 \times \exp\left[\frac{(18.678 - T/234.5) \times T}{257.14 + T}\right].$$

The temperature scale would be plotted to the inverse function of that equation.

While this works with "read only" instrumentation, *process control instruments* have their output tailored to the requirements of the system they monitor. For instance, conditions in a boiler might be controlled by steam pressure commanded adjustments of fuel supply to the burner. This so-called proportional control, once appropriately set up, functions as long as changes in operating conditions, such as arbitrary variation in demand for steam, are small enough to be compensated by the controller's action. Such limits are known as the proportional band. Beyond, manual or automatic reset is called for in order to reestablish the same pressure as before. To this end, control instruments are available with automatic setpoint recovery, which can be integral or, for processes with long time delays, derivative.

Since it would be a formidable task to develop for every particular control system an appropriate instrument, sensors of all kind are designed to output voltage, amperage, or pressure in bands of usually 0 to 5 volt, 4 to 20 mA, or 3 to 15 psi. Within this context, sensors do more than just react to system conditions, such as temperature, pressure, humidity, etc. They linearize and scale their output to one of the standards above. This became possible with the consistent downsizing of the incredible shrinking CPU[4] in size and price to the point where these chips fit into the sensors housing. Only in hostile (e.g., overheated, acidic) environment, sensor signals are remotely processed.

Pressure sensors with electrical output resemble those in mechanical stress

4 central processing unit.

measurement: One of the resistors of a Wheatstone bridge is made to change its ohmic resistance in response to pressure, generating an error voltage in the formerly balanced bridge circuit, which the CPU converts into a linear 0 to 5 V or 4 to 20 mA output.

While bridge circuits depend on a proper electric power source to function, *piezoelectric pressure transducers* generate their proper voltage in response to pressure changes. However, the minute size of piezoelectric crystals makes the area a exposed to gas or liquid pressure exceedingly small. The force generated by a certain pressure p, being ap, may not suffice for generating a precisely measurable voltage. For that reason, a mechanical amplifier, including a plunger mounted at the center of a pressure-sensing membrane, is used to apply pressure on the crystal. If the area of the membrane is, say, 100 times the exposed area of the crystal, this device amplifies signal strength 100 times.

The beauty of signal standardization is that identical sets of instruments can be used in the most diverse applications, from simple boilers to 14,000 psi autoclaves in gasoline synthesis.

The control center of a satellite launching facility and a nuclear power plant may employ identical instruments. And get ready to see them in the space station, next time you decide to spend some 20 million of your allowance on a ticket to get there.

The atmosphere and the cosmos

Uncontested record holder in the stampede toward lowest vacuum is intergalactic space, thought of holding no more than a few molecules per cubic meter. But potential investors in an intergalactic CRT industry – beware! Even if you traveled at the velocity of light, it would take you a million years to reach the point halfway to one of the closest of our galactic neighbors, the Andromeda Nebula. A long-term investment indeed!

Closer to home, our atmosphere may be seen as the prototype of a "world without borders," since there is no clear separation between "some air" and "no air" zones in space. But if we tentatively define its boundaries by the core of the Van Allen belt, where the Earth's magnetic field lines lock into closed loops, the atmosphere would be confined into an egg-shaped volume of 220,000/60,000 km radii, respectively.

Atmospheric pressure for up to 1000 km above sea level is listed in Table 8.1.

Table 8.1. Atmospheric pressure

H (km)	p (atm)	H (km)	p (atm)
0	1.000	50	0.000788
5	0.534	75	23.6×10^{-6}
10	0.262	100	0.316×10^{-6}
15	0.120	200	0.836×10^{-9}
20	0.0546	400	0.0143×10^{-9}
25	0.0252	600	0.811×10^{-12}
30	0.0118	800	0.175×10^{-12}
40	0.00283	1000	0.0742×10^{-12}

[a] H, altitude; p, barometric pressure

Northern Lights occur from 400 km down to 80 km of altitude, while the ionospheric E, F1, and F2 layers, best known from their reflectivity to radio signals in the shortwave band, locate between 100 and 300 km. Around 80 km, luminescent night clouds have been observed.

The stratosphere reaches from 45 km down to 10 km, with cirrus and stratus ice crystal clouds in its lower layers. It conducts sound. Commercial airliners, flying typically at 13,000 m of altitude, thus cruise in the stratosphere.

Only from 10 km downward lies the troposphere, which provides conditions for life on Earth, regardless of its playing host to such unpleasantries as hurricanes, tornados, and electric discharges of unsurpassed energy.

And – oh yes – I nearly forgot to mention the wish-fulfilling shooting stars, also known as meteorites, that burn off between 50 and 100 km of altitude, though the specters of the wishes they receive from us earthlings within the split seconds of their existence are thought to linger for much longer periods over the shelves of Victoria's Secret and Home Depot.

9 Density of solids, liquids, and gases

Quick: What's heavier, a pound of lead or a pound of feathers? Ha, ha, ha. Ho, ho! Did I hear lead? Gotcha! – A pound is a pound … See?

In a courtroom, such questioning would be objected to on grounds of leading the respondent on to think of *density* rather than weight. The density of a substance is – according to the International System of Units (SI) – the mass of substance per cubic meter of volume (kg/m^3). DIN 1306 add as "legal units for density" the *gram per milliliter* (g/cm^3); further the *kilogram per liter* (kg/ℓ or kg/dm^3), and *tons per cubic meter* (t/m^3), with all those DIN units equal to 1 for water. The mass of 1 m^3 of pure water at 4 °C (water's point of highest density) equals 1000 kg or 4535.9237 pounds.

Oops! Here comes the caveat! That's what it was for the founding fathers of the metric system and remained standard from 1901 to 1964. But refined methods of measurement led to corrections. To begin with, the temperature of the water's highest density is 3.984 °C, rather than a straight 4 °C; and second, the cubic meter of water at that temperature weighs 999.972 kg, rather than that appealingly round figure of 1000 kg.

But we shouldn't try to be more papal than the Pope; let's good enough be good enough and take the easy way out of that dilemma by sticking with our forefathers' 1000 kilogram for the cubic meter of water, unless exceptional accuracy is called for, such as – for instance – in the calibration of density standards. After all, it is this whole-number relation what makes computing with metric units easier than with ours. And we are still free to use the figures of highest accuracy where such a degree of precision is really needed.

Within this context, the term *relative density*, formerly *specific gravity*, stands for the ratio of the mass of a volume of a substance to the mass of the same volume of deaerated water at 4 °C under 760 mmHg of atmospheric pressure. Numerically, this equals gram per cubic centimeter; however, specific gravity is a dimensionless figure. For instance, commercial steel with the density 7.85 g/cm^3 is 7.85 times heavier than water, which gives 7.85 for its specific gravity.

In more inclusive terms, the concept of *density* includes all the expressions for the magnitude of a physical entity per unit of volume, such as: *impulse density*, *energy density*, *(electric) charge density*, etc., and in statistical mathematics, *probability density*. *Electric current density* – determinant for the ohmic heating of an electrical conductor – is defined as the quotient of amperage over the cross-sectional area of the cable.

The older term *specific weight* fell from grace because centrifugal forces combined with the gravitational effects of the oblateness of the globe make an object weightier near the poles than at the equator. This led to the concept of *standard gravitation* for the gravitational force at a place with *standard gravitational acceleration*, defined by $g = 9.80665\ \mathrm{m/s^2}$. Thus, the spring scales in Fig. 9.4 must be set up on that kind of place in order to show the true specific weight of the 1 dm^3 cubes on their pans. By contrast, two pan balances are immune to gravitational variations which affect equally the gravitational force on the weights in one pan *and* the load on the other – and cancel out.

Density and atomic mass

Early in the development of the modern concept of atoms, the weight of the atom of a given substance was defined in multiples of the weight of the hydrogen atom. But the hydrogen nucleus differs from the nuclei of all other atoms by its lack of neutrons. Since the mass of the proton is not identical with that of the neutron (1.007276 vs. 1.008665), an all-encompassing atomic mass unit (u) has ultimately been defined as $1/12$ of the atomic mass of the element carbon $\left({}^{12}_{6}\mathrm{C}\right)$, which nucleus contains equal numbers of protons and neutrons (6 each). In that new scale, hydrogen weighs in with 1.008 u, oxygen with 15.999 u, and nitrogen with 14.007 u.

Intuitively, one would expect the density of matter to rise and fall with the number of protons and neutrons in its nucleus. But only for gases exists such a correlation insofar as the density of any gas can be estimated as 0.0446 times the molecular mass of that gas, for instance:

Element	Predicted density	True density
hydrogen, H_2	$0.0446 \times 2 = 0.0892$	0.08988
oxygen, O_2	$0.0446 \times 16 \times 2 = 1.427$	1.429
nitrogen, N_2	$0.0446 \times 2 \times 14 = 1.2488$	1.2505

The small differences between predicted and true density values are caused by the presence of isotopes in the samples.

For monoatomic (called rare or noble) gases, such as helium and neon, molecular mass equals atomic mass, which gives for their densities:

helium, He	$0.0446 \times 4 = 0.1784$	0.1786
neon, Ne	$0.0446 \times 20 = 0.892$	0.9002

Measurements of gas density

Like we did for solids, the density of gases too can be deduced from the weight of a given volume. A 1-liter spherical container (124.07 mm internal diameter), made of, say, 0.5 mm thick duralumin sheet, would weigh about 64 gram and could thus be weighed on standard analytical scales of typically ± 0.0001 gram of accuracy, rated usually for up to 200 gram. Weighing the container first evacuated, then with air let in, allows for figuring the difference as the weight of 1 dm^3 of air – read density – under the prevailing atmospheric conditions.

For still better precision, the gas in the container can be compressed by a factor of 10, to 10 atm absolute or 9.034 kgf/cm^2 (128 psi) above atmospheric pressure without loading the tank's walls with more than 1/3 of the rupture stress of duralumin. Working with ten times the amount of gas in the tank adds another decimal to the results.

But all that presupposes the availability of laboratory equipment and trained operators to handle it. In industrial environment, in particular gasworks and coking plants, gas density is routinely measured with the effusiometer.

Effusiometer

The effusiometer derives the density of gases from the time a predetermined volume of gas takes to escape through a small opening or nozzle.

An apparatus (Fig. 9.1) built to this principle consists of a cylindrical container with a calibrated glass tube, the metering cylinder, reaching down from the cover. The tube's upper end is piped to a three-way cock, which can be selectively switched between charge and exhaust through a nozzle of typically 0.5 to 1.0 mm bore.

Operation of the setup starts with the cock in charge position and the gas under test entering the tube from above and bubbling out at the lower end until the tube has been purged of all residues from earlier use and filled with the new charge.

Next, the cock is turned into its exhaust position and the raising of the water level in the tube being timed. The time interval from the low to the high mark on the walls of the metering cylinder is then proportional to the square root of gas density.

Inversely, the ratio of gas density ρ_G to air density ρ_A follows from the raise-times t_G and t_A for gas and air respectively as $\rho_G/\rho_A = (t_G/t_A)^2$.

If it took with air, say, 10.00 s for the liquid level to rise from the low to the high mark, and for a 60/40% propane-butane mixture 13.64 s, the ratio of densities

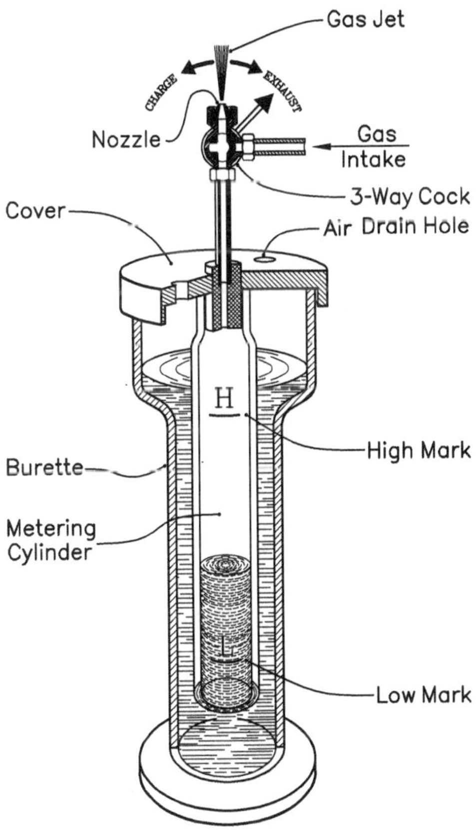

Fig. 9.1. Effusiometer for gas density measurements

becomes $\rho_G/\rho_A = (13.64/10)^2 = 1.860$, which gives the density of the mixture as $1.860 \times 1.293 = 2.406$.

With 2.670 for the density of pure propane and 2.01 for butane, this checks neatly with the *rule of blends* for computing the density of gas mixtures, in the present case: $0.60 \times 2.670 + 0.40 \times 2.010 = 2.406\ \mathrm{g/cm^3}$.

Hydrogen and helium

Although helium accounts for 23% of the mass of the universe, it used to be so rare on our home planet that its discovery in 1868 by J. Norman Lockyear happened in the spectrum of a solar eclipse rather than in a laboratory on Earth. Only the rise in production of natural gas, which contains up to 7.8% of helium, made helium available in quantities sufficient for the replacement of hydrogen in balloons and airships.

Previously, hydrogen-filled toy balloons had been playthings – even for grown-ups, who occasionally misused their cigarettes as igniters for balloon bombs.

First to fly a hydrogen filled-balloon was the French physicist Charles Jacques Alexandre César (1746–1823), who in 1783 reached 3200 meter of altitude. As recent as 1931, the Swiss scientists Auguste Piccard and Paul Kipfer flew a hydrogen-filled balloon 52,000 feet high into the stratosphere. Fear of fire was then a way of life for ballooners, and Piccard reportedly struck a lit cigarette from between the lips of a bystander who had placidly been waiting to watch the take-off.

Likewise, the hydrogen-inflated airship *Graf Zeppelin*[1] circled the globe in 1929 and made 144 ocean crossings between 1928 and 1938. And in the period from 4 March 1936 to 6 May 1937, the rigid airship *Hindenburg* made 10 scheduled cruises from Ludwigshafen, Germany, to New York and back. But a fire at the landing in Lakehurst, New Jersey, marked the end of hydrogen-filled "lighter than air" flying devices, and helium became the norm.

It seems surprising that helium, of atomic mass 4, could so easily supplant hydrogen, of atomic mass 1, without significant loss of lift. One reason is that helium, from the group of noble gases, exists in the atomic state, while the highly reactive hydrogen atom, known in chemistry as *hydrogenium nascendi*, hardly survives for more than 0.5 seconds before it binds with atoms next door, of principally its own kind. Therefore, the chemical formula of the *gas hydrogen* is not H, but H_2, which makes the ratio of the *molecular masses* of helium to hydrogen 2:1, rather than 4:1.

The lift obtained by supplanting 1 cubic meter of air, weighing 1.293 kg at 0 °C, by 1 cubic meter of hydrogen, that weighs 0.090 kg, is $1.293 - 0.090 = 1.203\ \mathrm{kg/m^3}$, while the lift from replacing air by helium, of $0.180\ \mathrm{kg/m^3}$ of density, is $1.293 - 0.180 = 1.113\ \mathrm{kg/m^3}$ – a mere 7.5% less –, which shows that it wasn't the higher density of helium that made lighter than air craft obsolete, but rather the spectacular progress of commercial airplanes previous and beyond the introduction of the jet engine.

1 named after Count Ferdinand von Zeppelin (1838–1917).

Since the state of gases changes with environmental factors, such as barometric pressure and temperature, their density is usually defined at 0 °C and 760 mmHg, and the unit *normal-cubic-meter* (Nm^3) is based on these conditions. Rather than adding the apodosis "at 760 mmHg and 0 °C" to every density figure, density is expressed in kg/Nm^3. The density of air thus becomes conveniently expressed as 1.293 kg/Nm^3; of carbon dioxide (CO_2), 1.600 kg/Nm^3; and of carbon monoxide (CO), 0.789 kg/Nm^3. That shows CO_2 heavier and CO lighter than air, and explains why carbon dioxide gas lures in the depths of vine cellars, while a leakage of the highly toxic carbon monoxide endangers principally those above the source.

Among the heavier gases, sulfur dioxide, an air pollutant, weighs in with 2.93 kg/Nm^3, while the rare (noble) gas xenon tops them all with 5.894 kg/Nm^3. The average density of the Sun, essentially a compressed ball of hot gases, reaches 1410 kg/m^3.

Acoustic gas density meter

The relation of gas density and the speed of propagation of sound waves lends itself for gas density measurements.

The speed of sound in an ideal gas is independent of the frequency and amplitude (pitch and loudness) of the acoustic wave, and varies only slightly with humidity. But it depends on temperature T. For air, this relation is

$$c_{\text{air}} = 331.5 + 0.6 \times T$$

with T expressed in °C. For instance, the speed of sound in air at 20 °C, or 68 °F, is $c_{20} = 331.5 + 0.6 \times 20 = 343.5$ meter/second.

The general expression for the speed of sound in gas reads $c = \sqrt{\kappa p/\rho}$ and κ stands for the ratio of specific heat capacity at constant pressure to specific heat capacity at constant volume, wherein specific heat capacity of a substance is defined by the heat energy to raise the temperature of 1 kg of said substance by 1 °C, or better: from 14.5 to 15.5 °C.

While 0.2385 Cal is required to heat 1 kg of atmospheric air by 1 °C, it takes only 0.1701 Cal for an equal amount of air trapped in a tank. This gives $\kappa = 0.2385/0.1701 = 1.402$ for air, a figure that varies only little for most other biatomic gases, such as hydrogen, oxygen, nitrogen, etc. For triatomic gases, the values for κ group around 1.3.

With $\kappa = 1.402$, $p = 1$ atm $= 101325$ Pa, and $\rho = 1.293$ kg/m^3, the formula above gives the velocity of sound in air at 0 °C as

$$c = \sqrt{1.402 \times 101325/1.293} = 331.5 \text{ m/s},$$

which checks with the value obtained from the temperature formula. In all that, the *pascal* (Pa) equals 1 $newton/m^2$ or 1/9.80665 kgf/m^2.

Speed of sound is thus inversely proportional to the square root of gas density; and gas density is inversely proportional to the speed of sound squared, a relation that became the design principle of the acoustic gas density meter in Fig. 9.2, an

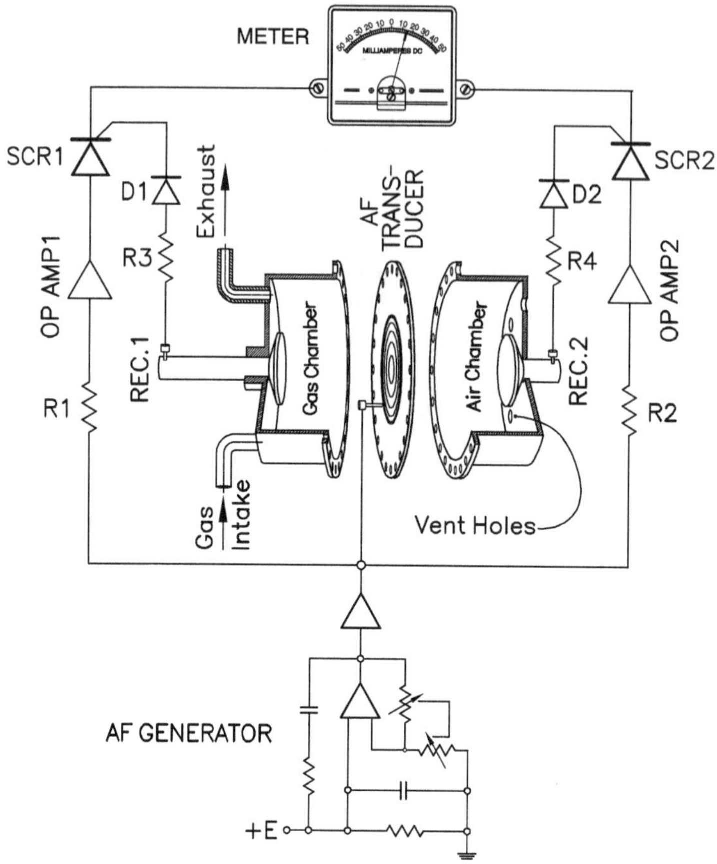

Fig. 9.2. Acoustic density meter

instrument that measures the ratio of the speed of sound in the gas under test and the speed of sound in air.

At the core of the instrument are face-to-face mounted flanged bins, bolted together over a sheet-metal bulkhead with a central opening that houses the membrane, dimensioned to resonate at some predetermined frequency between 1000 and 5000 Hz.

Audio frequency (AF) pick-ups locate at the bottom of the chambers at identical distances from the oscillating membrane. Much like in conventional speakers, oscillations of the membrane are induced by sinusoidal alternating current.

If both chambers were filled with air, the times for the sound waves to travel from source to receiver would be the same on either side. But with gases of different density, sound waves in the gas of higher density travel slower, which causes a phase shift between waves from the left and right side sound receiver that shows on the galvanometer in the bridge circuit, sketched in Fig. 9.2, top.

For example, let one chamber be filled with air ($c_{\mathrm{air}} = 331.5\ \mathrm{m/s}$), and the other with carbon dioxide, in which sound travels at $c_{\mathrm{CO_2}} = 259\ \mathrm{m/s}$.

For, say, 50 mm (0.050 meter) spacing between sound source and sound receptors, the times for the signals to travel that distance are

$$t_{air} = 0.050/331.5 = 0.1508 \times 10^{-3}\ \text{s},\ \text{or } 150.8\ \mu\text{s},$$
$$t_{CO_2} = 0.050/259 = 0.1930 \times 10^{-3}\ \text{s},\ \text{or } 193.0\ \mu\text{s}.$$

For that case, the gates of the silicon controlled rectifiers SCR1 and SCR2 become positive 150.8 μs and, respectively, 193.0 μs after the start of the membrane's oscillations, which frequency we assume set at 1 kHz.

Since the cycle time of a 1000 Hz sound source is $T = 1/f = 1/1000\ \text{s} = 1000\ \mu\text{s}$, or 500 μs per semicycle, the 150.8 μs time delay in air makes for $150.8/500 = 30.16\%$ loss of ON-time on the air side, while the 193.0 μs time delay in CO_2 causes a loss of $193.0/500 = 38.6\%$ of ON-time on the carbon dioxide side. With π radians or 180° for the length of the half-wave, these percentages convert to $0.3016 \times 180 = 54.29°$ and $0.386 \times 180 = 69.48°$, respectively.

The related voltage drop can be figured from the area under the sinusoid $y = \sin x$, which is $\int_0^x \sin x\, dx = -\cos x\big|_0^x = 1 - \cos 54.29° = 0.416$ for air and $1 - \cos 69.48° = 0.649$ for CO_2. Since the area under half of a sinusoid equals 2, the remaining voltages are $(2 - 0.416)/2 = 79.2\%$ and $(2 - 0.649)/2 = 67.6\%$ of the supply voltage. For an operating voltage of 10 Vpp (peak to peak), this amounts to 7.92 V average on one terminal of the meter, and 6.76 V on the other, resulting in a 1.16 V readout, directly measurable with a standard galvanometer or – better still – a bridge circuit.

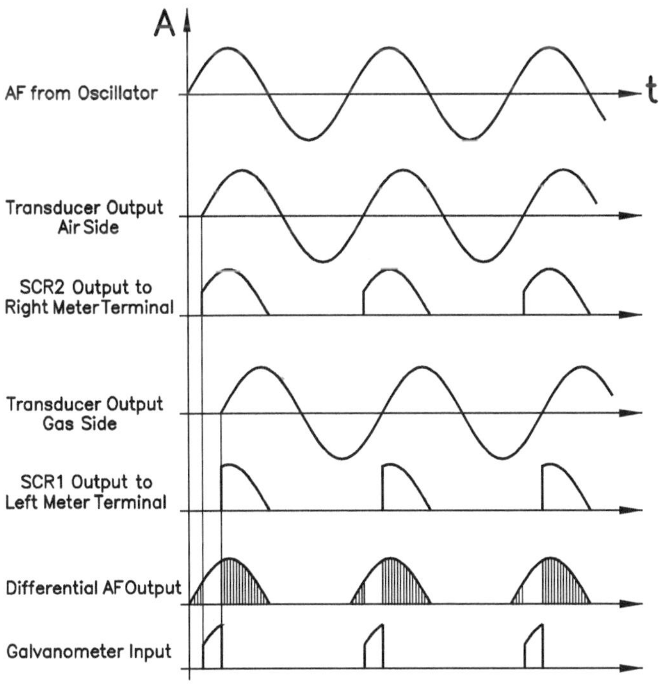

Fig. 9.3. Waveforms in the acoustic gas density meter

Figure 9.3 shows the AC waveforms generated in the process, starting in the first row with the membrane's drive voltage. The identical, yet phase-shifted AC waveform in the second row stems from the air-side sound transducer and starts 150.8 μs later, its rise triggering the silicon-controlled rectifier SCR1, which outputs the wave shown in the third row – still sinusoidal but clipped at the start and devoid of the negative semicycles.

The fourth and fifth rows repeat the previous two rows for 193.0 μs of time delay on the CO_2 side. When the clipped voltages of the third and fifth rows are connected to the terminals of a galvanometer, they subtract from each other (sixth row). The instrument then measures the pulsating direct current[2] in the seventh row.

For correct results, the peak voltages from driver and receivers must be identical. Therefore, automatic gain control (known from radio circuits under the acronym AGC) is a must for the circuit in Fig. 9.2. The respective error voltages are taken from a capacitor, connected through a diode that will charge to the peak-to-peak voltage of the source.

Instead of the traditional Wien Bridge oscillator in Fig. 9.2, bottom, a *direct digital synthesizer* (DDS) chip is lately being used as the sinusoidal AC generator. Driven by an external or internal clock, the DDS generates the sine wave mathematically point per point and subsequently uses a digital-to-analog converter to create the related AC output.

The instrument's heavy housing, typical for *acoustic gas density meters*, protects it from the effects of occasional impacts and internal pressure differentials. Like yesterday's microphones, the density meter hangs suspended between tension springs in a ring on a tripod with heavy baseplate.

The effects of environmental temperature changes can be prevented by wrapping the housing with heater plates, kept somewhat above ambient temperature (e.g., at 35 °C).

Density of solids

The density of solids depends on their atomic mass and the spacing between atoms in their crystalline or amorphous molecular pattern. There are no such simple relations between density and atomic mass as in gases.

The definition $density = weight/volume$ makes us think of weighing machined test cubes of 1 decimeter (100 mm) side length, as shown in Fig. 9.4. A less costly (but probably less precise) process is to submerge an arbitrarily shaped test piece in a graduated measuring cylinder and get its volume from the raise of liquid level.

Higher accuracy is achieved by finding the volume of a probe from its buoyancy in a liquid, such as water, which equals the weight of the amount of liquid replaced by the probe. *Archimedes Scale* measures that lift as the difference between the weight of the test specimen in air and in water, which – as long as we work with metric units (kg) – equals volume in liters or dm^3.

Among metals, lithium is the lightest with density of 0.534 g/cm^3, followed by magnesium with 1.74, and aluminum with 2.70. Technical steels group around 7.85, lead weighs in with 11.34, gold with 19.30, and platinum with 21.40. The specific heaviest metal is osmium with 22.48, and the heaviest element so far is

2 abbr. DC

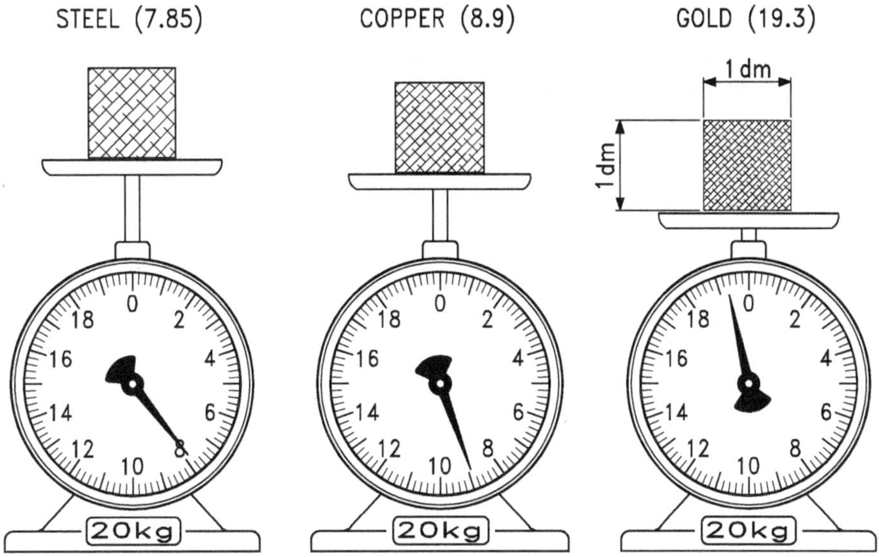

Fig. 9.4. Weight per cubic decimeter of steel, copper, and gold

iridium, which density is 22.65. Oh yes, as not to forget our home planet, Earth, which iron core brings its average density up to 5.52 g/cm^3, making it the specific heaviest planet in the Solar System.

The average density of 0.5 of timber makes that logs float on water. During the 19th century, many of the country's waterways became traffic roads for river logging until rail transport took over. Entire communities, such as Muskegon, Traverse City, and Menominee in the state of Michigan, sprang into existence as the hubs on the way of timber into the sawmills.

Density of liquids

The group of "lighter than water fluids" includes gasoline with density of 0.68 g/cm^3, kerosene (0.80 g/cm^3), pyridine (one of the most radical organic solvents, 0.982 g/cm^3), while glycerin (1.261 g/cm^3), hydrogen peroxide (1.442 g/cm^3), nitroglycerin (1.596 g/cm^3), sulfuric acid (1.831 g/cm^3), and bromine (3.119 g/cm^3) are among the heavyweights, outdone only by mercury (13.6 g/cm^3) if we classify it as a liquid rather than as molten metal.

Density of mixtures of various liquids can be figured by multiplying their respective densities with their concentrations and add them up. For instance, a 40% sulfuric acid solution weighs $0.40 \times 1.831 + 0.60 \times 1 = 1.332$ g/cm^3.

Pycnometer

The pycnometer in Fig. 9.5, a simple and accurate device for measuring the density of liquids, also known under the term *density bottle*, is a glass flask shaped like a sphere or a truncated cone. Here again, measurement is based on the definition of density ρ as the quotient of mass m over volume V:

$$\rho = m/V .$$

The capacity of a pycnometer flask or can (Fig. 9.5) is derived from the weight of a filling with deaerated water, which density is taken from pertinent tables, such as those from Wagenbreth and Blanke (Table 9.1).

For other temperatures (t in °C), the density ρ of water in g/cm^3 is given by the empirical polynomial

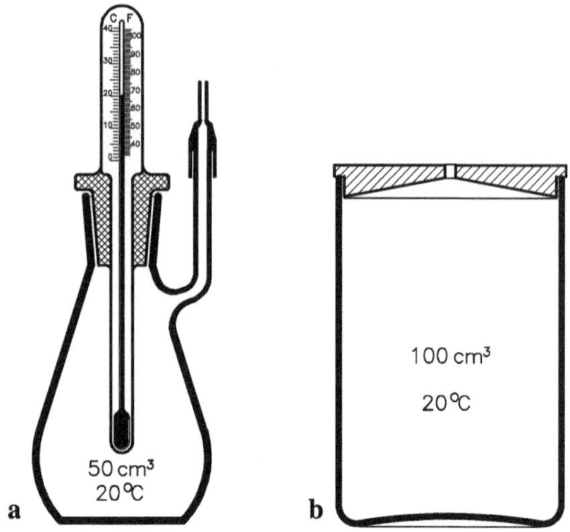

Fig. 9.5. a Pycnometer and **b** specific gravity weighing can

$$\rho = a + bt + ct^2 + dt^3 + et^4 + ft^5 ,$$

where the constants are

$$a = 0.999839564,\ b = 6.7998613 \times 10^{-5},\ c = -9.1101468 \times 10^{-6},$$
$$d = 1.0058299 \times 10^{-7},\ e = -1.1275659 \times 10^{-9},\ f = 6.5985371 \times 10^{-12}.$$

The volume of the water in the flask is then found as the quotient of the water's weight and its density at the temperature shown on the thermometer reaching through the pycnometer's plug. The result should come out close but not necessarily identical with the manufacturer's standard marked on the flask, including such sizes as 5 cm³, 10 cm³, 25 cm³, 50 cm³, and 100 cm³.

The degree of filling must be such that the liquid drains through the capillary when the ground glass stopper is put in place. While the Gay-Lussac pycnometer

Table 9.1. Density of water at various temperatures and 1 atm

t (°C)	ρ (g/cm³)	t (°C)	ρ (g/cm³)	t (°C)	ρ (g/cm³)
10	0.999699	17	0.998772	24	0.997293
11	0.999604	18	0.998593	25	0.997041
12	0.999496	19	0.998402	26	0.996780
13	0.999375	20	0.998201	27	0.996509
14	0.999242	21	0.997989	28	0.996230
15	0.999097	22	0.997767	29	0.995941
16	0.998940	23	0.997535	30	0.996643

drains through a narrow channel in the stopper, the Jaulmes pycnometer (Fig. 9.5 a) sports a lateral discharge tube and nozzle. The rationale for expelling excess liquid through a narrow opening is that, like in the mercury thermometer, the raising liquid level in a capillary makes very small changes in the volume of the overall filling readily evident. For instance, a cubic millimeter of liquid would hardly show in the liquid level at the neck of an open bottle, but in a capillary of, say, a 0.5 square millimeter opening, would cause a 2 mm raise.

Previous to final weighing, the container should be held in vertical position and wiped clean. Its weight minus the weight of the empty pycnometer flask (including the plug) gives the net weight of the contents. Some models of high precision pycnometers are accurate to 1/100,000 gram per cubic centimeter of volume.

In industrial environment, the *gravity weighing can* (Fig. 9.5 b) makes for a robust variety of the pycnometer, though the principles of measurement are the same. While the British *pint weighing can* has been an offspring of the Imperial System of Weights and Measures, metric weighing cans usually come in sizes of 100 cm^3, 500 cm^3, and 1000 cm^3 (1 liter). Their finely machined, straight walls allow for a close fit of the lid, which must be pressed down until it settles snuggly on the brink of the vessel. Excess liquid is made to escape through the lid's center bore, of typically 2 mm diameter. Here again, the container must be wiped rigorously clean before taking its weight.

Unlike the blown glass containers of laboratory type pycnometers, metal containers are machined to make their true volume match the standard stamped on the lid. Additional identification numbers on lid and body make sure that similar looking lids aren't accidentally interchanged from can to can.

Both types of pycnometer can also be used to determine the density of powdered or granulated solids, such as chips from machining operations, by mixing a weighed quantity of them into the water fill of a pycnometric can.

With w_S for the weight of the solid, and w_{W+S} the weight of the blend of solid and water together, the ratio of the densities of solids (ρ_S) and water (ρ_W) is

$$\frac{\rho_S}{\rho_W} = \frac{w_S}{w_S + w_W - w_{W+S}}.$$

The weight of water in the voids left by the solids, w_W, is found as the difference between the water in the can without the solids and the weight of the water being spilled when the solids are being added.

To take this (not quite obvious) formula to the test, let the volume of the pycnometer be 100 cm^3, and the dry weight of the grains of solids (e.g., metal chips) 500 g. Considering the density of water as equal to 1, and the density of the solids 8.00 g/cm^3, we get the volume of the solids as $500/8 = 62.5$ cm^3, and that of the water surrounding them as $100 - 62.5 = 37.5$ cm^3. The weight of solids plus water becomes $500 + 37.5 = 537.5$ gram.

With these figures, the formula we set out to test, yields (not surprisingly, I hope):

$$\rho_S/\rho_W = 500/(500 + 100 - 537.5) = 500/62.5 = 8.$$

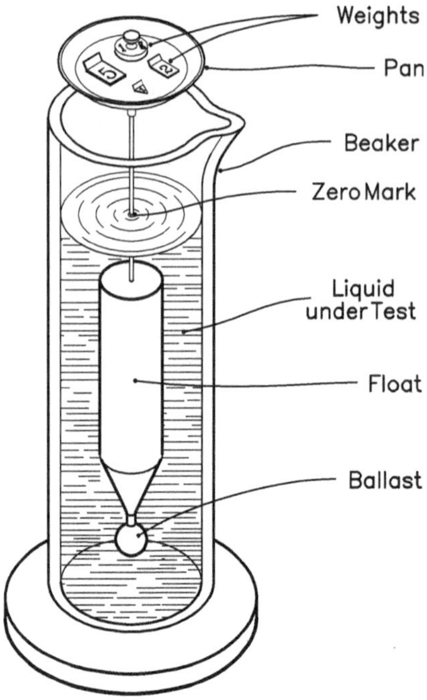

Fig. 9.6. Nicholson's hydrometer

Nicholson's hydrometer

The hydrometer introduced in 1790 by the English chemist William Nicholson (1753–1815) allows for faster and less labor-intensive testing of fluid density than the pycnometer. While standard hydrometers dip only partially into the liquid under test, the float of the Nicholson hydrometer is always fully submerged. The one in Fig. 9.6 consists of a plummet from which upper end a thin metal post, supporting a tray, projects out of the liquid. A notch on the stem marks the point to which the unit should be made to submerge by placing the appropriate weights on the tray.

Before use, the device is calibrated with deaerated water. With W_o for the weight of the hydrometer, and W_W for the total of ballast that makes it sink to the mark on the stem, the lift on the submerged portion of the device becomes $W_o + W_W$. In water, this equals V, the volume of the float plus that of the submerged portion of the post, multiplied with the density ρ_W of water under test conditions, that is, $V\rho_w = W_o + W_w$.

Conversely, in tests with a fluid of density ρ_L, a ballast W_L would be needed to make the hydrometer dip once again down to the zero mark, so that we further get $V\rho_L = W_o + W_L$.

Division of those equations by each other yields for the ratio of the densities of the fluid to that of water

$$\rho_L/\rho_w = (W_o + W_L)/(W_o + W_W).$$

For instance, if a 100 gram heavy Nicholson hydrometer had to be weighed down with 20 gram to make it submerge in water to the zero mark on the stem, and does the same with 51.68 g while submerged in the liquid under test, the equation yields $\rho_L/\rho_w = (100+51.68)/(100+20) = 1.264$. Thus, the liquid could be glycerin, which density happens to be 1.264.

Mohr (Westphal) balance

The Mohr, or Westphal, balance, a precision instrument for measuring the density of liquids, could be seen as the fluid counterpart of Archimedes' density balance for solids. Also known as the *hydrostatic balance*, this instrument derives the density of liquids from the *degree of buoyancy* of an immersed solid body of known weight and volume.

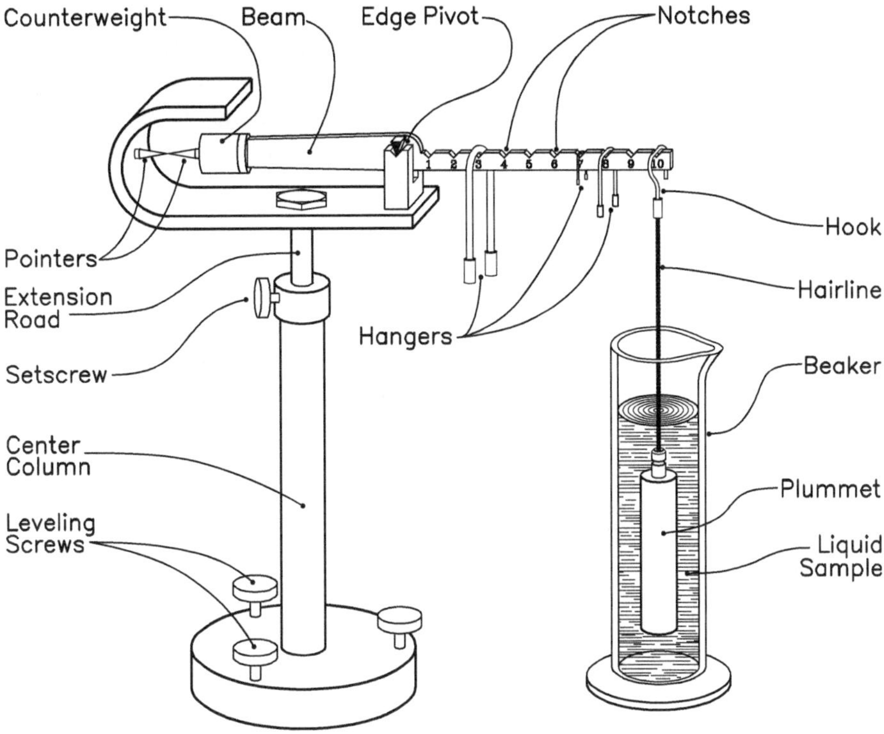

Fig. 9.7. Mohr (hydrostatic) balance

True to Archimedes' principle, lift equals the weight of the volume of liquid replaced by the solid. For instance, a solid glass plummet of 100 cm^3 of volume and 260 gram of mass weighs when immersed in water of unit density (1 g/cm^3) only $260 - 100 = 160$ gram. Submersed in gasoline rather than water, its apparent weight would fall to 192 gram, which makes the density of gasoline $(260 - 192)/100 = 0.68$.

The Mohr balance shown in Fig. 9.7 resembles a sliding weight balance with an adjustable counterweight at the left end of the beam and the float suspended on the right. The space between the fulcrum and the suspension point of the float is divided into ten equal parts, earmarked by numbered notches for the placement of hanging weights.

For instance, a set of 5 gram, 0.5 gram, and 0.05 gram, used with a 5 cm^3 plummet, allows for reading the density of the liquid probe in the beaker directly from the positions of the hanging weights. If the first, of 5 gram, is placed in, say, notch number 6, and the second, of 0.5 gram, in notch number 8, while the lightest isn't used, we read the density of the liquid under test as 0.680 g/cm^3, identifying it once again as likely (but not necessarily) gasoline.

The three leveling screws in the base plate allow for making sure of the vertical position of the instrument's center column previous to setting the counterweight.

If the volume of the float is not known with the required precision, it can be found by using the hydrostatic balance with pure water of known temperature and its precise density from Table 9.1 or other pertinent sources, rather than setting it equal to 1. For instance, a float of nominal volume V_o dipped into water of 20 °C should give 0.998201 for the water's density. If the result of the weighing is different, say, 0.900, then the plummet is bigger than assumed by a factor of 0.998201/0.9. The true volume, V, of the plummet then is $V = 0.9\,V_o/0.998201 = 0.901622\,V_o$.

Hydrometer

The hydrometer, also known under the term *areometer* (Fig. 9.8), is the instrument of choice for speedy checks of liquid density. The idea to derive liquid density from the depth of immersion of a float reaches back to Archimedes' treatise *On Gravity and Buoyancy.* Present-day hydrometers are hollow cylinders of glass or clear plastic, e.g., polymethylpentane, divided into a thinner section with the upside down scale on top of the far thicker bottom section, that ends in a bulb filled with a measured amount of heavy granulate (generally lead shot).

The farther down the hydrometer sags, the more liquid it displaces. Here again, Archimedes' law applies insofar as the hydrometer will sink until the weight of the liquid replaced, ρV, equals the weight W of the instrument.

Let V_o stand for the volume of the bulge of the hydrometer, including the part holding the ballast, and D for the neck diameter. $A = D^2\pi/4$ is then the cross-sectional area of the neck, and y, the depth of submersion, measured from the shoulder of the neck upwards, follows from the relation of submerged volume $V = V_o + Ay$, which yields with $\rho = W/V$ the liquid density as

$$\rho = W/(V_o + Ay)\,.$$

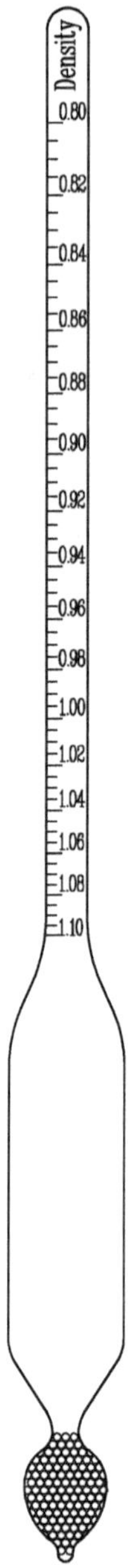

Fig. 9.8. Hydrometer

The equation explains the inverted, non regular scale in Fig. 9.8. And the term Ay in the denominator shows that the slimmer the neck, the higher the instrument's sensitivity. But slimming the neck goes at the expense of range.

For ready temperature correction, hydrometers often have a thermometer in the bulge section, so that density and temperature of the liquid under test are found simultaneously.

Since hydrometers show the specific gravity of the liquid probe relative to water, users should be aware that

Table 9.2. Hydrometer readings and electrolyte concentration in function of charge

State of charge (% of full charge)	Open circuit battery voltage (V)	Hydrometer reading (g/cm^3)	Mass concentration of electrolyte (%)
100	12.65	1.265	35.59
75	12.45	1.225	30.73
50	12.24	1.190	26.35
25	12.06	1.155	21.95
0	11.89	1.120	17.40

their instrument might not have been calibrated with 4 °C water, but to different standards, "double distilled water at 15.6 °C" being one of them.

The familiar battery syringe for checking the acid in car batteries has a hydrometer housed in the cylinder of the syringe, showing the density of the sulfuric battery acid. The electrolyte's density rises with charge, as shown in Table 9.2, which gives the relation of battery voltage, specific gravity, and concentration of the electrolyte for various degrees of charge at 70 °F. However, those figures may vary somewhat from supplier to supplier.

For tests performed at lower temperatures, add 0.012 volt for each degree Fahrenheit below 70 °F.

Highest accuracy (i.e., 0.001 g/cm^3) is achieved with application-oriented models of hydrometers, such as *alcohol hydrometers* with scales for direct readout of the alcohol content of aqueous alcohol solutions; *acidity hydrometers*, showing the concentration of acetic, hydrochloric, nitric, sulfuric, and phosphoric acids as mass-percentage (formerly % of weight); same for *alkalis*, including sodium hydroxide, hydrogen peroxide, and potash lye.

Further, there are *lactometers* for milk. *Brix hydrometers* for the percentage of sugar in aqueous sucrose solutions. *Balling hydrometers* showing the sugar content of beer and wort (mash). And last but not least, *Baumé hydrometers*, measuring the specific gravity of all kinds of aqueous solutions in a special unit, fittingly called *degrees Baumé*; for instance, U.S. Grade A honey must have at least 42.49 °Bé at 60 °F for FDA approval, and the concentration of hydrochloric acid for balancing the pH value of swimming pool water is specified as 20 °Bé.

Two particular Baumé scales are in use, one for liquids lighter than water, and one for liquids heavier than water. The relation between specific gravity G and degrees Baumé is given by the formulas

$$G = 140/(130 + {}^\circ\text{Bé}) \quad \text{for liquids lighter than water,}$$

$$G = 145/(145 - {}^\circ\text{Bé}) \quad \text{for liquids heavier than water.}$$

Special purpose hydrometers can be crafted from standard models by varying the mass of lead shot in the instrument's bulb.

Previous to density measurements, the liquids and fluids under test must be purged of air bubbles and, if more than one component is involved, be thoroughly mixed. Finally, the uppermost divvy of the liquid should be discarded in order to clear the surface of impurities, such as floating dirt or dusk particles.

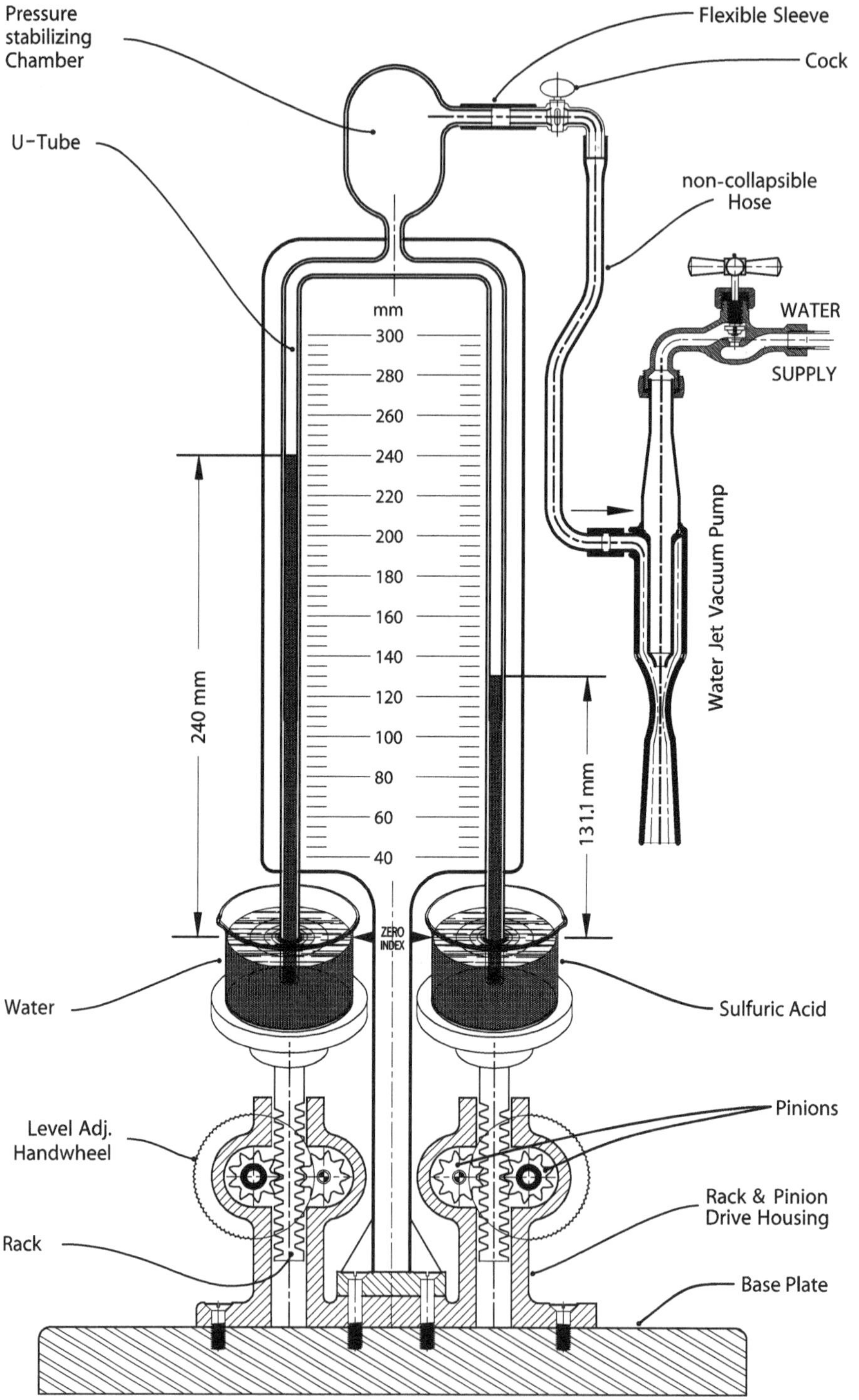

Fig. 9.9. Hare's apparatus

For best readout precision, the hydrometer should be dipped into the liquid about 1/8 inch beyond the point to which it would sink if left alone, and only then been released to *lift* itself into its final position – rather than letting it *sink* toward it. To avoid parallactic reading errors, observations should be done at eye level with the surface of the probe – the point where the apparent shape of the liquid surface changes from oval to straight.

Hare's apparatus

While pycnometer and hydrometer are built to deduce density from the weight of a given volume of liquid, Hare's apparatus (Fig. 9.9) uses a matched pair of Torricelli's barometers to the same end.

As we recall from the preceding chapter, atmospheric pressure makes a mercury column rise to 0.760 meter, while an equivalent water column is 10 meters high. Gasoline as barometric fluid would reach even higher, to a whopping $10/0.680 = 14.70$ meter.

Obviously, the height of the liquid column would be indicative for the density of the liquid, yet out of range for most practical applications; therefore, Hare's apparatus does not fully evacuate the space above the columns of barometric fluids, but just enough to make their height compatible with the overall dimensions of the instrument. The partial vacuum in Hare's apparatus can be established with a ventury vacuum pump available in most physical and chemical laboratories.

For instance, to set up Hare's apparatus shown in Fig. 9.9 for comparative measurements of the densities of water and sulfuric acid, the air in the upper sector of the manometer tubes has been thinned just enough for lifting the water column to a convenient height, such as 240 mm. Hereby, the rack-and-pinion drives are used to match the instrument scale's zero-index with the liquid surface of the water sump as well as the level of the sulfuric acid in the other bin.

Due to the equal weights of the columns of water and sulfuric acid, the equation $\rho_W H_W = \rho H_{H_2SO_4}$ or $\rho/\rho_W = H_W/H_{H_2SO_4}$ applies, where H_W and $H_{H_2SO_4}$ stand for the respective heights of the liquid columns, and ρ_W and ρ for the densities of water and of sulfuric acid.

For instance, in a test with water at 20 °C temperature and therefore $0.998203\ \mathrm{g/cm^3}$ of density, the sulfuric acid column raises to 130.8 mm of height, indicating the density of the acid as

$$\rho = \rho_W \times \frac{H_W}{H_{H_2SO_4}} = 0.998203 \times \frac{240}{130.8} = 1.832\,.$$

The scale of the instrument is regular and can be drawn in any units (e.g., inches, millimeters), though millimeter scales are the most convenient, because the size of the millimeter suits best the capacity of resolution of the human eyesight[3].

3 Hebra: *Measure for Measure*, Johns Hopkins University Press, 2003.

Oscillating tube densitometer

The oscillating tube densitometer in Fig. 9.10 is an instrument for "in-line" density tests on industrial products, such as gases and liquids. It derives density from the resonance frequency of an oscillating U-tube which is inversely proportional to the summed up masses of the tube and the product it is charged with.

The instrument includes a cantilever mounted U-tube, usually of glass or stainless steel, secured in sturdy sockets on a massive base. The first node of the oscillation lies thus at the boundary line between U-tube and socket.

If k and m, the spring constant and the mass of the free-swinging portion of the tube, are known, the tube's eigenfrequency is computed from

$$f = \frac{1}{2\pi}\sqrt{\frac{k}{m}}.$$

Herein, the spring constant k of a cantilever is given by the ratio of the applied force F (in newton) and the deflection induced by that force at the lever's outer end; with 1000 millimeter to the meter, F equals 1000 times the force causing 1 mm of deflection.

The important part of all this is that it doesn't matter if the product contained in the tube stems from a one-time fill, or is continuously circulating. In both cases, the unit's resonance frequency remains inversely proportional to the square root of the summed up mass of the tube and its contents from the clamping points outward.

Though k and m of a U-tube can be computed from its dimensional and mechanical characteristics, such as internal and external diameter, overall length, and elastic modulus of the tube's material, calibration of vibrating tube densimeters is

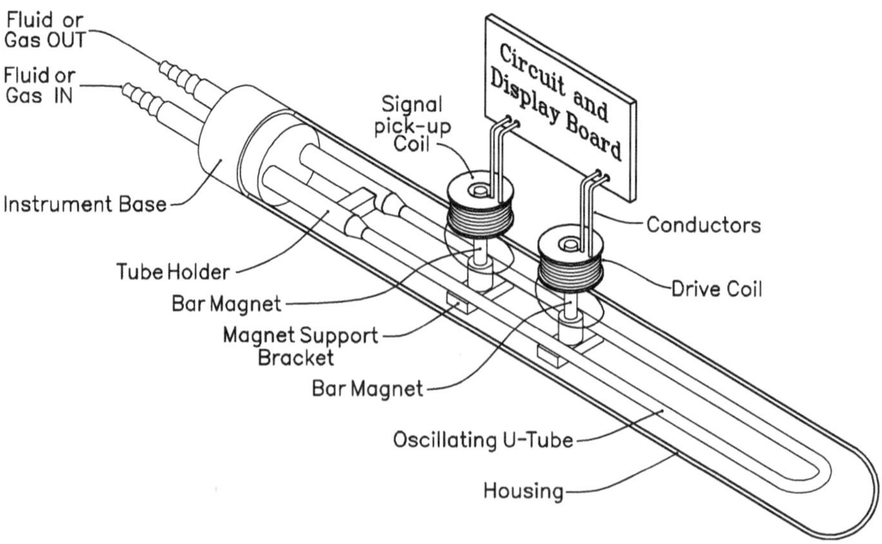

Fig. 9.10. Oscillating densimeter

done by measuring the instruments' frequency response with two different substances of accurately known densities, such as water, of density $\rho_1 = 0.999097$ at 15 °C, and, say, selenic acid, of density $\rho_2 = 2.602$ at that same temperature. If the frequencies with these samples are f_1 and f_2, the unknown density ρ of a probe, swinging at its own resonance frequency f, can be computed from the formula

$$\rho = \rho_1 + (\rho_2 - \rho_1)\left(\frac{f_2}{f}\right)^2 \frac{f_1^2 - f^2}{f_1^2 - f_2^2}\,.$$

For instance, if a U-tube resonated with water filling at $f_1 = 100\,\text{Hz}$, and at $f_2 = 61.966\,\text{Hz}$ with selenic acid, and resonates at 79.1 Hz while filled with an unknown probe, the probe's density is:

$$\rho = 0.999097 + (2.602 - 0.999097)\left(\frac{61.966}{79.1}\right)^2 \frac{100^2 - 79.1^2}{100^2 - 61.966^2} = 1.597\,,$$

equaling $\rho = 1.597$ of nitroglycerin at 15 °C.

Bracketed to the U-tube is a pair of permanent magnets, which poles reach into fixed inductor coils (Fig. 9.10). Alternating current (AC) from the instrument's power source is fed to one of the coils, causing its magnetic plunger to swing with the second magnet in tow. The alternating voltage induced in the second coil by its swinging core is amplified and sustains and stabilizes the oscillations. In short, one coil acts as the signal generator, the other as the driver of the tubes' oscillations. At the U-tubes' mechanical resonance frequency, oscillations build up into well defined and accurately measurable peaks.

Radio buffs remember a similar principle from the classical *Hartley* one-tube receivers of the 1920s, a time when vacuum tubes still counted as luxury items. Signal regeneration, similar to that in the oscillating tube densitometer, happened there in a few extra windings on the antenna coil, charged with an adjustable fraction of the anode voltage of the amplifier tube.

The degree of that positive feedback depended on the settings of a variable coupling capacitor, doubling as "volume control", though excessive volume was the least of those times radios troubles. But this brute force approach at getting a little something from near to nothing was curtailed by the allowable degree of feedback before the circuit began oscillating on its own at its proper frequency, snobbishly ignoring the antenna signals. This made that the loudness of reception depended on the listener's aptitude in setting the feedback control knob as close as possible to that "whistle/non-whistle" position.

To obtain sustained oscillations in the oscillating tube densitometer, one could manually set the frequency of the driving voltage close to the equivalent of the whistle point in Hartley receivers. But in our days, closed loop circuitry as we know from auto-tuning modern radios and TVs, does all that by itself.

Influence of the environmental temperature on the tubes' frequency can be avoided by heating the entire device to a constant level somewhat above the usual temperature of the surroundings. Peltier thermostat controlled heating coils are a good choice. In the absence of such luxuries, measurements should be done in an environment similar to that of the calibration.

The effects of viscosity on densitometer measurements are negligible below $\eta = 1\,\mathrm{mPa} \times \mathrm{s}$, but need to be considered with liquids of higher viscosity. For instance, density measurements on a 65% aqueous sugar solution come out 0.4% too high and would need correction by a factor k, compounded from the formula $k = -0.018 + 0.058\sqrt{\eta} - 0.00124\eta$. k comes in $\mathrm{kg/m^3}$, and η in $\mathrm{mPa \times s}$, a unit equivalent to 1/100 poise. In digital instruments, the CPU does the computing for us, and outputs viscosity-corrected results.

Oscillating densitometers with typically 0.1% precision are employed for material quality control in general, such as testing of battery acid, milk control, spot-checks on photo processing liquids, and probing the SO_2 content in exhaust gases.

Precision instruments of 0.01% accuracy are found in sugar refineries, beverage and cosmetics industries, electrolytic metal plating plants, and mineral-oil refineries.

Still higher precision (0.001%) is demanded in the pharmaceutical industry, alcoholic beverage producers, nuclear plants, and principally research laboratories.

The relatively high price of oscillating density meters is compensated for by the instruments' potential for automated operation and the ease of handling sample quantities as small as 1–2 cm^3.

The weakest point of the principle is its susceptibility to the presence of solid particles in suspension and entrapped microscopic gas bubbles.

Density of extrasolar objects

If our Earth emerged victoriously from the Solar System's *Miss Density pageant*, with Venus and Mars in second and third place, our chances tumble if we set foot into extra sidereal space.

First to unknowingly contesting Earth' density privileges was the elusive companion of the brightest star on our skies, Sirius. F. W. Bessel in 1834 suspected Sirius' double star nature from observations of its proper motion, but it took Alvan G. Clark 11 years with the world's then largest refracting telescope to actually *see* the Sirius-companion, *Sirius B*, nicknamed then *the Pup* (Fig. 9.11).

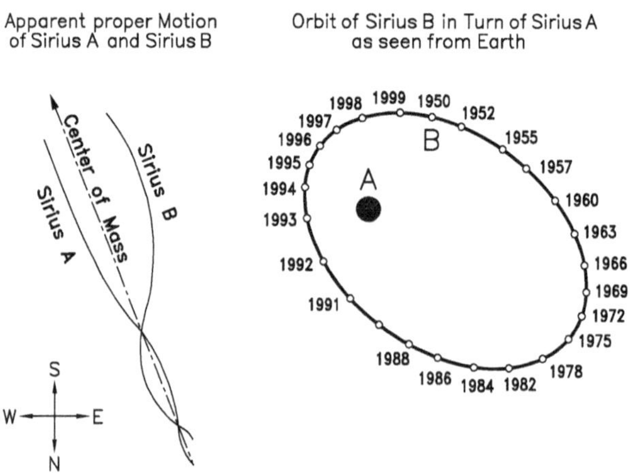

Fig. 9.11. Sirius' massive companion

While the mass of the Sirius companion was deduced from its orbital parameters, the star's angular diameter couldn't be measured directly, because unlike such objects as sun, moon, and solar planets, fixed stars are too far away to be seen as disks. They remain punctiform, even at highest magnification. So the Pup's size had to be deduced from its brightness, distance, and temperature. Of spectral type A5, Sirius B is hotter than our Sun, itself a spectral type G2 star, and the respective differences in surface temperature make the intensity of light from Sirius B four times that of light from the Sun.

Sirius locates 8.7 light-years (ly) from the Sun, and so does its companion, Sirius B, which apparent brightness is 8.65^{m}. Seen from that same distance of 8.7 ly, our Sun would shine with brightness 1.98^{m}, that is $8.65 - 1.98 = 6.67^{\mathrm{m}}$ classes brighter than Sirius B. That would make the Sun (according to the Pogson logarithmic Magnitude Scale of star brightness) $2.512^{6.67} = 465$ times brighter than Sirius B.

In other words, if both, the Sun and the Pup, were of identical surface temperature, the apparent area of the Pup would show with $1/465$ of the apparent area of the sun disk, seen from the same distance. But the Pup's radiation intensity, 4 times that of the Sun, brings that down to $1/(4 \times 465) = 1/1860$, making the size of Sirius B $\sqrt{1/1860} = 0.023$ sun diameters. With that, the ratio of the volumes of Sun and Sirius B becomes $(1/0.023)^3 \approx 82200$. Considering that the mass of Sirius B is 0.98 that of the sun, the ratio of the densities of Sun and Sirius B follows as $1 : (82200/0.98) \approx 1 : 83900$. With 1.409 for the average density of the sun, that brings the density of Sirius B to $1.409 \times 83900 \approx 120{,}000\ \mathrm{g/cm^3}$. A gallon of material of such density would weigh some 450 metric tons. Imagine how much easier Henry Cavendish's measurements of the gravitational constant would have been, had he had access to that sort of matter to work with!

Which brings us to the question if gravitation alone could have compressed stellar matter to such a degree of compactness. According to Newton's inverse square law, gravitational forces between two bodies, placed at distance R from each other, increase proportional to $1/R^2$. Mathematically speaking, gravitation would thus reach infinity for $R = 0$, but such a condition is a physical impossibility because two bodies remain always separated by at least the sum of their respective radii, $R = R_1 + R_2$, even in collapsing stars.

Nevertheless, when a star the size of the Sun abruptly shrinks from the status of "red giant star" to "dwarf star," gravitational forces get high enough to compress its mass into a sphere of approximately the size of the Earth.

Conversely, heavier stars with 4 to 8 sun masses shed most of their mass in the process of becoming supernovas. But in their central regions, worth about 1.4 times the mass of the Sun, gravitational and inertial forces grow to the point of destruction of the regular atomic structure of matter. The atoms' orbiting electrons interact with the positive charges from the nuclei, and only neutrons remain. Unlike protons, which violently repulse each other, the uncharged neutrons pack closely together into a neutron star, the most compacted kind of matter in the universe – except for black holes.

The density of neutron stars, typically $1.4 \times 10^{14}\ \mathrm{g/cm^3}$, makes surface gravitation so intense that the equivalent of our Mt. Everest on one of those stars would tower no higher than half an inch. And yet, a lifetime of a human body's energy production would not suffice for climbing one of those half inch neutronic mountains.

So where in our density pageant fits Sirius B? Its material is flimsy in comparison to that of neutron stars, but some 6000 times denser than gold. Which brings to mind the question if Sirius B still consists of ordinary matter, comprising a nucleus surrounded by a cloud of orbiting electrons. Intuitively, one would expect that atoms could be brought close together without damage as long as their outermost electron orbits aren't made to overlap. In other words, as long as atom spacing exceeds the atomic diameter, estimated between 0.1 and 0.5 nanometer.

Intuitively, one would expect heavier atoms as the bigger ones, but the higher attraction between the greater number of nuclei and their orbiting electrons in heavy elements makes for tighter packed atoms. Plutonium, for instance, weighs 200 times more than hydrogen, but its atomic diameter is just about 3 times that of the hydrogen atom.

Spacing of atoms in ordinary matter can be deduced from the number of atoms per mol (Avogadro's number), $A = 6.02 \times 10^{23}$. 1 mol of hydrogen H_2 molecules thus weigh 2.016 gram, and 1 gram of helium atoms 4.003 gram. 1 mol of material from the Sun, consisting of approximately 75% hydrogen and 25% helium, would thus weigh $0.75 \times 2.016 + 0.25 \times 4.003 = 2.513$gram.

On the other hand, the hydrogen's density is 0.0899×10^{-3}, and that of helium 0.1785×10^{-3}, which gives for the Sun's material (if cooled down to 0 °C and put under 760 mmHg of pressure) $0.75 \times 0.0899 \times 10^{-3} + 0.25 \times 0.1785 \times 10^{-3} = 0.1120 \times 10^{-3}$.

The volume of 1 mol or 2.513 g of solar material thus becomes $V = 2.513/(0.1120 \times 10^{-3}) = 22438\ \text{cm}^3$, which, according to Avogadro, is also the room for 6.02×10^{23} atoms. The space taken by each atom thus becomes $V_o = 22438/(6.02 \times 10^{23}) = 3.727 \times 10^{-20}\ \text{cm}^3$.

Taking the average diameter of an atom as 0.84×10^{-8} cm, V_o could be compressed by $3.727 \times 10^{-20}/(0.84 \times 10^{-8})^3 = 62881$ times before the atoms "touch." But Johannes Kepler's formula for the packing density $\pi/3\sqrt{2} = 0.74048$ of globes in cubic or hexagonal patterns signals a way to $1/0.74 = 1.35$ times higher compression, amounting to $62881 \times 1.35 \approx 84890$ times.

With $1.411\ \text{g/cm}^3$ for the density of the Sun, such a degree of compression would result in material of the density $84890 \times 1.411 \approx 119780\ \text{g/cm}^3$.

That's close enough to the density of 120,000 of Sirius B, as to assume the star's matter still consisting of intact atoms, but it would convert fully or in part into "degenerate matter" if density went any higher.

10 Light and radiation

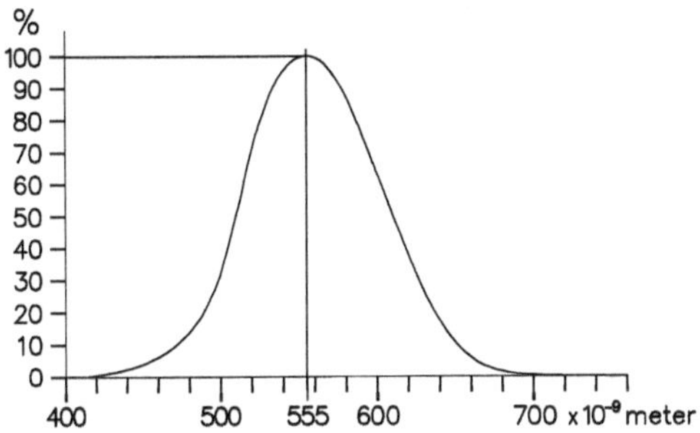

Fig. 10.1. Sensitivity of the human vision vs. wavelength of perceived light

Opposed to radiometry, which includes the full spectrum of electromagnetic radiation, photometry is restricted to the visible portion, including wavelengths from 360 to 830 nanometer, of frequencies between 360×10^{12} and 830×10^{12} Hz.

A first attempt toward measurement of light in the early eighteen hundreds was the creation of a standard light source, the *candlepower*, defined as a 7/8 inch diameter spermaceti wax candle consuming itself at the rate of 120 grains[1] per hour.

Spermaceti is a waxlike product extracted from the head substance of Sperm Whales and yields flames more luminous than those from candles of stearin or beeswax. A similar standard in Europe, called Hefnerkerze, equals 0.903 American candles.

Unlike such empirical definitions of luminous intensity, the modern unit of candela has been defined in line with the units of mechanics and electricity. Equaling approximately 1.02 candles, the candela is a light source emitting the equivalent of 1/683 W per steradian, or 18.3988 mW over a complete sphere centered at the light source. Left as a stand-alone definition, this would include the infrared and the ultraviolet bands of spectral wavelengths, which the eye cannot see. But even within the visual part of the spectrum, the relation of light perception to light energy (Fig. 10.1) varies widely and is highest at the wavelength of 555 nm or the frequency of 540×10^{12} Hz, and drops to zero below the wavelengths of 400 nm and above 700 nm. Noteworthy, the point of our vision's highest sensitivity coincides with the peak intensity of sunlight – an example of the adaptability of living beings to environmental conditions.

Vanishing grease blot

Such were the humble beginnings of *photometry*, the science of measuring the intensity of light. True to the definition of *measurement* as the comparison of a given

1 1 grain = 1/7000 pounds.

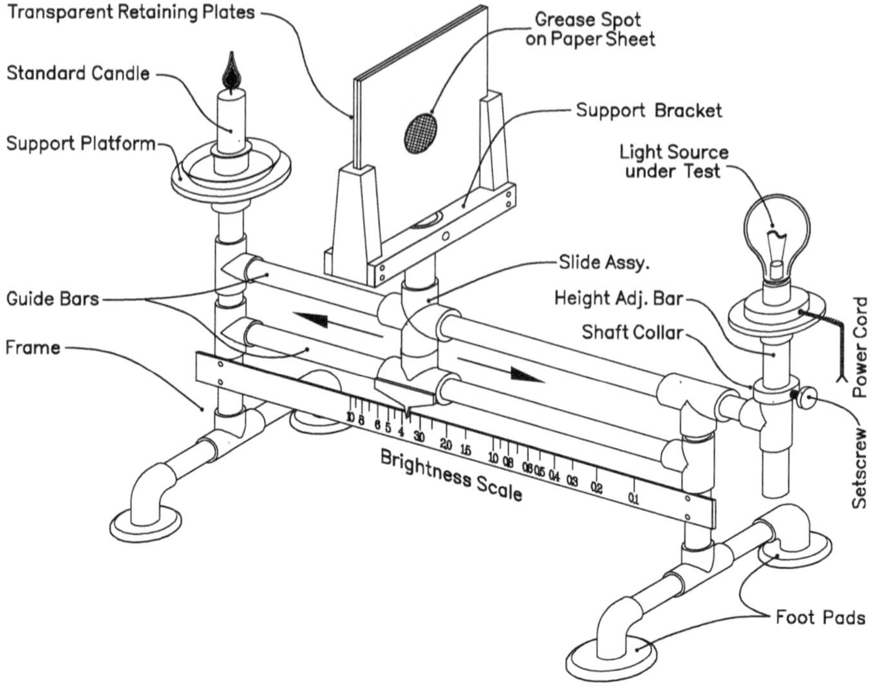

Fig. 10.2. Bunsen photometer

magnitude to an established standard, the *Bunsen grease spot photometer* (1843) does just that by means of a semitransparent sheet of paper with a central grease spot as detector. The spot appears dark on bright background if head-on illumination on the sheet exceeds that from behind, and bright on a grayish background in the reversed situation. Likewise, the spot becomes virtually invisible with equal incidence of light from both sides.

The simplicity of the Bunsen photometer in Fig. 10.2 makes it a tempting high school project your youngster could tackle with readily available PVC pipes and fittings, a can of CPVC cement, and some pieces of timber from Home Depot's scrap box.

The sheet of paper with the grease spot is mounted between a pair of acrylic panels in a wooden frame supported on a bracket slidable on a pair of guideposts. If the latters are 1/2 inch or 3/4 inch, the fittings for the bracket itself must be of the next higher gauge. With 1/2 inch Schedule #40 PVC pipes as guideposts, 3/4 inch fittings are used for the slide bracket and will need internal reaming by at least 0.016 inch in diameter. 1 inch fittings, used with 3/4 inch guideposts, need opening up by only 0.001 inch plus some clearance to allow for smooth riding.

Besides, the guideposts support the reference light source left, and the sample tray on the right. The latter's support base is vertically adjustable for best alignment between lights and grease blot.

Although the reference light source is shown as a candle, symbolizing the venerable candlepower, few of us will find spermaceti wax candles on the shelves of our favorite supermarket; and further, most of the lights we test these days would be bright enough to outshine a humble standard candle too much for good measurement. Rather, a bulb with its ratings in watts and in candelas marked on the package would make a convenient – if not absolutely precise – light standard.

If the grease spot (actually a flattened drop of oil absorbed into the paper) seems to disappear with the slide at x inches from the reference light source, the inverse square law gives the intensity of the probe as $[(d-x)/x]^2$, where d stands for the distance between probe and reference light source. In case, for instance, that the grease spot vanishes when the slide gets to $d/4$ from the standard light, the probe's intensity would be

$$\left(\frac{d-d/4}{d/4}\right)^2 = \left(\frac{3d/4}{d/4}\right)^2 = 9 \text{ times}$$

that of the reference light source. The scale on the photometer in Fig. 10.2, plotted to this formula, allows to read that ratio directly.

Tests should happen in a darkened room as to eliminate the effects of stray light from the surroundings. Still, a cardboard shield – the bigger the better – with a cutout for the support bracket, is helpful in preventing diffuse reflections to interfere with measuring.

The voltage for the tests should match the one specified for the lamp used as reference, because not all types of incandescent lamps react equally to deviations from the nominal 120 VAC.

The Bunsen photometer, a child of an era when "illumination" meant "candles", works fine with equally colored lights on both sides of the screen, but cannot reliably compare the intensity of differently colored lights. For instance, our vision perceives the luminous intensity of a yellowish flame as half that of a greenish one, though the energy of radiation is the same for both (Fig. 10.1).

Photometry aside, the Bunsen photometer allows for convincing demonstrations of the inverse square law. Put, for instance, one candle on the left, and 4 identical candles on the right, and you have to place the slide $1/3$ from the single candle for the grease-blot to vanish, because the square of $((2/3)/(1/3))$ equals 4.

Clap hands – here comes the semiconductor!

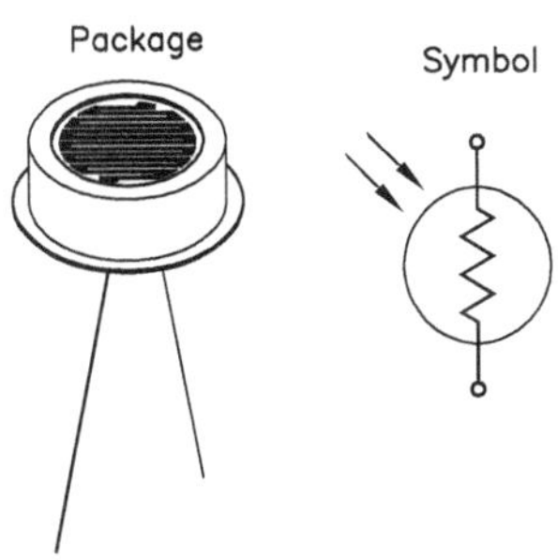

Fig. 10.3. Photoresistor

Although each of the three groups of light-sensitive semiconductors – photoresistors (LDR), photodiodes, and phototransistors – has its proper field of applications, photoresistors might be seen as the handiest of the lot. They are not polarized, work over a wide range of inputs, and respond to light of various wavelengths (colors) much like the human vision.

As shown in Fig. 10.3, most photoresistors rely on a meandering strip of cadmium sulfide (CdS), vacuum-deposited on a base of ceramic substrate and encapsulated in clear plastic. Their overall resistance varies with the brightness of incident light from about 100 ohm for

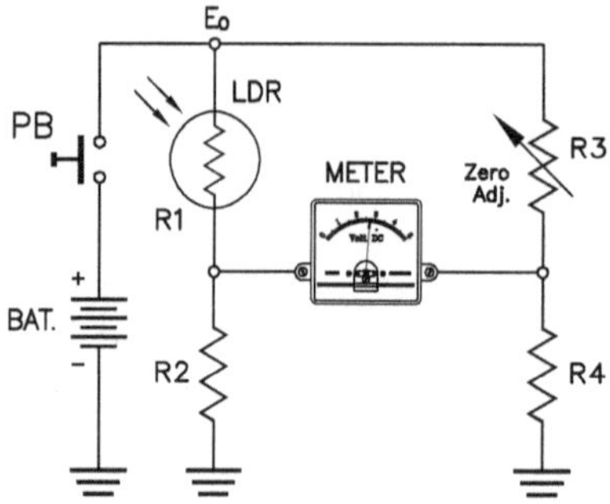

Fig. 10.4. Light meter circuit

fully illuminated to several megaohms in darkness. But their response times are about 20 ms (milliseconds) for rise and 30 ms for decay.

The wide field of applications of photoresistors includes commercial light meters, light activated switching devices, on/off controllers of street lights, shaft encoders, paper and coin handlers, counting systems, etc.

The basic light meter circuit in Fig. 10.4 has the LDR serial connected with a fixed resistor R_2 between the poles of a battery of nominal voltage E_o. Photoresistor R_1 and the fixed resistor R_2 constitute a voltage splitter, outputting at the R_1/R_2 junction the voltage $E = E_o R_2/(R_1 + R_2)$, from which we derive:

$$R_1 = R_2\left(E_o/E - 1\right)$$

Manufacturers' data sheets are the best sources for the conversion of R_1, the resistance of the photoresistor, into lumen.

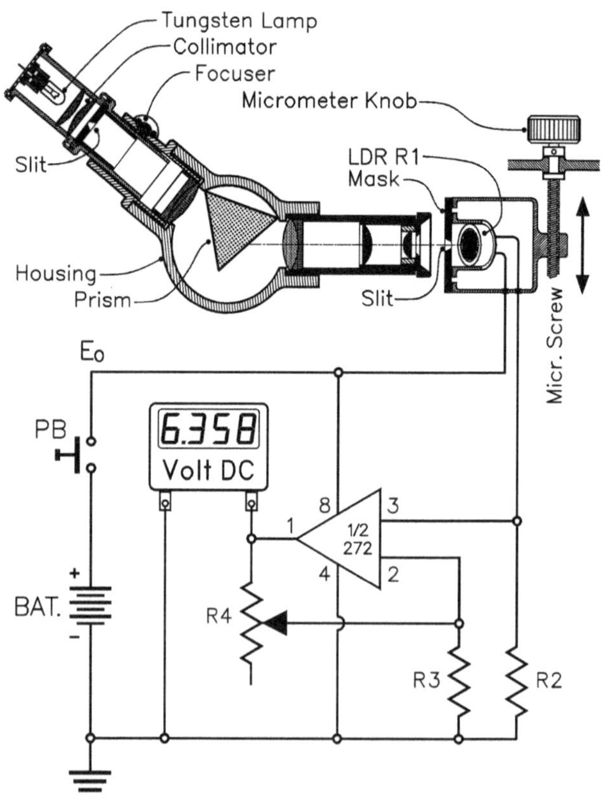

Fig. 10.5. Sensitivity measurement setup

The potentiometer R_3 allows for nulling the meter at *dark* or any other desired minimum light condition by setting it at $R_3 = R_4(R_1/R_2)$. That same circuit becomes a supply voltage independent bridge circuit if R_3 is made a scaled potentiometer for nulling the meter. The resistance R_1 of the photoresistor and with it the degree of illumination can then be derived from the position of the rotor of R_3, using once again the bridge equation $R_1 = R_2(R_3/R_4)$.

But radiation does not come as a uniform package of energy, but rather a host of discrete streams, each at a certain level of energy. These levels do not melt continuously into each other, but go in steps, defined by Planck's constant. That is the reason why Planck's law of energy distribution in the spectrum of black body radiation needed quantum mechanics for its deduction.

In applications where wavelength gets into the picture, such as the plotting of the sensitivity curve of human vision in Fig. 10.1, the setup shown in Fig. 10.5 is the solution. Here, a spectroscope projects the spectrum of light from a source of known filament temperature on a mask with a narrow slit, that shields the photoresistor except for the portion of the spectrum coincident with the slit's location.

An infinitely narrow slit, if such a thing existed, would pass light of only one single wavelength, while a slit of finite width covers an entire – if narrow – frequency band.

The term *spectral brightness* at a given wavelength thus refers to the integral light output from a host of wavelengths which bandwidth depends on the opening of the slit. The ratio of intensity vs. bandwidth is known as *radiant energy density*.

All said, the instrument in Fig. 10.5 measures the integral of radiation within a certain band of wavelengths, which width is given by the width of the slit in the darkening mask and the allover extent of the spectrum projected by the spectrometer. Turning the micrometer screw shifts the position of that slit gradually from one

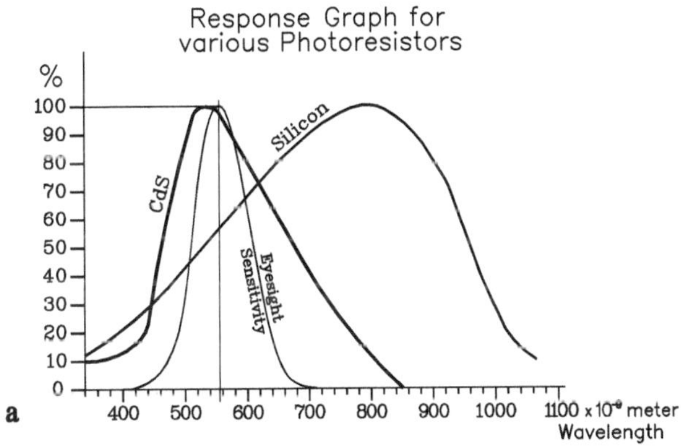

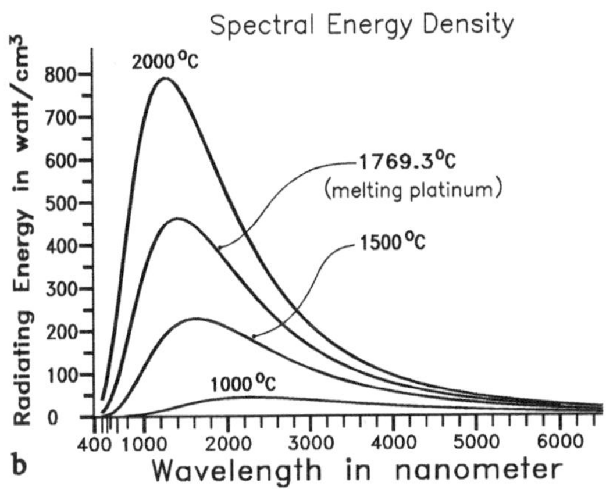

Fig. 10.6. **a** Sensitivity and **b** spectral energy graphs

end of the spectrum to the other, allowing for any number of readouts in between. The resulting curves in Fig. 10.6a have been plotted for typical CdS and silicon photoresistors. Details may vary between products from different suppliers.

Voltage gain of the Op-Amp is determined by the feedback resistors R_4 and R_3. If we set the potentiometer R_4 at, say, 840 kΩ and use 840 Ω for R_3, the circuit amplifies by a factor of $840000/840 = 1000$. R_4 can thus be used to boost or, respectively, restrict the gain of the device.

The graph for the relation of radiant energy and source temperature in Fig. 10.6b shows radiant energy far from constant over the band of spectral wavelengths. But the output we get from our spectral measurements represents the product of radiation intensity and the sensitivity of the photoresistor used as detector.

Since the degree of radiation depends on the temperature of the light source, a preliminary calibrated light bulb of known filament temperature must be used in the spectroscope. The relative sensitivity figures plotted in Fig. 10.6a thus represent the ratio of the instrument's readout and the radiant intensity at the wavelength under investigation (Fig. 10.6b) divided by the highest value among the results.

Because a standard galvanometer with, say, 20,000 Ω per volt, would practically short circuit the 1 MΩ potentiometer to ground, a >10 MΩ impedance (digital) instrument as shown in Fig. 10.5 must be used.

Using an outfit similar to Fig. 10.5, however with a punctured shadow mask instead of the slit, photographs of the Sun's surface in the light of one single wavelength or frequency can be obtained. With a CCD or a photographic plate in place of the photoresistor and the shadow mask set at the point of interest within the spectrum of the Sun, such as the Hα line of hydrogen, scanning the sun disk yields an image of solar hydrogen and nothing else. Unlike regular photographs that picture the Sun as a nearly uniform luminous disk with a few sunspots at best, spectroheliographic pictures reveal a surprising degree of details by focusing on radiation from one chemical element at a time.

Photodiodes, on the other hand, are used for scientific light intensity measurements, in medical applications such as detectors for computed tomography, and in remote control receivers of TVs and VCRs.

While common diodes conduct with the cathode on negative and the anode on positive, but block reverse currents, photodiodes lower their reverse resistance with the incidence of light on the junction. In that sense, they resemble photoresistors, but for their far shorter response times.

Silicon diodes excel by their wide range of working frequencies, from ultraviolet into the infrared, or 290–1100 nm of wavelength (1 nm equals 10 angstrom units), while gallium arsenide (GaAs) diodes operate in the range of 800–2000 nm.

PIN photodiodes consist of a thick base layer of N-type semiconductor with a thin layer of P-type material on top, separated by a thin deposit of intrinsic semiconductor material – hence the name. They excel by their frequency response of up to 10^{10} Hz.

In the absence of an applied voltage, photodiodes exposed to bright light develop a voltage of their own across the junction. Mind me, solar cells are basically arrays of oversized photodiodes.

Phototransistors are more sensitive than photodiodes, because light targeting the base-emitter junction polarizes the base much like the input signal in transistor amplifiers. The resulting emitter/collector current is then the base current times the gain of the transistor, of typically 100 and up. The low noise levels of phototransistors make high amplifications practical. On the other hand, their response is slower than that of photodiodes, and they conduct a small current even in the total absence of light.

Like standard transistors, early phototransistors were based on silicon (NPN) and germanium (PNP), while modern devices employ more exotic materials, such as gallium arsenide. Gain is typically from 50 to a few 100, but can be brought up to several thousand. However, the size of the base-collector junction, which as the light gathering element is much larger than in common transistors, leads to heightened parasitic capacity that limits the frequency response of phototransistors to approximately 250 kHz.

Darlington transistors, consisting of a pair of transistors connected in ways that feed the emitter-collector current of the input transistor through the emitter-base junction of the output transistor, thus multiplying the gain of both, also have their equivalent of phototransistors, but their frequency response is often as low as 20 kHz.

The photo-electric version of the *field effect transistor* (FET) is of interest where high input impedance is mandatory. Much as transistors conduct even in the absence of any emitter-base current, phototransistors conduct to a certain degree even in the total absence of light. This so-called dark current, amplified by transistor action, can be adverse to circuit performance.

Photomultipliers resemble traditional vacuum tubes if you replace their cathode by a photocathode and instead of one single anode use a series of ten or so, then called dynodes. Respective anode voltages come from a voltage divider, consisting of a cascaded chain of serial connected resistors.

Electrons emitted by the cathode bounce from dynode to dynode, gaining speed in function of the successively increased voltage of each. The release of new electrons with each impact amounts to an avalanche effect which makes that even a small number of photons at the input may trigger a measurable output. In combination with powerful astronomical telescopes, this has allowed to spot the faint galaxies populating the outer fringes of the known universe.

The finite velocity of propagation of light

In the early days of broadcast television, viewers were plagued by an annoying phenomenon called ghosts, or double images, caused by the simultaneous reception of signals directly from the transmission towers along with their reflections from nearby hills, buildings, or other obstacles. Although radio signals travel at the speed of light (300,000 km/s), the difference between the zigzag path of the reflected signal and the straight line from transmitter to receiver accounts for a time lag between the two signals of $1/300000 = 3.33 \times 10^{-6}$ or 3.33 microseconds per kilometer.

The 15750 Hz frequency of a TV's horizontal scan makes that the image-composing luminous spot takes $1/15750 = 63.5 \times 10^{-6}$ seconds to sweep the length of one line on the screen, so that 3.33 microseconds account for

$3.33 \times 10^{-6}/63.5 \times 10^{-6} = 0.052$ of a line. On a 21 inch TV screen with 17.5 inch line length, 1 kilometer of path-difference between the two inputs would have left $0.052 \times 17.5 = 0.91$ inch, or nearly 1 inch between Lucy and Ricky and the ghosts tracking them.

Annoying or not, this phenomenon is a firsthand demonstration of the finite velocity of propagation of electromagnetic waves, of which light makes part. Light transmission had been assumed instantaneous until the Danish astronomer Olaf C. Römer (1644–1710) figured the speed of light from observations of the Galilean moons of Jupiter. This largest planet of the Solar System circles the Sun in close to 12 years (4333 days), and thus wanders in each year approximately from one constellation of the zodiac to the next. At Römer's times, when mechanical timepieces lacked the precision called for in maritime navigation, celestial motions were used as worldwide time standards. The period of the eclipses of Jupiter's innermost moon, Io, tabulated by Giovanni Cassini, was one of them. In an attempt at refining Cassini's tables, Römer proved the finite velocity of light almost by accident.

Io orbits close enough to Jupiter to get eclipsed by the planet's shadow in every single one of its orbital revolutions. From Cassini's tables, Römer deduced that Io went through 225 eclipses in the 1.091 years of Jupiter's synodic period. From that, Io's average period of eclipses resulted in $1.091 \times 365.25/225 = 1.7711$ days or 42.50 hours. The 103 revolutions Römer could actually observe before the planet vanished in the glare of the Sun should thus have taken $103 \times 42.5 = 4377.5$ hours, but to Römer's surprise, Io's final observable eclipse was delayed by some 1450 seconds, or about 24 minutes.

At a time when the diameter of the orbit of Earth was thought to be 311 million kilometers, Römer took the 1450 seconds of delay as the time that light or the absence thereof from the eclipsing Io had taken to cover that distance, and figured the speed of light as $c = 311 \times 10^6/1450 \approx 214{,}500$ km/s. Though about 30% below the correct value, this was pretty close for a first attempt and, most important, obsoleted the idea of instantaneous signal transfer through space.

Over a century had to pass until direct measurements of the speed of light were attempted in 1849 by Hippolyte Fizeau (1819–1896) in cooperation with Léon Foucault. They sent the intense light from a hydrogen-oxygen lantern through the gaps between the 720 teeth of a rotating gear to a mirror on a hilltop, 8633 meters away. With the gear at a standstill, the mirror was carefully adjusted so that light from the lantern got reflected back through the same tooth gap it had passed through on its way out, and Fizeau could view its reflection in a 45° semi-silvered mirror through a telescope.

At slow rotation of the gear, the image would go on and off as the teeth successively eclipsed the light from the mirror. Speeding up the gear's rotation made that the eye could no longer separate the flashes until further acceleration made the gear turn by half a tooth-spacing in the time interval the light-beam took to travel from the lamp to the mirror and back. At that stage, gear teeth would successively block the returning beam, and the view through the telescope would darken. At double that critical rotation, the returning ray arrives just when the gear has turned by one full tooth spacing, bringing the image back to full glare.

Fizeau turned his toothed wheel either by hand over a speedup geardrive or by a weight on a rope slung around a drum. In 28 experiments, he recorded the first eclipse at 12.6 turns per second, which gave the average speed of light as 313300 km/s. The error was likely caused by lack of precision in timing the gear's revolutions.

Since then, atomic clocks and radar technology helped refining these early results to $c = 299{,}792{,}458$ meter per second for the speed of propagation of light and for this matter, electromagnetic waves in general; proven by the special theory of relativity as an invariable cosmic constant, it became the basis of a new definition of the meter, the length unit of the International System of Units. In 1983, the meter's former definition as the *ten millionth of the meridian quadrant* changed into *1/299,792,458 of the space light travels in vacuum during the time interval of one second.*

I see the light!

It all began in 1690 with Christian Huygens *Treatise on light*, which explained light as a series of waves in an omnipresent, yet ideally penetrable medium, called the aether. Though a contradiction by its very definition as matter with nonmaterial qualities, that lofty carrier of light persisted under the name of ether until the late eighteen hundreds, when the Michelson–Morley experiment disproved its existence by showing that the 30 km/s orbital velocity of Earth through the supposed ether did neither ad to nor subtract from the measured speed of light, which remained what it was in all orientations of the setup.

Incidentally, had the experiment come out according to the ether theory, it still wouldn't have shown the expected 30 km/s up and down from the average, but rather 250 km/s, which is the tangential component of our Milky Way galaxy's rotation at our sun's 30,000 light-years distance from its center.

The reason for the longevity of such a far-fetched idea as the world ether was that Isaac Newton's particle theory, though readily explaining light propagation through cosmic space, failed on diffraction, scatter, and polarization. On the other hand, Maxwell's theory, *Electricity and Magnetism* (published in 1873), supported the wave nature of light by encompassing it into the wider field of electromagnetic radiation, for which the *telegraph equation* mathematically predicted the formation of radio waves. Of universal importance however is the theory's expression for c, the speed of propagation of electromagnetic waves, as the inverse product of electric permittivity ε_o and the magnetic permeability μ_o of empty space, by the equation

$$c^2 = 1/(\varepsilon_o \mu_o)$$

Used with $\varepsilon_o = 8.85419 \times 10^{-12}$ and $\mu_o = 1.25664 \times 10^{-6}$, the equation yields $c = 2.99793 \times 10^8$ m/s, matching experimental results for the speed of light exceedingly well. Not enough that this is numerical proof of the convertibility of electromagnetic and light waves, the equation's independence of a frame of reference already suggested c as a universal constant. Thus was the birth of the tenet that led Einstein to postulate his special theory of relativity.

Previously, Albert Einstein's discovery of the photoelectric effect, namely, the interaction of light with electrons, suggested once again the particle structure of light, which led him in 1905 to assign wave as well as particle properties to light quanta, read photons. But Einstein didn't stop there. In his analysis of the laws of physics in the light of relativity, he developed a numerical expression for the relation between mass (like in particle) and energy (like in wave energy), the well-known equation: $E = mc^2$, which became one of the pillars of atomistics.

While photons traveling through cosmic space are elementary energy packages, they convert into wavetrains while interacting with matter. With $c = 3 \times 10^8$ m/s, the length of typical wavetrains, known as *coherence length,* follows from their lifespan of approximately 10^{-8} seconds as $10^{-8} \times 3 \times 10^8 \approx 3$ meter.

That makes "light as we know it" a sequence of flashes of various lengths, too fast for our eyes to separate, and yet, unable to interact with each other because no two flashes happen at precisely the same point in time. Unlike the Hertzian (radio) waves and the coherent light rays from lasers, everyday light waves do not produce interference patterns unless one and the same ray is made to interact with its own image. Good for us. Would it be otherwise, interference would make our morning paper's crisp letters into psychedelic pictures framed by the colors of the rainbow.

The *interferometer* in Fig. 10.7 uses a semi-silvered mirror, tilted by 45°, to split each beam from a light source into two, as shown individually in details 1 and 2 of Fig. 10.7. A ray from the light source gets partially reflected by the semi-coated, tilted mirror up to mirror 1 and from there into the observation telescope. The remainder passes straight through the tilted mirror onto mirror 2, gets reflected back on the same tilted mirror and from there into the telescope. Both components arrive simultaneously if mirrors 1 and 2 are equidistant from the center of the tilted mirror, and interference occurs over the full length of the wavetrains.

Detail 3 in Fig. 10.7 shows what happens if mirror 2 is shifted just enough to make one beam's path longer than the other by the waivetrain's coherence length. In that case, the waivetrains arrive at the telescope one after the other and cannot interfere with each other.

The light beam's coherence length is thus determined as the displacement of mirror 2 that makes all interference rings disappear.

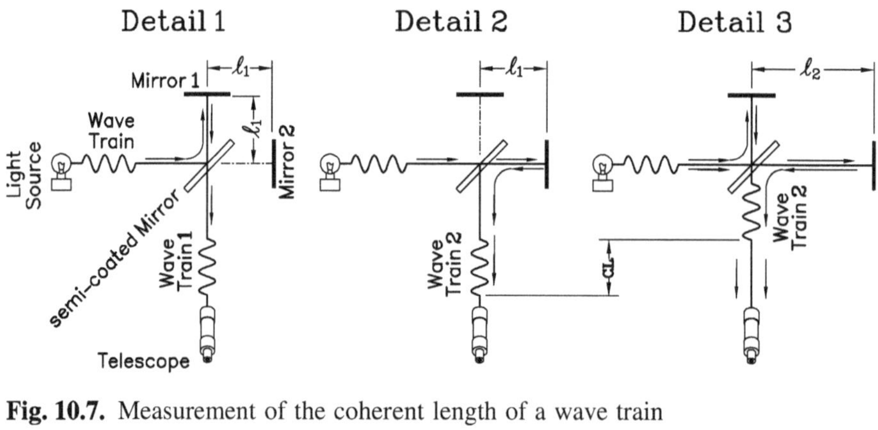

Fig. 10.7. Measurement of the coherent length of a wave train

The enchanted land of lenses

If you acknowledged the first view through your offspring's astronomical telescope with a yawn, don't despair. You fell victim to the Users Manual syndrome. At the rated 200 times or so of the instrument's highest magnification, you expected to see the dust lanes in the Andromeda galaxy and the rest of the wonders of the heavens much like on high-gloss NASA photographs, though chances are that no dust lanes, and for that matter, not even an Andromeda galaxy will show up in your telescope's field of vision.

It always takes a portion of good fortune to spot a celestial object at the instrument's highest magnification. Rather, one should begin with the weakest eyepiece and gradually shift to stronger ones without letting the object of observation out of the telescope's field of vision. This is not an easy task, first because the field of vision becomes smaller with every step-up of magnification, and second because not all tripods are steady like the Khéops pyramid; and last but not least, the rotation of Earth constantly shifts your celestial object away from the point last seen, unless you have a clockwork-driven parallactic mount on your tripod or, better still, a computerized star tracking system.

In times when astronomical telescopes, such as Fig. 10.8, weren't so incredibly cheap as today, many amateurs were building their own. Even lenses of outdated picture projectors had to do as objectives, while eyepieces were rescued from surplus binoculars from the great wars.

The mechanical simplicity of the typical astronomical telescope is a sure invitation for the "do it yourselfers",

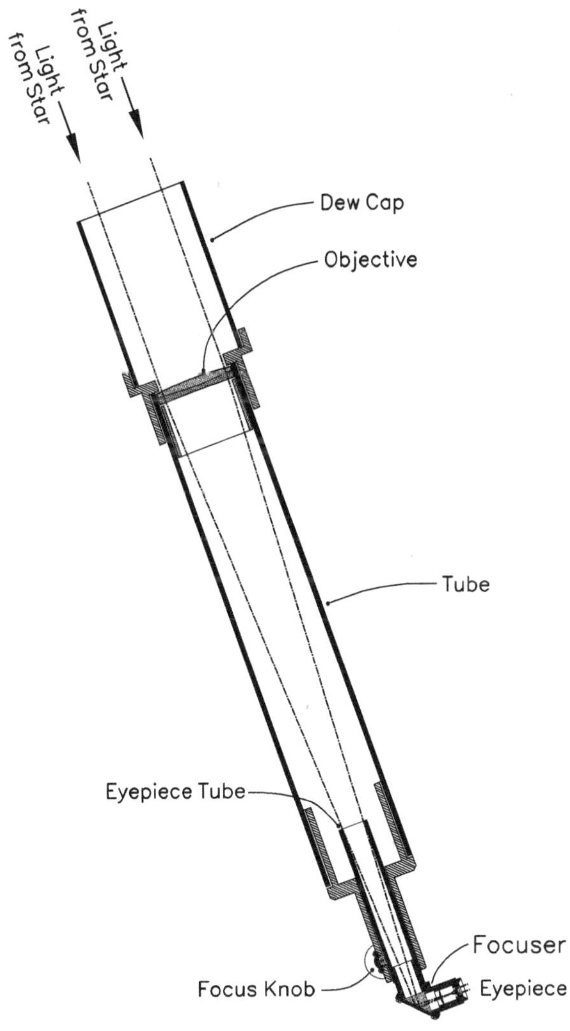

Fig. 10.8. Refracting astronomical telescope

but unless the builder has access to a not too small engine lathe and some other hard to get outfits of a mechanical workshop, lack of accurate machining becomes often the main reason for the poor performance of many first-time telescopes.

Within this concept, an interesting idea was presented in 1937 by the out of print book *Astronomie, endlich einmal praktisch* (The practical side of Astronomy), suggesting that builders boldly switch to square tubes of well dried, planed timber boards. The resulting boxlike instruments may not have evoked the usual outcries of admiration from innocent bystanders, but photographs of moon craters and sunspots taken with those edgy instruments competed successfully with those from far more costly equipment.

Building a telescope should start with an estimate of its expected magnification M from the expression $M = F/f$, where F is the focal length of the objective lens, and f that of the eyepiece. A slightly above 1 meter long telescope, equipped with an objective lens of 100 cm focal length and a 2 cm eyepiece thus gets you $100/2 = 50$ times magnification; enough to resolve planet Jupiter into a disk and to show its four Galilean moons as a row of goose-stepping little stars; and likewise the phases of Venus, Saturn's rings, and of course the craters on the moon. Stronger eyepieces of, say, 1 cm focal length provide larger images, but often at the expense of acuity – read sharpness.

Next step up the learning curve comes with the law of the light gathering power of lenses. Divide the diameter of the objective lens by the diameter of the eye's dark-extended pupil, of typically 8 mm (5/16 inch), and square the result. Even a modest 3 inch objective thus provides a luminosity gain of

$$\left(\frac{3}{5/16}\right)^2 = \left(\frac{3 \times 16}{5}\right)^2 = 9.6^2 \approx 92\,,$$

theoretically showing nearly 5 orders of magnitude fainter stars than the naked eye. Wow! From points of observation with really clear skies, where stars of 5th and sometimes even 6th order of magnitude are visible to the unaided eye, a 3 inch telescope should thus show everything down to 10th or even 11th magnitude.

For comparison, 9.5^{m} had been the limit magnitude in the famous *Bonner Durchmusterung*[2], which after 11 years of observations got published in 1859. It showed the positions of 324,188 stars on the northern hemisphere with an accuracy of 0.1 s in right ascension, and 0.1 minutes of arc in declination, and gave magnitudes with 0.1 mag tolerance.

For many decades, the Bonner Durchmusterung remained the number one and only accurate net of star positions on the northern sky and was basic for research on fixed stars' proper motions and the changes in magnitude of variable stars. It got periodically updated and reissued into our days.

But there is a limiting factor to the gain of luminosity a telescope can provide. I'm not talking the suspended dust in the atmosphere that lately cut the number of astronomer-friendly locations down to a precious few, the summit of Mt. Everest

2 sky survey of the observatory in Bonn, Germany.

among them. It is that the real big gain in brilliance is peculiar to fixed stars, because they remain punctiform at whatever magnification. In other words, a star is a star, whether seen with the naked eye or through a fabulously expensive telescope, and presently achievable magnifications don't get it beyond that status.

Even the closest fixed star, Alpha Centauri, which two main components are about as big as our Sun, would not be an exception. The star's distance from us is 4.34 light-years, or $4.34 \times 365 \times 24 \times 60 = 2{,}281{,}000$ light-minutes, while light from the Sun reaches Earth within 8.3 minutes. From that and the 30 minutes of arc apparent diameter of the sun as seen from Earth, we figure the apparent diameter of an Alpha Centauri star as $30 \times 8.3/2281000 \approx 0.00011$ minutes of arc, or 0.00655 arc-seconds.

Through a 1000× telescope, this would be seen as 6.55 arc-seconds or about $1/10$ of an arc-minute, while the acuity of human vision lies at best around 0.5 to 1.0 arc-minutes, as every observer – amateur or professional – knows from proper (and sometimes disheartening) experience. Thus, a 10,000 times magnification telescope would have to be built to make Alpha Centauri's image into a disk, provided we get perfectly sharp images from such a hypothetical mammoth instrument. If not, the star might still resemble nothing better than a spread-out point of light.

While the image of a fixed star accumulates all the light there is, objects of disklike appearance, such as planets and nebulae, lose brilliance to the tune of enlargement. A 50× telescope, sized for 100 times light gain, would boost the luminosity of planets and nebulas by only about $100/50 = 2$ times, with part of that gain still lost through the less than 100% transparency of the lens system.

That's what makes it an art to spot some of the simmering blobs of nebulas, galaxies, and star clusters, known under the collective noun *Messier Objects*. Experienced Messier hunters often use "averted vision," based on the idea that a glow – too faint for the color-sensitive cones at the center of our retina (fovea centralis) – might still be enough to excite the monochromatic rods on the margins.

Light-gathering power is not the only factor for the trend towards tubes of ever greater diameter. So far, we considered the light beams in optical instruments as straight lines, but that too is a simplification. When light rays skim the edge of the frontal opening of a telescope, they give birth to new wavelets which propagate at diverting angles from the mainstream and cause the omnipresent halo around star images even in top of the line telescopes. For an instrument of diameter D and focal length f of the objective lens, the spread d in the size of the image is given by $d = 2.44\lambda f/D$. For instance, an instrument of 10 meter focal length and a 1 meter diameter objective lens, focused on a star shining at a wavelength λ of principally 555 nm, is expected to spread out the star's image from punctiform to an *airy disk* of $d = 2.44 \times 555 \times 10^{-9} \times 10/1 = 13.5 \times 10^{-6}$ meter or 13.5 μm of diameter.

The simple telescope Galileo Galilei built in 1609 was supposedly 30 to 40 inches long and had a concave eye-lens of 2 inches virtual focus. Unlike the upside down images you get from telescopes with convex lenses on both ends, the virtual images seen through concave eye-lenses stand upright. But such telescopes' principal shortcoming is their extremely narrow field of vision – only $1/4$ degree of arc in Galilei's telescope. Nevertheless, it showed him the four inner moons of Jupiter (collectively named after him), the rings of Saturn, the craters on the moon, sunspots, and 2 years later, the phases of the inner planet Venus.

Most important, it allowed him to resolve the continuous glow of the Milky Way into the concerted light-output of a great number of individual stars.

The possible use of convex lenses in eyepieces is first mentioned in Johannes Kepler's thesis *Dioptrice* (1611), which also predicts image inversion; but only in 1630 Christoph Scheiner used throughout convex telescopes for projecting images of the Sun on a screen.

As far as astronomical observations go, top-down images are acceptable, because the concepts of up and down are strictly earthbound. That north is upwards on our globes and maps stems from conventions rather than the laws of nature. Frequent observers get so used to inverted imagery, that it ceases to bother them even on terrestrial objects. For the rest of us perishables, Kepler already suggested the use of an additional convex lens between objective and eyepiece, set for projecting the image from the objective on the focal plane of the eyepiece. This doubling of inversions makes for an upright image seen through the eyepiece.

The 90 degree roof prisms in right angle eyepieces too provide upright images, but sideways inverted. For a truly upright image, two such prisms, installed in the so-called Porro configuration, must be used, as it is the case with binoculars.

Binoculars

Binoculars show things the way we see them – only much bigger. The simplest form of binoculars is an assembly of two Galilean telescopes in parallel, providing

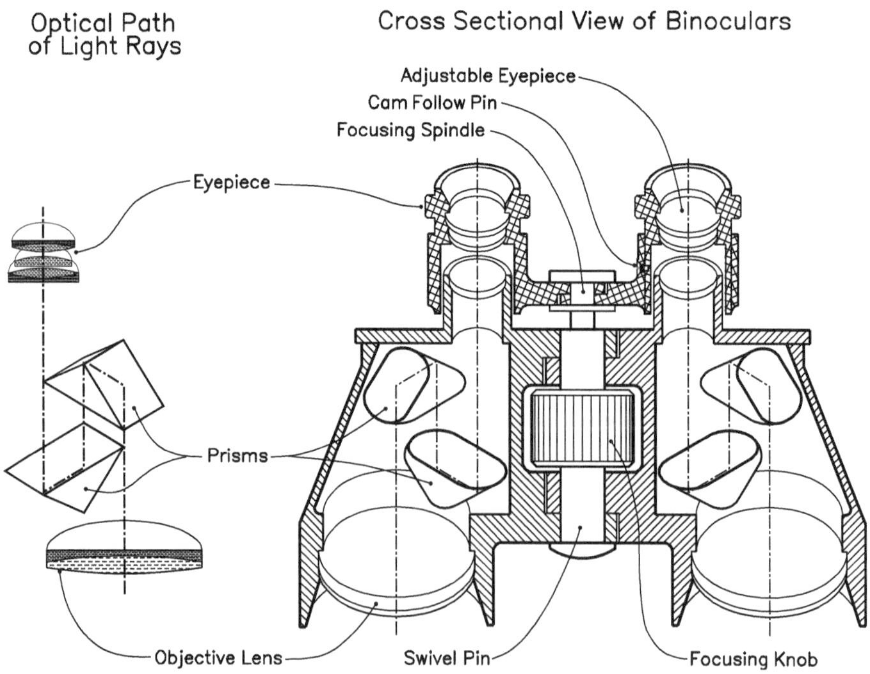

Fig. 10.9. Prismatic binoculars

an upright stereoscopic image within the characteristically narrow field of vision of that kind of instrument. Magnification seldom exceeds 2 times, since anything higher would show reality in bits too small for comfort. Named opera glasses for their frequent use in classical theaters and opera houses, they couldn't keep up with the sprawling movie theaters, whose broadwand projections couldn't be outdone – even if you brought the Mt. Palomar telescope with you.

To enhance the in depth perception of human vision, binoculars use a pair of right-angle (Porro) prisms (Fig. 10.9) to distance the objective lenses beyond the spacing of our eyes. The image inversion inherent to such prisms allows for positive eye lenses and yet shorter tubing than their direct view cousins of equal power, because binoculars' twice broken light path adds length to the optical distance between objective and eye lenses.

For adjustments of the eye distance, the two halves of binoculars are hinged with each other, wherein the linchpin doubles as micrometer screw for concerted focusing of both eyepieces. Additionally, one eyepiece, usually the right one, is guided in a helically grooved socket, which makes it slide for- and backwards as you rotate it. This allows for correction of possible differences between the user's left and right eyesight.

The ratio of objective diameter to magnification gives the diameter of the *exit pupil*, which should be kept somewhat below the diameter of the eye lens, of 8 mm while dark adapted, and 3 mm at daylight. Lately, 5 mm exit pupils, such as in 10×50 or 8×40 binoculars, became popular, while the older 7×50 models fell out of grace. For daytime observations, like bird watching, a 3 mm exit pupil is sufficient.

Readying a pair of quality binoculars should start with adjustment of the eye distance while watching a piece of sky or any other clueless sight until the two fields merge into one. Next the central focusing knob is rotated to bring the left eye image in sharp focus. That should automatically focus the right side image as well, but if it remains less than sharp, the socket of the right eye lens must be rotated to correct the discrepancy. From then on, further focusing on different targets should be possible with the central focuser alone.

The roots of white light

Stories abound on Isaac Newton's use of prisms to dissect sunlight into the colors of the rainbow, and the ensuing theories, which in later, not always totally rational times, had to compete with Johann Wolfgang von Goethe's philosophy-bordering *Farbenlehre* (Theory of Colors).

A prism, such as in Newton's experiments, generates a spectrum because of the peculiar property of light rays to follow the path of *shortest travel time* rather *shortest distance* (Fermat's Law of least time). The situation resembles that of a swimmer who jogs every morning from her inland tent to her boat, anchored in deep water some way down the beach. Since she swims slower than she jogs, following a straight line would not be her best bet, because it leads over too much water. Another option, jogging to a place opposite the boat and swim straight out, would add too much mileage to the distance "as the crow flies." The path of shortest time for reaching the boat from the tent lies somewhere between those two extremes. As it turns out, the angles α and β that the legs of such a "shortest time path" subtend with a line perpendicular to the coastline, follow the simple relation $\sin\alpha/\sin\beta = c_1/c_2$. If you take c_1 as the camper's speed of jogging, and c_2 as the speed at which she swims, $(90-\alpha)$ and $(90-\beta)$ are the angles between the coastline and the directions of her jogging and, respectively, swimming.

Applied to the passage of light through transparent material, this same formula yields the material's refractive index n as the ratio of the speed of light in said material to c, the speed of light in vacuum.

While the velocity of photons, the carriers of light in empty space, is a universal constant, light waves, generated in the interaction of photons with matter, travel at only a fraction of c, namely, c/n. With c lowest for red and highest for blue, this explains why Newton's prism deflected the blue portion of sunlight the most, and reddish-yellow the least, with green somewhere in between. Note that pure red and purple do not make part of the spectrum of white light.

An easy way of measuring the refractive index of transparent material, such as various kinds of glasses, is to insert a probe of thickness d into the light path of a microscope, previously focused on an easily definable object, such as a micrometric scale, etched on glass. Seen through the glass sample, the image appears blurred, but can be brought back into sharp focus by lifting up the stage or lowering the objective by Δh. The refractive index of the inserted glass plate is then $n = 1 + \Delta h/d$. As the late Professor Heinrich Mache put it: "The passage of light in glass is weightier than in air."

With 1 for the refractive index of empty space, it is 1.00029 in air. For water, it is 1.33, while crown glass and windowpanes weigh in with 1.52. For heaviest flint glass, n equals 1.89, and for diamond 2.42. Therefore, even the glint of a crystal of heaviest flint glass makes a deplorable substitute for the sparkle of an $n = 2.42$ diamond, as eternalized in the Brazilian folksong:

O anel que tu me deste
Era vidro – se quebrou.
O amor que prometeste
Era assim – se acabou.
(But the ring, to me bestowed,
Was just glass – and cracked asunder.
Oh – the love for me, you vowed,
Fared no better – and went under.)

However, the different n-values of flint and crown glass are vital in the design of objectives for photographic cameras, telescopes, and microscopes, because similar to a prism, the angled surfaces of a lens split the white light into its constituents. Telescopes with a single lens objective generate images bordered by the colors of the rainbow, and single-lens cameras fatally produced washed out babyfotos.

Therefore, objectives of cameras, telescopes, and microscopes consist of at least two lenses: a crown glass biconvex lens, cemented with optical gelatin to a plane-concave lens of flint glass. Due to their different refractive indexes of 1.52 vs. 1.89, the color-selective properties of the concave lens counteract those of the

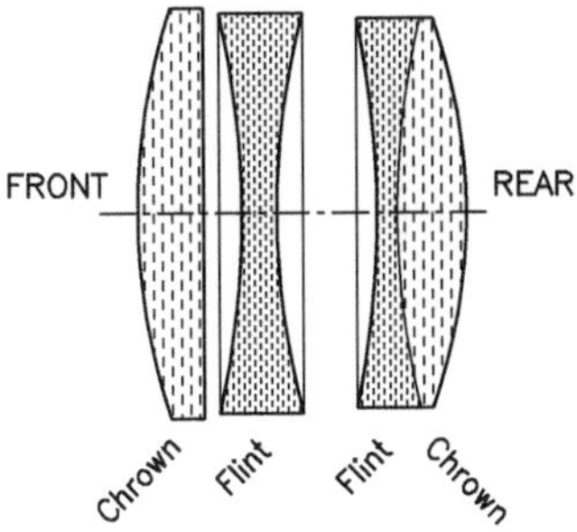

Fig. 10.10. Lens arrangement in a Zeiss Tessar photographic objective

biconvex lens, generating images surprisingly free of parasitic colors.

Critically speaking however, the focal points of no more than two wavelengths of light can be made coincident that way, while the colors of the remaining wavelengths are still showing to different degrees. In times when sensitivity of photographic films and plates peaked in cyan and blue, objective lenses for cameras were corrected at shorter wavelength than the green centered of visual instruments, such as telescopes and binoculars. Likewise, the objective lenses of instruments for astro-photography were of different design from those of visual telescopes.

Adding a third lens allows for gathering three wavelengths, normally red, green, and blue, in a common focal point where they recombine into white. Such lens arrangements, called apochromatic, were the predecessors of the famous *Zeiss Tessar* photographic objective lens (Fig. 10.10), consisting of four lenses in three groups: A positive crown glass lens up front, followed by a biconcave (negative) flint glass lens in the middle, and a cemented pair of biconcave and biconvex flint and crown glass lenses facing the photographic film or plate. Designed in 1902 by Paul Rudolf, Tessar lenses are still going strong in quality photographic equipment, including digital cameras, where 6 or 8 MB of resolution would have little effect unless equaled by the acuity of the camera's objective. After all, a washed out image can't be sharpened by projecting it on a plethora of megabytes.

But that's not the end of lens problems. Since the focal length of convex lenses remains essentially the same in all directions, a well focused image is spread over a spherical surface, not unlike the maps on a globe, while the image sensors in digital cameras and the film surface in the others are, of course, strictly planar. Therefore, a well focused photograph may lose sharpness from the center outward, unless produced by an anastigmatic objective.

Photography aside, anastigmatic lenses are used as wide field (over 2 inch diameter) magnifying glasses.

Altogether, $x + 1$ lenses are afforded to correct x types of defects, but that shouldn't make us rate an instrument's objective by its number of lenses. We could reach the point where *transmission losses* exceed our gains. As it happens, a small fraction of the light striking a lens is being reflected rather than transmitted and is invariably lost. According to Fresnel's law, the intensity of light transiting from a media of refractive index n_1 into another of refractive index n_2 gets reduced by

$$r = \left(\frac{n_\lambda - 1}{n_\lambda + 1}\right)^2,$$

where n_λ stands for the fraction n_2/n_1.

Minimizing such losses is the reason for cementing lenses together, such as the popular pair of crown-glass and flint-glass in achromatic objectives. With $n = 1.52$

for crown-glass and $n = 1.89$ for flint-glass, we get

$$n_2/n_1 = 1.89/1.52 = 1.24, \quad \text{and} \quad r = \left(\frac{1.24 - 1}{1.24 + 1}\right)^2 = 0.011\,,$$

or 1.1% of reflection losses for a glued pair of lenses.

If the same lenses were used separately, we would get the loss crown-glass/air and subsequently that of air/flint-glass, namely, $1.52/1.00 = 1.52$ and $1.89/1 = 1.89$, resulting in $(0.52/2.52)^2 = 0.0425$ and, respectively, $(0.89/2.89)^2 = 0.095$, a total of $0.0425 + 0.0950 = 0.1375$. That is 13.8% – or more than 12 times the losses of a cemented pair!

Reflecting telescopes

In the light of the complexity of lens design, the use of curved mirrors instead could be seen as the panacea to end all our troubles. Mirrors treat light of all wavelengths equally und therefore are inherently achromatic – or apochromatic if you will – and relatively cheap.

Mathematical proof that light-rays parallel to the center-axis of a parabola get reflected into a single focal point is easy to do. But the step from mathematics down on the shop floor is surprisingly tough. To begin with, most concave mirrors, such as we know them for their ability of vilifying our faces while we shave, are not parabolic but spherical, because only spherical shapes can be produced simply and inexpensively by pairing convex and concave blanks to grind each other into spherical shape.

The convex blank is hereby sealed to a "dop stick," which other end the operator – human or machine – guides through an epicycloidal path, including a number of small loops in the course of slowly moving through a larger circle. With the help of emery paste, the two halves grind and smoothen each other into nearly perfect spherical shapes. Early eyeglasses, magnifiers, microscopes, and even telescopes testify the precision inherent to this rudimentary manufacturing process.

Primary telescope mirrors – on the other hand – might still be pre-ground that way, but need finishing into a paraboloid with a degree of precision measured in fractions of the wavelength of light. Blanks for real big mirrors are being cast in rotating molds, which spin just fast enough for the melt to settle into a parabolic surface of the desired depth; success depends on the smoothness of the mold's rotation. Still, final grinding and polishing to typically 0.05 arc-seconds acutance has to follow.

But *getting* an image is not the only design challenge of reflecting telescopes. *Seeing* that image without blocking the telescope's mouth with your head is the other. The biggies among reflecting telescopes used to place the observing astronomer like a power-line repairman in a basket and crane him into the instrument's focal point.

Newtonian reflectors

Different from big reflecting telescopes, the more modestly sized Newtonian reflectors (Fig. 10.11) employ an elliptical 45° diagonal mirror to reflect the image sideways into the eyepiece at the front end of the tube. This placement often leads to the

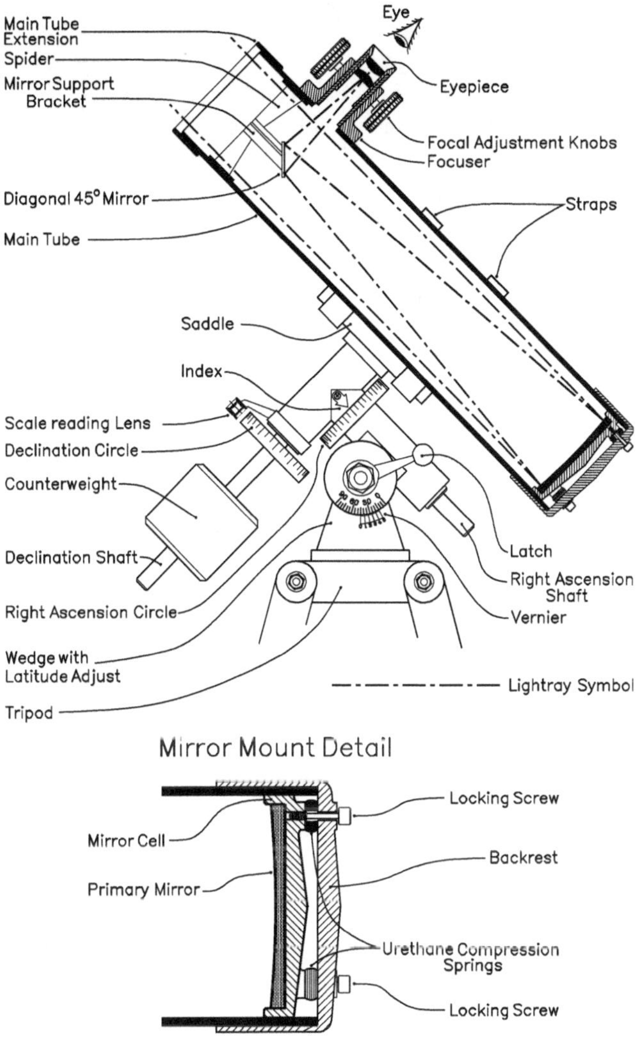

Fig. 10.11. Newtonian telescope

need for stepstools, ladders, and similar equipment for "comfortable" observing. Occasionally, even the tube must be rotated in its straps to keep the eyepiece from pointing straight up into the night.

Not surprisingly, the diagonal mirror and its mounting hardware block some of the incoming light but locate too close to the primary mirror for showing up in the eyepiece. Likewise, we don't see the mirror support bracket and the *spider*, the diagonal's three- or four-legged cross-support inside the tube. For minimal obstruction of incident light, the legs of the spider are S-shaped, made from over-the-edge-mounted flat steel strips.

The parabolic mirror is cemented into the mirror cell, which itself is bolted to the rear-end cover of the main tube with compression springs or thick rubber or Urethane washers as spacers. Prestressed, they allow for adjustments in the mirror's angular position by tightening or untightening the bolts during alignment of the primary mirror, the diagonal, and the eyepiece. In a well collimated telescope, a laser beam aimed through the eye lens into the telescope should get reflected by primary and secondary mirrors back into its point of origin.

The least demanding way of installing the telescope on a tripod is the *alt-azimuth mount* we remember from terrestrial telescopes, such as the ones charging quarters for a view at New York's skyline or Alaskan glaciers. The shortcoming of this economic mount is the need for changing the horizontal as well as the vertical pointing of the telescope to follow the daily motion of the skies which, boosted by

100 or 200 times magnifications, makes your images run out of the field of vision under your very eyes.

Equatorial or parallactic mounts as in Fig. 10.11 let you follow the daily rotation of the sky by swinging the instrument in turn of one single axis, called the hour-axis. In principle, they resemble an alt-azimuth mount tilted to the point where one of its shafts parallels the axis of rotation of the Earth and thus points at the center of the apparent orbits of all stars and planets, the *celestial pole*, marked in northern latitudes by the close-by Polaris star in the Little Dipper constellation. Powering the telescope's rotation in turn of the hour-axis by a clock motor freezes your objects permanently into the field of vision.

The other shaft, mounted at right angles to the hour-axis, is the *declination axis*, used to set the angular distance (declination) of the object of observation from the celestial equator. Like latitude, declination is given in degrees and fractions thereof. That makes the setting of the declination circle a straightforward operation, while the hour circle must be set to the hour of observation in *sidereal time* minus the object's right ascension, which is the reason for division of that circle in hours, minutes, and seconds rather than degrees. The conversion factor between the two scales is 15, that is 1 hour equals 15°, 1 minute equals 15 minutes of arc (15′), and 1 second equals 15 seconds of arc (15″).

The sidereal timescale is based on the uniform rotation of the Earth rather than the apparent positions of the Sun; its relation to local time changes by location and throughout the year and must be looked up in periodically published astronomical tables.

Several sidelines of alt-azimuth and parallactic mounts are available. Among the former, the *Dobsonian mount* replaces the horizontal axis of the alt-azimuth mount by a pair of studs installed at the center of gravity of the telescope, which makes counterweights unnecessary. Bearings for sustaining the instrument consist of circular wooden disks sliding on Teflon pads, a combination that provides smooth tilt and yet keeps the instrument firmly in position if left alone.

A variety of the equatorial mount puts it all – instrument and observer as well – on a platform supported on two shaft stubs mounted in line with each other and the celestial pole. Powered by a hefty clock motor, the platform keeps track with the motion of the stars over the sky, but tilts with the passing of time with everything on it – tacked and nailed or not – and must be reset from time to time unless it becomes a toboggan. Back in the horizontal, refocusing of the telescope on your object of observation is needed.

Schmidt–Cassegrain telescopes

Short length instruments, such as the Schmidt–Cassegrain telescopes (Fig. 10.12) rest usually in a furcated bearing-support for the declination axis. Their overall length is less than half that of equivalent Newtonians, because light-rays from the object of observation pass through the tube twice. Reflected by the primary mirror, they strike the secondary mirror, a slightly convex mirror set a small distance inwards from the primary mirror's focal point, and from there get reflected back through a central orifice at the center of the primary mirror into the eyepiece.

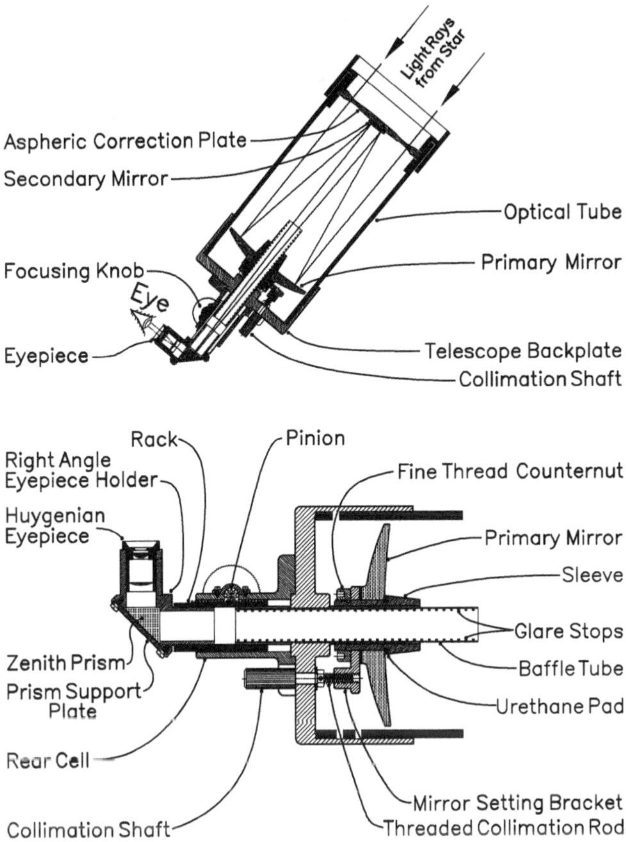

Fig. 10.12. Schmidt–Cassegrain reflector

A flat mirror as secondary would measure about half of the tube's diameter, but a convex mirror can be smaller and becomes less of an obstruction to incident light. Moreover, its convex profile brings the virtual focal length of the instrument up to the value commensurate with the vertex angle of the cone of light from that mirror to the eyepiece, making the optical length of Cassegrain telescopes exceed the length of the tube by a factor of 2. Magnification of Cassegrains is therefore far above that of Newtonian reflectors of comparable size.

All those extras and the convenience of observing from the rear end of the instrument must be paid for with the need for optical components of higher precision. Cassegrains could thus been dubbed "the rich guy's choice", and even more so the Schmidt–Cassegrain telescopes, which have a full-size lens of complex profile mounted at the open end of the tube, specifically designed to make up for the sins of the instrument's other optical components.

From extreme to extreme

Just like your first glance through a telescope may have been disappointing, the view of a "world too small to be seen" through an affordable microscope may not fare better, though the problem could be that you bought a Ford while hoping for a Learjet. Spoiled by images of microcosmical creatures in books and magazines, we tend to blame the instrument at hand for not seeing what, in reality, it cannot show; because even the best brands of microscopes cannot generate an image of higher resolution than about half the wavelength of the light illuminating the probe.

Anton van Leeuwenhoek (1632–1723) is known as the designer and builder of magnifying lenses, by then called microscopes, but the compound microscope as we know it still took many centuries of technological development, and only in 1880 presented Ernst Abbe (1840–1905) an instrument with usable magnifications of up to 2000 times. And that's about as far as it can get: The 555 nm length of the light wave of highest intensity in the spectrum of sunlight – green – enlarged 2000-fold, becomes $2000 \times 555 \times 10^{-9} = 1.11 \times 10^{-3}$ meter $\approx$ 1 mm long, a degree of resolution you would call blurred on photographs of your wife leaning on the new (and not yet paid for) family car.

Opposed to one-lens magnifiers, the compound microscope consists – much like the telescope – of a system of at least two lenses: The objective and the eyepiece. Owners of autofocusing cameras are familiar with seeing the objective lens popping out the farthest while aimed at objects close by. Veteran photographers remember the 6 by 9 or 9 by 12 cm direct view cameras with double length expansion bellows that could show objects at their true size and beyond. Microscopes aim still higher than that und employ short-focus objectives to throw highly enlarged images of the probe toward the eyepiece, which then multiplies overall magnification by its particular magnifying power.

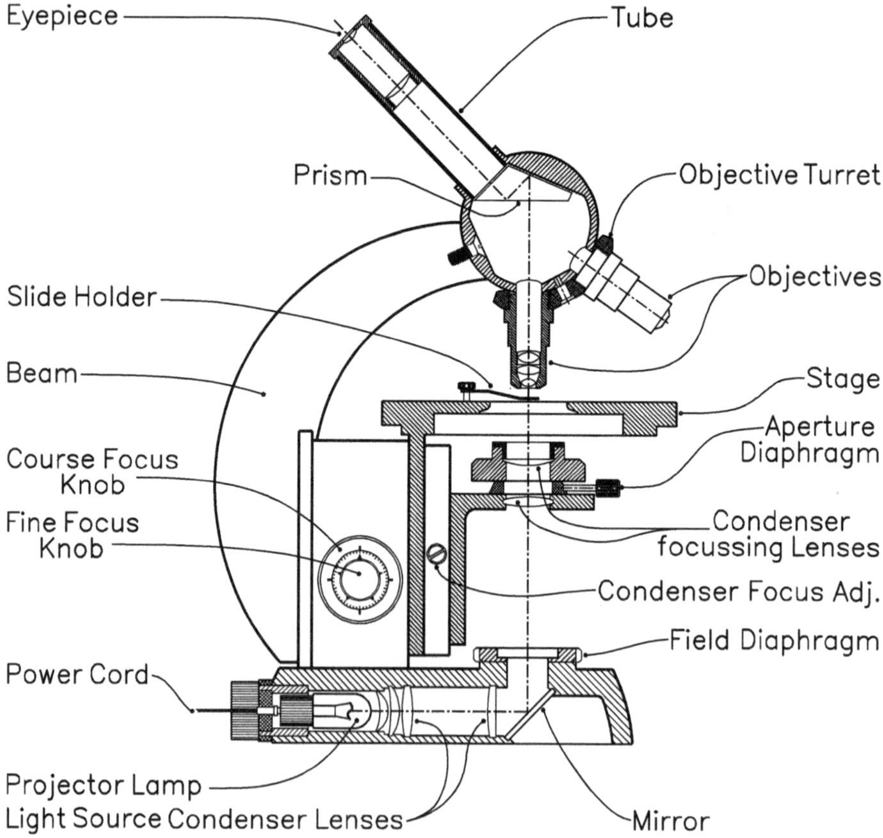

Fig. 10.13. Optical microscope with 3-objective turret

Like telescopes, microscopes employ multilens objectives and eyepieces. The microscope in Fig. 10.13 has a two-lens eyepiece and a set of four and, respectively, five-lens objectives, mounted on a turret for easy exchange. Certain instruments come with objectives and eyepieces in matched sets, designed in ways that make the lateral chromatic aberration of the eyepiece to compensate for that of the objective in a technique developed in the late eighteen hundreds by Ernst Abbe. An associate of Carl Zeiss, Abbe became known as the first to base his lens designs exclusively on mathematical theory.

Electron microscope

The growing need for much higher magnification and better resolution led to attempts at using beta rays, actually streams of electrons, in ways the classical microscope uses light. But it is far from self-evident that the image-forming power of the virtually massless photons could be outdone by streams of the much heavier electrons; intuition points rather the other way. But that didn't deter Ernst Ruska and Max Knoll from building a prototype electron microscope in 1932 by replacing the conventional glass lenses with "magnetic lenses," actually coils acting on the same principles as the yoke on the neck of television picture tubes.

The magnetic fields of a set of coils in an electron microscope focus the flow of electrons on a phosphorescent screen or, more recently, a CCD (charge-coupled device) similar to those in digital cameras. However, the effect of *magnetic* fields on a stream of electrons is far more complex than that of the *electric* fields. In the traditional cathode rays tube (CRT) of the oscilloscope, an electron beam passes between electrically charged plates. Attracted by the plate with the positive charge and repelled by the negative one, the beam deflects proportional to field strength, given by the voltage gradient from plate to plate. With a sawtooth AC voltage on the vertical plates and the signal current on the horizontal plates, a graph of the signal input appears on the screen.

Using that simple design in TV sets would afford exceedingly high voltages to sustain the far greater deflections called for, while the magnetic field induced by a few amperes of electric current through the deflection coil do the job effortlessly – so to say.

What complicates matters is that deflection of electrons in a magnetic field happens not along but at right angles to the field. Envisioned in a spatial x, y, z coordinate system, a magnetic field oriented in the direction of the z-axis makes that a stream of electrons in the direction of the y-axis deflects parallel to the x-axis. For instance, electrons in Wilson's fog chamber move in circles when a magnetic field is applied. Likewise, the highly accelerated electrons in an electron microscope spiral through the instrument's magnetic lenses (Fig. 10.14), driven by the coil's magnetic field, which strength within the airgap amounts to several hundred times that in the pole pieces.

But the anatomy of winding such coils for best performance has taken the efforts of teams of engineers throughout the 20th century, which makes it the more surprising that two lone inventers, such as Ruska and Knoll, succeeded in coming up with an electron microscope that triggered a rush into design and perfection of

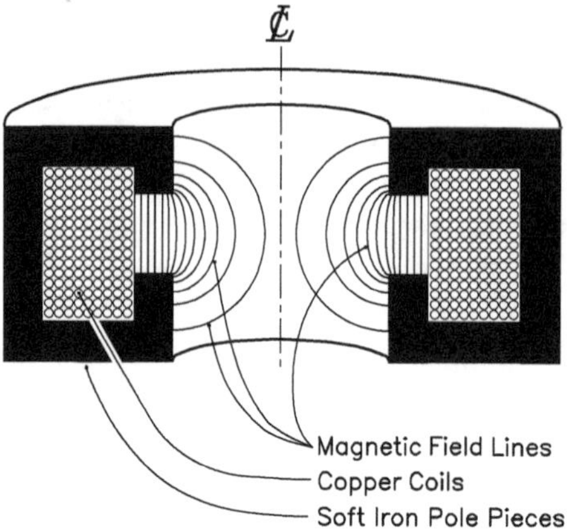

Fig. 10.14. Magnetic lens

such instruments; more important still, that first electron microscope became living proof for *Albert Einstein*'s wave–particle duality thesis of light, which assigns to light rays what in human society would be called a double life: wavetrains in contact with matter, particle beams in empty space. In a generalization of this concept, which earned *Louis de Broglie* the Nobel Prize in Physics in 1929, Einstein's duality concept is extended to *any* kind of moving particle, including such macrocosmic objects as speeding bullets. According to the theory, the wavelength λ assigned to a moving particle of mass m and velocity v is $\lambda = h/(mv)$. As before, h stands for Plank's constant of 6.62618×10^{-34} joule $\times$ seconds.

The length of the wave equivalent of a moving particle is thus inversely proportional to its mass, which makes it too short for physical detection on objects weightier than the elements of particle beams. For electrons of mass 9.11×10^{-31} kg and an electric charge of 1.60×10^{-19} coulombs, the kinetic energy K has been measured as 54 eV, and the electron's de Broglie wavelength, $\lambda = h/\sqrt{2mK}$, results as

$$\lambda = 6.62 \times 10^{-34}/\sqrt{2 \times 9.11 \times 10^{-31} \times 54 \times 1.60 \times 10^{-19}} = 1.67 \times 10^{-10} \text{ meter}.$$

Since 10^{-10} meter equals 1 angstrom unit, the Broglie wavelength of an electron is 1.67 angstrom, compared to 5550 angstrom of the green section of sunlight. This makes us expect the power of electron microscopes to top that of light microscopes by a factor of several thousands. However, magnetic lenses cannot be expected to control the path of electron beams with the same precision as glass lenses guide the course of light rays. Several types of electron microscopes have been developed over time to overcome such difficulties.

The *transmission electron microscope*, the closest cousin of Ruska and Knoll's prototypes, works with very thin slices of the materials which internal structure is being researched. Transmission microscopes have shown things formerly confined to human imagination, such as the spacing of the atoms in carbon and silicon foils of 0.89 and 0.78 angstrom, respectively.

The *scanning electron microscope* (SEM) makes a focused electron beam hit the sample point by point in a raster resembling the scan on a TV screen. The image is generated by secondary electrons emitted by the sample in response to the impact of the primary ones.

With its higher resolution and greater depth of view, the SEM returns impressive in-depth images, but the impact of the concentrated electron beam that helped generating them sometimes endangers the integrity of the investigated samples and can make it hard to distinguish between the sample's "natural" and "man-made" features.

Spectrometry

Few branches of science have opened as many doors to research as the study of spectra. Our knowledge of the structure of the atom stems from studies of the location of the spectral lines of the elements. And much of our knowledge of the universe resides in studies of the spectra of stars and nebulas. Not enough that the spectra of celestial objects help identify the materials these bodies are made from, they make it possible to predict their radial velocity with surprising precision through the phenomenon of *redshift*. Spectroscopy was thus the precondition for *Hubble's law* of the expansion of the universe with all its implications, such as the idea of the *Big Bang*. Figure 10.15 exemplifies the redshift on the most prominent lines of the hydrogen spectrum (first row), and in second and third row, their shift if emitted by sources speeding away from, or respectively, toward the observer with $v = 0.002c = 600\,\mathrm{km/s}$. In the fourth row in Fig. 10.15, the difference is readily detectable in the superposition of the red- and the blueshift.

While torrid solids, liquids, and dense gases create the continuous spectra we know from the rainbow and Isaac Newton's observations of sunlight refracted by glass prisms, gases excited by heat or high-voltage electricity emit light not continuously, but in certain well-defined wavelengths, showing up in their spectra as narrow lines. Wavelengths of those so-called *emission lines* can be derived from their positions within the spectrum and are the footprints of their elements of origin.

When emission lines from a given element pass through an unheated gaseous layer of that same element, the bright emission lines change into dark *absorption lines*, which allow to search the chemical composition of planetary atmospheres.

The link between today's atomistic ideas and the wavelengths of spectral lines can be seen in Albert Einstein's postulation of the energy E of light quanta, a.k.a. photons, as proportional to their frequency f: $E \propto f$. Introduction of the Plank's proportionality

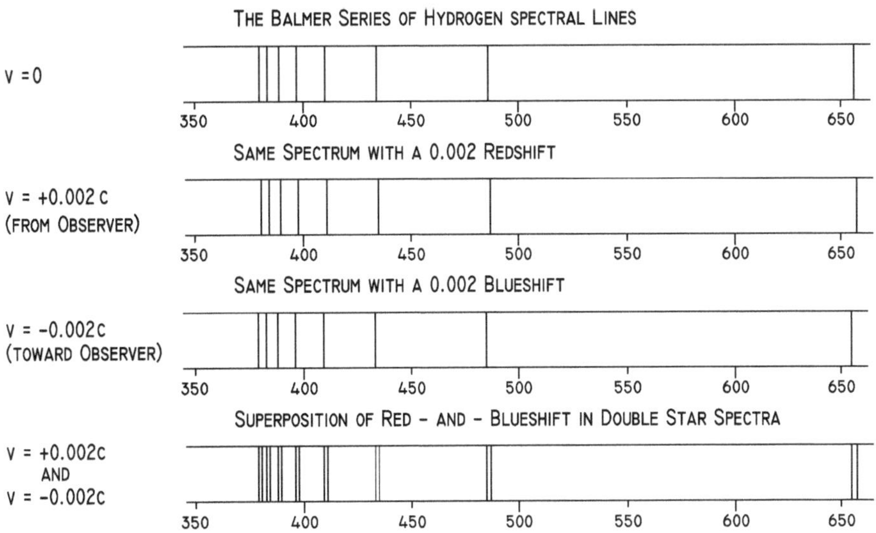

Fig. 10.15. Typical sectors of the hydrogen spectrum

constant h of 6.63×10^{-34} joule × seconds, makes this into the familiar equation $E = hf$, which assigns a certain level of energy to each spectral line of the elements.

The assumption that this energy stems from processes within the atom, such as the displacement of electrons from one orbit into a different one, inspired *Niels Bohr*'s model of the hydrogen atom, which ended three millennia of belief in indivisible atoms.

Historians trace the roots of that concept back to *Mochus of Sidon* in the days of the Trojan war. About half a millennium later, in the 5th century BC, *Leucippus* and *Democritus* suggested the atom as a particle too small to allow further splitting. Since the Grecian adjective *atomos* means "uncuttable," it actually referred to *molecules*, not atoms. Only the birth of chemical processing taught us to split molecules into their atomic constituents.

In the atomic scale, energy is expressed in electron volt (eV), where 1 eV = 1.602×10^{-19} joule, but maintains the characteristics we know from our macroscopic environment, save for the definition of the zero level of energy. Taking it as the surface of the Earth makes the potential energy of 1 kilogram of water in a reservoir 100 meter above the place where it's being used, $1 \times 100 = +100$ kg × m. But the electrostatic energy that binds electrons to the nucleus lacks such an obvious level of reference, which led to assigning zero potential energy to an electron locating very, very far out (infinity). Electron orbits of highest energy are thus the ones closest to the atomic nucleus, and vice versa.

In the hydrogen atom, energy levels of the sole electron lie between zero for hydrogen in the unexcited state, and 13.60 eV for a hydrogen atom heated or electrically charged (ionized) to the point of losing its electron. Lesser excitation can make the atom's energy level 10.20, 12.09, 12.75, and 13.06 eV, but nothing in between, because those levels are not accidental, but represent the energy of the electron's possible orbits. The energy of photons, emitted in the process of electrons jumping between orbits, can thus be the difference of any two of those orbital energy levels, such as: $12.09 - 10.20 = 1.89$ eV, or $12.75 - 10.20 = 2.55$ eV, or still $13.06 - 10.20 = 2.86$ eV for respectively the Hα, Hβ, and Hγ spectral lines, collectively known as the *Balmer Series*.

Previous to Niels Bohr, the idea of negative elementary charges, such as electrons, spinning around a positively charged nucleus, had been rejected on grounds that such an arrangement would fatally spend its energy in the generation of electromagnetic waves, much like a radio transmitter consummates electrical energy for powering its oscillator–antenna circuit. By circulating the nucleus like electric currents circulating in a coil, orbiting electrons – each an elementary carrier of electricity – were expected to radiate their energy away.

But the de Broglie formula $\lambda = h/\sqrt{2mK}$ allows to describe the electron as a wave phenomenon and to estimate a set of specific orbits whose circumferences amount to an integer multiple of the electron's associated wavelength. Those are the orbits where the electron waves become standing waves and, instead of radiating their energy into space, keep it confined at home – so to say.

And that was the idea that ultimately opened ways to explaining the structure of atoms in general. In the case of hydrogen, Bohr's atomistic model postulates the positions of 3 groups of lines in the hydrogen spectrum, commonly known as the *Lyman* series, *Balmer* series (Fig. 10.13), and *Paschen* series, which wavelength follow simple progressions, such as $\left(n^2/(n^2 - 2^2)\right)$, with $n = 3, 4, 5, \ldots$ for the Balmer lines.

Prism spectroscope

The prism spectroscope comes in a variety of models, from hand-held straight look-through sets to laboratory instruments like that shown in Fig. 10.16, a three-arm model that shows spectra along with a superimposed scale in nanometers (10^{-9} meter) or angstrom units (10^{-10} meter). Figure 10.16 shows a probe, such as a cube of sodium, being vaporized and subsequently ionized by the heat of a Bunsen flame so that the spectral lines of ionized sodium vapors show up in the eyepiece at the far right of the picture.

The instrument has three optical arms arranged around the central prism. The latter does the essential work of fanning out the light from the collimator tube (left in Fig. 10.16) into a spectrum, and direct it into the observing telescope (right). Both have achromatic objective lenses facing the prism, which is installed for shortest light path in the glass. For an equilateral prism, this means that the rays of light traverse the prism parallel to its base.

The angular position of the collimator tube with respect to the telescope depends on the refractive index of the glass the prism is made from. The scale projector tube is set in ways that make the reflection of its light beam on the lower surface of the prism enter the telescope.

The fine slit in the aperture plate on the left of the collimator tube is what actually accounts for the shape and width of the spectral lines you see in the eyepiece. Due to the slit's position at the focal point of the collimator lens, all the

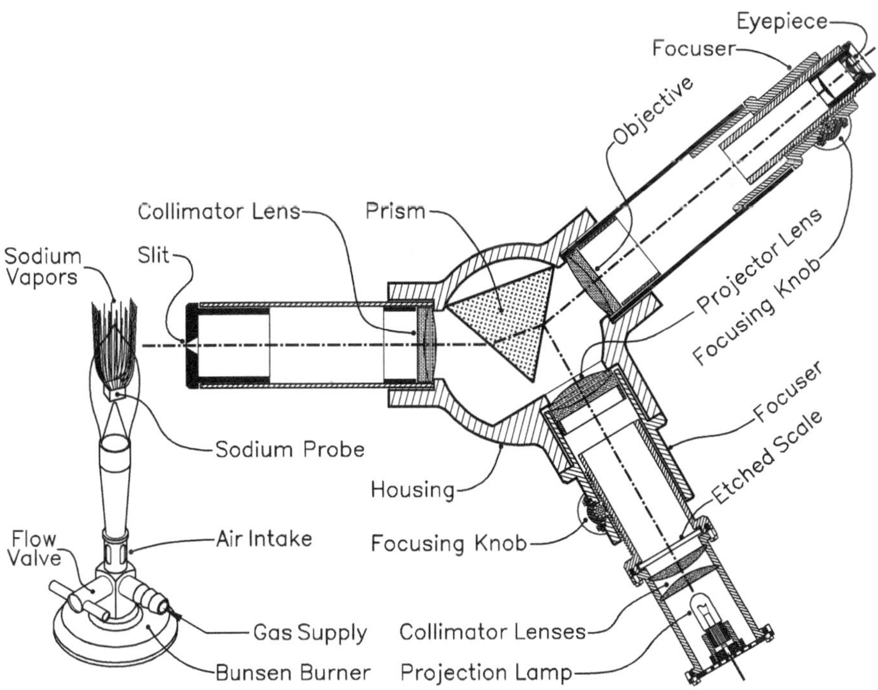

Fig. 10.16. Prism spectroscope

light entering the instrument exits the collimator tube as a bundle of parallel rays, which pass through the prism as such and continue parallel into the telescope. Monochromatic light would be seen by the observer as an amplified image of the slit, but light composed of several wavelengths shows the slit once for every specific wavelength. Since white light consists of a great number of waves of different lengths, they overlap and generate the familiar band spectrum. By contrast, light from ionized gases shows itself as easily discernable spectral lines, indicative for the atomic structure of the probe.

The scale projector tube (Fig. 10.16, bottom) has on its lower end a conventional light source with collimator lenses that homogeneously illuminate the frosted glass with the etched on scale.

While the distance between slit and objective in the collimator tube is fixed, the telescope as well as the scale projector allow for manual focusing by rack and pinion drives. Initially, the telescope is set for best sharpness of the spectral image, and subsequently the projector's focuser is set to focus the scale.

Diffraction grating spectroscope

Spectra generated with gratings differ from prism spectra insofar as they are regular scaled and the sequence of colors is inverted. This because their mechanism of analyzing light by wavelength is a different one.

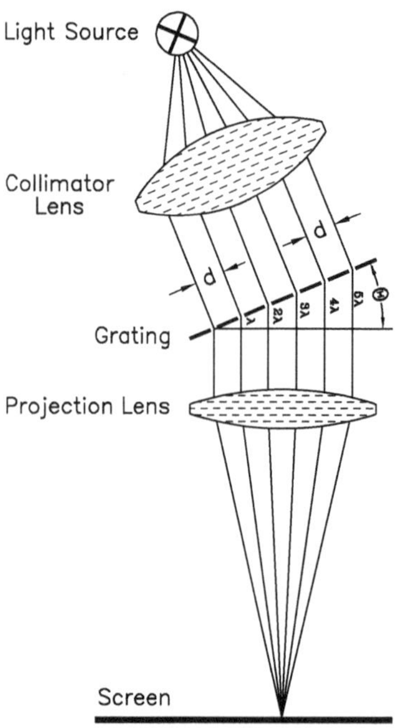

Fig. 10.17. Principle of grating spectroscopy

So far, we remember diffraction as the limiting factor of the resolution of astronomical telescopes, caused by parasitic wavelets generated at the inner edges of the front end of the tube. Such diffraction wavetrains settle in more regular patterns when they stem from a point source, such as an orifice. Their generation is best understood by comparison with ocean waves battering a small access channel to a harbor. Although one could expect a series of short "wave snippets" to pass through the entrance and proceed in straight procession, that's not what happens. True to the principles of Huygens' wave theory, the discharge of the access channel becomes the source of an entirely new family of waves which circular shape resembles the waves created by the impact of a rock landing on the surface of a pond. If two rocks rather than one hit the water simultaneously, the two resulting groups of waves interfere in a pattern which spacing relates directly to the distance between the points of impact.

By the same token, a bundle of parallel light rays from a collimator tube falling on a grating with a great number of slits, spaced by distance d (Fig. 10.17), makes that every slit becomes the source of light rays which spread out in all directions. A convex lens with its principal plane tilted by an angle Θ against the plane of the grating, singles out of that confusion those rays that emanate from the slits at that same angle Θ, focusing them on a screen at the focal distance of the lens. If the length of the light path from the lowermost slit to the principal plane of the lens is s_1, then the length of the light path from the next higher up slit becomes $s_1 + d \sin\theta$, and the length of the next higher up light path $s_1 + 2d \sin\theta$, etc. To obtain constructive interference, the difference $d \sin\theta$ must equal the wavelength λ of the incident light or an integer multiple thereof, as shown in Fig. 10.17.

Rewriting this relation into $\sin\theta = \lambda/d$ shows that the various components of light from the probe will spread out, each according to its wavelength λ. In other words, a spectral image will appear on the screen.

For a grating with 500 slits per millimeter, that is, a spacing of $d = 10^6/500 = 2000$ nm (nanometers) the relation $\sin\theta = \lambda/d$ yields for instance for the green light's wavelength of 555 nm the deflection angle of $\theta = \arcsin(\lambda/d) = \arcsin(555/2000) = 16.11°$. The visible light spectrum, including the wavelengths from 400 to 700 nm, will thus be stretched between the angles of $\arcsin(400/\ 2000) = 11.54°$ and $\arcsin(700/2000) = 20.49°$.

Since the conditions $\sin\theta = 2\lambda/d$, $\sin\theta = 3\lambda/d$, etc., likewise allow for constructive interference, secondary and tertiary spectra of successively lower luminosity also surge in the process but are cropped out by the instrument's internal apertures.

Resolution of grating spectra gets better with the number of slits per millimeter. For instance, a grating with 1800 slits per millimeter creates – at least theoretically – images with 0.003 nanometer of resolution.

Rather than slit gratings, reflecting gratings may be used to the same end. Their surface is covered by triangular grooves, sawtooth grooves, or sinusoidal grooves. The latter, also called holographic gratings, originate from engraving with a laser beam. They excel by their low level of stray light and the absence of ghosts in the images produced.

The *Fabry–Perot interferometer* makes part of the group of interference spectrometers, used for research into the structure of particular spectral lines, such as the Zeeman effect, the splitting of spectral lines into 2 or more narrowly spaced components if a strong magnetic field is applied to the light source.

These instruments do away with gratings altogether and use a pair of face-to-face mounted, partially silvered glass plates to the same effect. The narrow air space d between the plates makes that light entering at an angle Θ gets reflected in zigzag to and fro between the plates. Since the angle between incident and reflected light beams is small, its cosine can be considered equal to 1, and spacing for constructive interference becomes an integer multiple of $\lambda/2$.

Among many other applications, Fabry–Perot interferometers are used with the 1.7 m Antarctic Submillimeter Telescope and in the James Clerk Maxwell Telescope on Mauna Kea in Hawaii.

11 Acoustics

There were times when people planned their high fidelity sound systems long before they wondered what kind of house to build around them, hoping maybe that savings on opera tickets would some day pay for it all: The mighty power supplies, the delicate pickups and tuners, the vacuum tube spangled amplifiers, the mixers with their arrays of gain controls, the transformer congested output stages; and then the final, yet not weakest link in the stentorian chain: the solid wood speaker cabinets, housing an assembly of woofers, subwoofers, tweeters, and – you have it.

A truly intrepid audiophile of that era used the partition between his room and the attic as the baffle for a monster speaker, six foot across. The entire loft behind became a giant speakerbox, and the room up front his private auditorium. The editors of *The Guinness Book of Records* could have eternalized his name in print, but as things go, a hearing problem in later years might have become the most likely reward for his devotion to heavenly tunes.

The symphony orchestra in the living room

It all started quite modestly with those post World War I loudspeakers on an RCA Victor poster, featuring a foxterrier whose tilted head betrayed his intense listening into the black-hole shaped horn of one of those times speakers. But what eternalized this display was the capture: "*His Master's Voice.*"

Well, dogs are known for their capacity to single out the voices of their beloved ones from the environmental cacophony we got used to live in. Anyway, another world war had to come and go until industry came up with vacuum tubes capable of outputting something resembling life music. And with them came the rush into the 'high fidelity on every hearth' mindset. Pre- and output amplifiers were built as separate units and installed as far as possible from each other to forestall mutual interference. Entire bookshelves were cleared of Shakespeare's *Hamlet*, Dante's *Divina Comedia*, and Goethe's *Faust* – to no better end as to house all those electronic heavyweights and hide the speaker boxes in the space between – an ambitious aim at times when the quality of a speaker was judged by the size of its housing – the bigger the better, of course.

And the results? Well, let's call them far better than the fidelity of Edison's cylinder phonograph – if you remember the heartwarming story of the little poem the inventor played back by turning the crank. But something essential was missing. Maybe it was what the French call the "je non sais quoi," the "certain something" that still had to be paid for with the price of a concert ticket.

But then – when space travel led to the demand for smaller and lighter electronic components – came the transistor. Comparing its appearance with a tsunami would be an understatement. It just drowned the radio industry, challenged the makers of audio outfits, and united the throng of Heathkit builders into a worldwide community.

Mathematicians, not to be left behind, revived their matrix and Boolean algebra in search for infallible guidelines on counteracting the shortfalls of those early germanium PNP transistors by wicked design tricks. Industry – on the other hand – tried to put some order into the unpredictable characteristics of their creations by sorting them into never ending sets of categories, each with its own application sheets, which often seemed tailored to component properties rather than market forces. An abundance of circuits for transistorized Morse code practice oscillators and transistor testers swept circuit designers, who in reality needed only three things: sensitive amplifier transistors, high-current power transistors, and radio frequency transistors.

What sounded like a dream in the germanium age, the silicon NPN transistor delivered. Its negligible leakage currents and high gain made that transistorized audio gear soon swamped the market at a fraction of the price of tube amplifiers. And yet, it took years of tampering until, in the early 1960, some beaming electronic engineer showed a black box with the comments: "Operates like a length of copper wire – *but with amplification*."

The way to "high fidelity in every home" seemed wide open, and yet, an old problem still defied the developers: Emulating the sound effects of the concert hall in your living room. Sure enough – there was *Stereo*. Maybe the inventors were inspired by Grecian mythology, in particular the story of the one-eyed cyclop on the island of Youra in the Aegean Sea who threw boulders after the fleeing Odysseus' ships in hopes to sink them. Even if the cyclop's single eye had not been preemptively punched out by Odysseus wielding a pointed tree trunk, the cyclop would have missed, because his vision lacked the sense of depth that we twin-eyed creatures enjoy. Millennia later, 3-D movie strips with quite convincing spatial views through red–green or polarized eyeglasses became the wow of the day, and hadn't it been for the inconvenience of wearing such goggles and keep your head straight and motionless for two hours, that so-called anaglyph 3-D standard might just have made it. But, as so often in life, illusions got the upper hand on reality, and the quasi spatial effects of the curved broadwand in "Cinerama" theaters won out over the truly stereoscopic depth effects from the superposition of a pair of mutually offset images.

On similar principles, stereo sound is obtainable from signals off two properly spaced microphones in front of a life orchestra, but brought the need to transmit two distinct signals through a radio station's single channel. That's where multiplex systems, operating either on the principles of time division or frequency division, got into the picture.

Time division confines outputs from each microphone into a series of time slots. The receiver sorts out that blend of left and right signals with synchronized electronic switches into two channels – one for each ear to hear. Much like the series of separate pictures in a movie strip provides the illusion of uninterrupted motion, the series of multiplexed sound impulses is too fast paced for hearing them one by one, leaving the listeners with the impression of continuous sound from both sides.

Conversely, the frequency division system assigns a specific carrier frequency to each microphone. The idea surged in the days when we paid extra for the privilege of listening to certain programs, the so called background music among them. In standard radio reception, the radio frequency (RF) carrier-wave contains the audio information on two sidebands, which allows for detection of the radio wave's audio content by a diode. Encoded transmissions, such as background music, have the RF carrier modulated in two steps. For instance, a 1 MHz RF carrier may be modulated by a 100 kHz sine wave, which by itself has been modulated by up to 10 kHz audio signals. In the receiver, the 1 MHz carrier gets eliminated by superposition with a locally generated oscillation of the same frequency in counter phase. The remaining 10 kHz signal is then demodulated as usual, leaving only the audio signal to be heard. That's what happened in the black box you had to pay rent for in order to receive the soothing music in elevators, and sometimes even in offices, depending on the boss' opinion if the lure of music improved the employees' zeal or put them to sleep.

With these technologies in place, stereo sound depended mostly on users' readiness to pay extra for their two-channel audio systems, but once again, ended in disappointment. Regardless the undeniably spatial effects of sounds from two speakers, the "concert hall sound" still remained a delusion. Even the surround sound of four speakers, operating on four channels, or in a different approach, injected time delays, didn't help. So – what was going on?

Well, a Ford is a Ford and a Cadillac is a Cadillac. In other words, the sound you hear in a concert hall can only be reproduced in a physically identical concert hall, and nowhere else. Unlike light, which comes in well defined rays, sound propagates in fields of pressure surges from the source – be it a speaker, a single instrument, or an orchestra – and their reflections from everything around, which freely interfere with each other. That makes the sound wave pattern of a given place unique and unfit for reproduction somewhere else by whatever means, from the simplest to the most sophisticated.

But, investors in expensive sound systems, take heart! You in your living room still hear the orchestra sooner than the music enthusiast on his expensive seat in the theater. Radio waves travel about 1 million times faster than sound waves, and arrive in your home long before the heavenly tunes have traveled from the stage into the audience.

Sound propagation

Since the sounds we hear travel through air, we instinctively see the atmosphere as the predominant media for sound transmission, though all matter conducts sound to a higher or lesser degree. Actually, the velocity of sound propagation in air, of 331 m/s at 0 °C and 341 m/s under average conditions, is easily surpassed by other substances, such as hydrogen with 1300 m/s, helium with 971 m/s, water with 1407 m/s, and steel with a whopping 5100 m/s. Small wonder that an approaching train can be heard with your ear pressed to the rail long before the engine's whistle makes you jump.

But unlike light, which travels through transparent matter almost as readily as through the emptiness of cosmic space, sound waves are blocked by a vacuum like Superman at the sight of a krypton meteorite. To the point that he, who yearns for a quiet corner to read his morning paper, may want to consider a vacuum chamber, provided his wardrobe includes a surplus spacesuit from Ebay. While his reading lamp would still illuminate his paper unabashed, not the slightest traces of sound from lawnmowers and rock music boxes would reach his ears (Obs.: Forgive me, ladies, for skipping the "his or her" figure of speech, but only men can have such silly ideas as reading their paper in a vacuum chamber. Signed: the author).

Yet the experience would highlight the basic difference between energy transmission by light waves vs. sound waves. While we have come to know the former as transversal electromagnetic waves, the latter are longitudinal mechanical waves, meaning that a sound wave consists of a row of overpressured air pockets interlaced with an equal number of pockets of rarefied air. Sound intensity is proportional to the square of the highest momentary speed of the vibrating air particles.

However, pressures gradients in sound waves are surprisingly low. At the sound intensity of 1 W/m^2, which is about the noise output from a military jet-plane, the high- and low-pressure zones of the surging sound waves are a mere 0.0003 atm above and below ambient pressure. Since the 90 dB noise from a jet engine is about all the human ear can take without permanent damage, it shouldn't surprise that about 30 million Americans suffer from some degree of hearing loss from setting the volume control of their portable music players too high.

Which brings us to the unit of sound intensity, the *bel*, named after Alexander Graham Bell, best known as the inventor of the telephone. An increase by 1 bel in sound intensity means 10 times higher sound energy. This is a logarithmic scale of the form $\log(P_1/P_0)$ for comparison of P_1, the power of a source of sound, with the power P_0 of some sound standard. For convenience, the bel has been subdivided into the familiar 10 decibel (dB), which makes said formula: decibels $= 10 \times \log(P_1/P_0)$.

If the master sergeant teaches a battalion of 1000 soldiers to shout "Hail to the King" as loud as he does, the combined sound of the 1000 voices would amount to a scant $10 \times \log(1000/1) = 30$ dB above the loudness of the sergeant's voice alone.

At the low end of the sound scale are the faintest sounds our ear can detect, such as the impact of a fallen leaf with a sound power of typically 10^{-12} W/m^2. This is 1 billionth of the 90 dB jet noise.

The development of the telephone and later of audio technology brought the need for sound measurements in function of the sensitivity of the human ear, which led to the definition of the unit of *phon* as the decibel level at 1000 Hz of frequency. For instance, the sound from a 60 dB–1000 Hz audio generator has 60 phone of loudness.

NBC, CBS, and Bell Telephone Laboratories standardized an instrument they dubbed *VU-meter* that showed the volume of audio *as loud as we would hear it* in *volume units* (VU). The device accounts for the ear's varying sensitivity with a non-linear scale, marked in decibel. The readout instrument is a simple d'Arsonval gal-

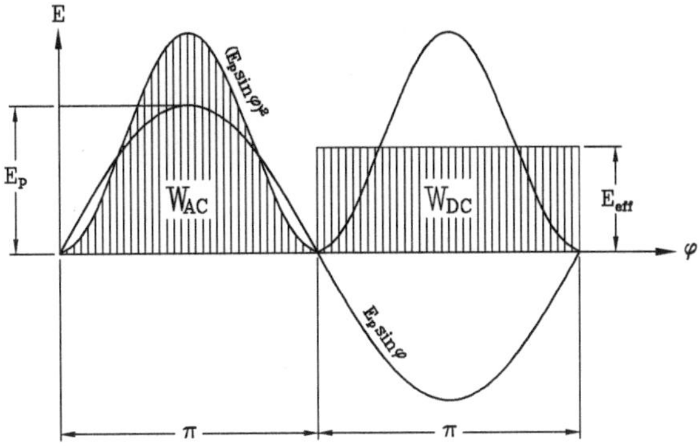

Fig. 11.1. AC characteristics

vanometer, while a small dynamic speaker from radios or other noise-producing gadgets can double as input transducer.

However, simple peak-to-peak measurements of sound intensity wouldn't do, because a few spikes could push the readings far above the de facto overall volume. Speech, for instance, comes in consecutive bursts of irregular alternating currents within the narrow frequency band of 300 to 3400 Hz. The significant magnitude here is the root-mean-square (RMS) of the sound-induced alternating current, whether sinusoidal or irregular.

The RMS standard stems from the early days of alternating current networks, when an AC equivalent to the voltage of the formerly dominant DC systems was created on the principle that equal voltages should produce equal power output in both systems. The graph in Fig. 11.1 shows the curve for sinusoidal AC and, above the x-axis, the related power output, which is proportional to the square of voltage and therefore always positive, even when E dips below the x-axis. The hatched area on the left is the integral of work per semicycle, which area equals the hatched rectangular area on the right.

This equivalence presupposes that a 100 watt light bulb should light equally on 115 V direct current (DC) and 115 V alternating current (AC). As it turned out, the power from 115 V direct current matched the power output of a 325 V peak-to-peak source of alternating current – a discrepancy that, combined with the evident danger of higher voltages, became Thomas Edison's principal argument against the Westinghouse-sponsored AC networks.

This AC/DC equivalence rests on two fundamental laws of electrical engineering. First that power P equals voltage E times current J: $P = EJ$; and second, Ohm's law, stating that voltage equals electric current J times resistance R: $E = JR$. Combining the two equations for elimination of J gets us $P = E^2/R$.

With E_p for the peak voltage of a sinusoidal alternating current (Fig. 11.1), the momentary voltage equals $E_p \times \sin\varphi$, and power becomes

$$P = \frac{{E_p}^2}{R} \int \sin^2 \varphi \times d\varphi \, .$$

Accordingly, the work W_{AC} from alternating current flowing through a resistor R during a single semicycle, comprising $\varphi = 0$ to $\varphi = \pi$, equals

$$W_{AC} = \frac{{E_p}^2}{R} \int_0^{\pi} \sin^2 \varphi \times d\varphi \, ,$$

which can be resolved into

$$W_{AC} = \frac{E_p^2}{R}\left[\frac{\varphi}{2} - \frac{1}{4}\sin 2\varphi\right]_0^\pi,$$

or

$$W_{AC} = \frac{E_p^2}{R}\left[\frac{\pi}{2} - \frac{1}{4}\times 0 - \left(0 - \frac{1}{4}\times 0\right)\right],$$

that is,

$$W_{AC} = \left(E_p^2/R\right)(\pi/2).$$

To find the DC voltage, E_{eff}, which under identical conditions yields an identical power output, we set

$$W_{DC} = W_{AC}$$

and get

$$\frac{E_{eff}^2}{R}\times\pi = \frac{E_p^2}{R}\times\frac{\pi}{2}$$

which boils down to $(E_{eff}/E_p)^2 = 1/2$.

Thus we finally arrive at the relation $E_p = E_{eff}\sqrt{2}$, and for the peak-to-peak AC voltage $E_{pp} = E_{eff}2\sqrt{2}$ or $E_{pp} = 2.828E_{eff}$.

E_{eff} is often referred to as *volt root-mean-square* (RMS), a term we remember from statistical mathematics as the square root of the mean of the squares of several variables. For instance, the root-mean-square of the numbers 4, 5, and −9, is $\sqrt{[4^2+5^2+(-9)^2]/3} = 6.38$, opposed to those numbers' average value of $(4+5-9)/3 = 0$. The latter is what we would get if we tried to measure our domestic AC voltage with a DC galvanometer. But don't try that with the 325 V peak to peak of a wall outlet. As expected, the needle would not deflect, but the instrument's moving coil would explode into flames and toxic fumes. And even if you manage to find a low-voltage power supply in your children's obsolete toys collection, a serial resistor is a must. For a 1 mA galvanometer to check a 6.3 VAC power supply, a serial connected resistor of at least $6.3/0.001 = 6300$ ohm is afforded, but 10 kΩ would be the safer choice. And remember to always unplug your handiwork before you call it a day.

In conclusion, the RMS voltage of the electrical signals from a microphone is what an AC instrument would read. In the absence of such, a galvanometer along with a rectifier is a good choice. A single diode would in principle suffice, but make us lose half of the signal strength, because a diode rectifies by clipping off either the positive or the negative halve-waves. Conversely, a 4-diode bridge rectifier, best known under the name of "full wave rectifier" (Fig. 11.2, bottom right) delivers the full DC equivalent of the AC input, safe for the diodes inherent voltage drop, 0.6 V for silicon and 0.2 V for germanium diodes. Thus, the combination of silicon diode rectifier and DC meter needs a voltage of 0.6 V to get the reading off the zero mark. That might be acceptable in measurements of, say, 115 VAC and above, but becomes a problem with low voltages.

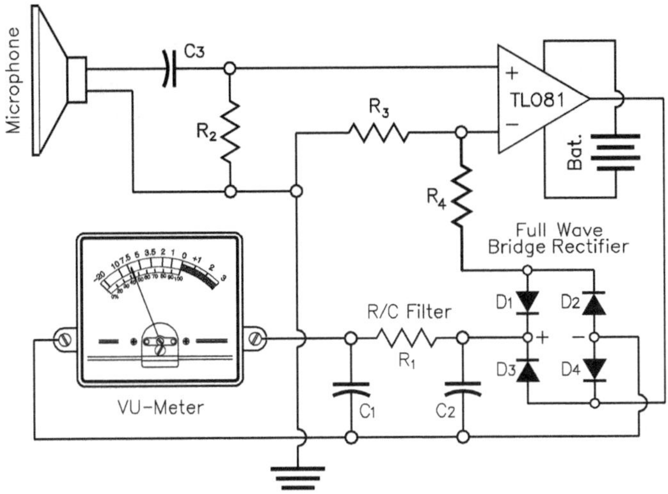

Fig. 11.2. Loudness meter

For this reason, the VU meter in Fig. 11.2 uses an operational amplifier (OpAmp) combined with a full wave rectifier bridge to compensate for the offset voltage by connecting one pair of diodes to the output of the OpAmp, and the other in the feedback loop. An RC filter in π configuration is there to smoothen the somewhat erratic signals captured by the microphone and dampen the meter needle's trepidations. The filter should be dimensioned for 300 milliseconds of rise time of the needle if loaded with a 1000 Hz signal of 0 VU loudness. The top scale on the dial reads VU, while the bottom scale refers to the degree of modulation.

According to 1939 standards of telephone companies, which we still use, the zero point on the VU dial should locate at the deflection caused by a voltage that would generate 1 milliwatt of power while fed into a 600 ohm impedance line. With $P = E^2/R$ and therefore $E = \sqrt{PR}$, the associated voltage becomes: $E = \sqrt{0.001 \times 600} = 0.775$ V. In other words, connected to a 0.775 VDC voltage source, the meter shows where the zero mark of the scale is to be located.

Present-day audio industry introduced a different standard with a 4 dB higher zero point. That's the equivalent of $0.001 \times 10^{0.4} = 2.511 \times 10^{-3}$ W of power, P, over a 600 Ω resistor. Here again, we derive from $P = E^2/R : E^2 = PR = 2.511 \times 10^{-3} \times 600 = 1.507$ and $E = \sqrt{1.507} = 1.227$ volt. An instrument using this standard would have the zero point of the scale at the deflection caused by 1.227 V.

The wave nature of sound

A handy demonstration tool for the longitudinal wave nature of sound and its conduction through solids is the sound-only based telephone in Fig. 11.3. Its simplicity makes one think that Jane could have used one to call Tarzan to lunch from her

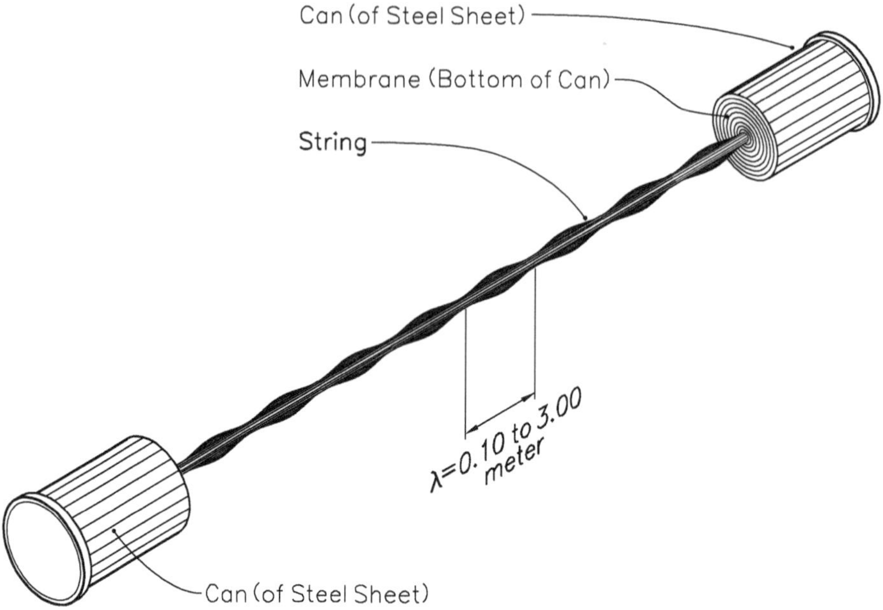

Fig. 11.3. Tin can telephone

tree house, had they remembered to pick up a pair of midsize empty soup cans from past expeditions, and find a thin, but not overly elastic liana string. The bottom of the cans, acting here as the membrane or diaphragm, should be of thin, zinc or tin-plated steel sheet, free of circular ribs. Soft, inelastic materials, such as pure aluminum, would unduly damp oscillations.

Somebody talking into the opening of one can is being heard by a guy listening into the other, because the oscillations of the talker's membrane propagate along the tightly extended string and impose similar oscillations in the listener's membrane.

In Fig. 11.3, the soundwave-generated bulges and dips wandering along the thread are zoomed in for the sake of demonstration; in reality, they are too small to be seen.

An alternative to the tin-can phone is a garden hose with plastic funnels taped to each end as mouth- and earpieces. If you have a long enough hose, you may even check the velocity of sound waves in air from the time delay between your sayings and the other party's hearing them, which should be about 1 second per 1000 feet of hose. A similar system was standard onboard ships for communication between bridge and machine center until electronic intercoms took over. Yet the *speech tube* has been preserved on board many ships as a fall-back system in emergencies and for its role in action movie slots, where the dramatic impact of a helmsman bellowing orders into a tube's mouthpiece dwarves that of a guy mumbling into a telephone.

An early antecessor of the speech tube was the *ear of Dionysius*, described as a wicked device of tyrant Dionysius of Syracuse to overhear the whispered escape schemes of brutally imprisoned members of his opposition party. But what present-day tourists mostly admire as the Orecchio di Dionisio is the arched entrance to a cave from where you can hear sounds from the cave's far end.

Most useful among the devices that transmit sound through a column of air is the stethoscope. Invented in 1816 by the French physician René-Théophile-Hyacinthe Laënnec, the idea was reportedly inspired by the doctor's wish for an alternative to checking symptomatic respiratory sounds by pressing his ear (and the head it belonged to) on the thorax of his blushing lady patients. A simple wooden tube had to do until Arthur Leared came up in 1851 with the binaural stethoscope with flexible tubes to both ears.

Electricity-based signal transmission started with the *carbon microphone* as transmitter and the *electromagnetic earphone* as the receiver. The carbon microphone's functioning is based on the changes in the electric resistance of an encapsulated pile of carbon granules with pressure. Rhythmically pressurized and depressurized by sound-induced vibrations of an electroconductive membrane, its electrical resistance mirrors the vibrations of the sound waves hitting the membrane. It works quite well within the bandwidth of about 300 to 3400 cycles per second (hertz), required for telephone conversations, but suffers from inherent distortion and the omnipresent background hiss caused by random alterations in the contact resistance between carbon particles.

Carbon microphones depend on a proper power supply to convert the carbon pile's sound-induced resistance variations, between 50 and 200 Ω, into the corresponding variations of electric current to feed the primary windings of the output transformer and induce some 15 volt peak to peak in the secondary. That's good enough to feed a few kilometers of transmission cable and to make the receiver's electromagnet actuate its membrane.

Audio systems

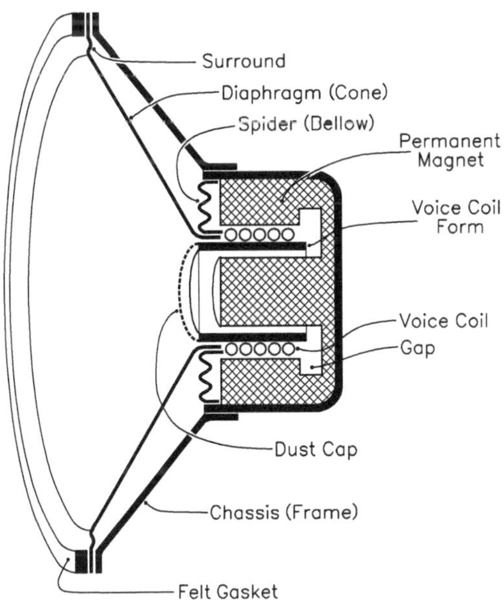

Fig. 11.4. Dynamic loudspeaker

Carbon microphones endured in telephones, but the much higher demands of audio systems on fidelity call for more sophisticated solutions, highlighted by the design of the *dynamic speaker* (Fig. 11.4), which principle could be described as that of a inversed solenoid. In a solenoid, the magnetic field of a fixed coil pulls the iron core into the coil, while the dynamic speaker has a coil of copper wire pulling itself lengthwise in and out of the static field of a permanent magnet. The coil must be light enough to oscillate at up to 20 kHz in Hi-Fi systems, and at about 10 kHz for general use. Voice coils capable of such rapid vibrations must be low mass and

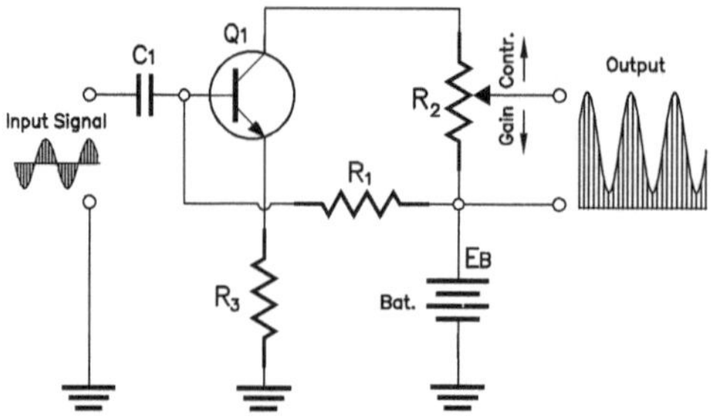

Fig. 11.5. Class A amplifier

yet stiff enough to withstand the resulting mechanical stress on their windings. That leads to coils of few turns of rather thick magnet wire with their inherent low impedance, typically 4 Ω, 8 Ω, and 16 Ω.

Matching that with the high output impedance of vacuum output tubes, such as the 50C5, called for audio transformers with primary windings of thousands of turns and only a few in the secondary.

In transistorized audio amplifiers, the primary windings of the output transformer act as the load resistor of the power transistor, but the inherently lower impedance of transistor circuits allows for far lower ratios of primary to secondary windings than in tube powered outputs. Here, impedance matching is only *one* function of the output transformer; DC to AC conversion is the other, because the output of a Class A amplifier (Fig. 11.5) is not a blueprint of the alternating current input from the microphone but consists of a direct current (DC) of periodically varying voltage and amperage. The DC component of this blend stems from the positive polarization of transistor Q_1 by resistor R_1, of typically 330 kΩ. Without this bias, the transistor would amplify the positive signal cycles but clip the negatives. Only proper bias gets the entirety of the signal transferred and amplified, yet as modulated DC rather than the desired amplified version of the alternating input voltage.

The alternating portion of that modulated DC induces in the primary of the output transformer the magnetic field inducing in the secondary the alternating output voltage. The output transformer thus performs impedance matching *and* DC blockage.

Supply voltage and component values in a circuit like Fig. 11.5 must be selected with an eye on these conditions. Emitter resistor R_3 should be made high enough to forestall random oscillations without unduly lowering gain, and generally ranges between 100 and 220 Ω. The impedance of the potentiometer R_2 should match the transistor's output impedance (typically 10 kΩ).

However, a transformer's operation is far from linear. The *impedance* of a coil, such as the windings of a transformer, actually consists of the vector sum of the coil's ohmic, inductive, and capacitive reactance. Inductance is what makes a transformer tick, while ohmic resistance shows as bothersome heating and sometimes (Heavens beware) the biting smell of roasting insulation.

A coil's capacitance C_p stems from the capacities between the windings of a given layer and the ones immediately above or below, and is also known as stray- or parasitic capacitance. It cannot be eliminated, but its effects can be lessened by winding the coil in ways that reduce the voltage gradient between adjacent layers. The standard wire-feed from left to right, followed by right to left, makes the voltage drop between the last wire loop of one layer and the one on top of it equal to 2 times the voltage drop ΔE per layer, e.g., $2\Delta E$. But if we wind from left to right,

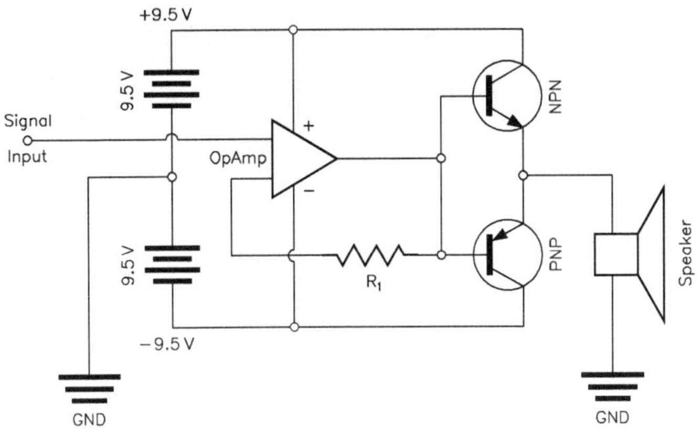

Fig. 11.6. Push-pull audio amplifier

pull a straight wire lengthwise back toward the left, and then again wind left to right, that same voltage difference gets reduced to ΔE.

Relative to the ohmic component of the transformer current, capacitive and inductive components are $\pm\pi/2$ or $\pm 90°$ out of phase. Only at resonant frequency, capacitive reactance $1/\omega C$ and inductive reactance ωL cancel out and leave a clean ohmic signal with E and J in phase. Otherwise, output voltage and amperage are out of phase with each other by the *phase angle* φ, and distortion results.

The event of power transistors of up to 10 ampere current carrying capacity, such as the 2N3055, allowed for direct coupled transformerless amplifier circuits, inherently free of all those complexities. In the push-pull system shown in Fig. 11.6, positive and negative halfwaves of the audio signal are amplified individually by a pair of serial connected NPN and PNP transistors. When the positive portion of the signal makes the NPN transistor conduct, current flows from the positive pole of the power supply through the speaker coil to ground, while the PNP transistor is turned off. Conversely, the negative halfwave makes that the PNP transistor conducts, channeling current from the negative pole of the power source through the speaker coil to ground, while the NPN transistor is off. Ground is the center tap of the power supply.

Direct coupling carries the risk of the buildup of bias currents to damaging levels, and thus calls for carefully balanced circuit design. Such "disaster waiting to happen" lost its dent with the use of operational amplifiers, which gain can be precisely controlled by selection of the feedback resistor (R_1 in Fig. 11.6).

Though transistorized audio amplifiers have since become the rule, certain audio enthusiasts out there still claim that the purity of sound of tube amplifiers cannot be doubled by their cousins in the transistor camp. Technically, the argument rests on the extremely high grid impedance of vacuum tubes, which makes that all a triode needs to control the flux of electrons from cathode to anode is voltage (opposed to current). Only the very small current through the grid-to-ground resistor is needed for the sake of circuit stability. Depending on the quality of shielding of the tubes and associated components, the grid-to-ground resistor can be

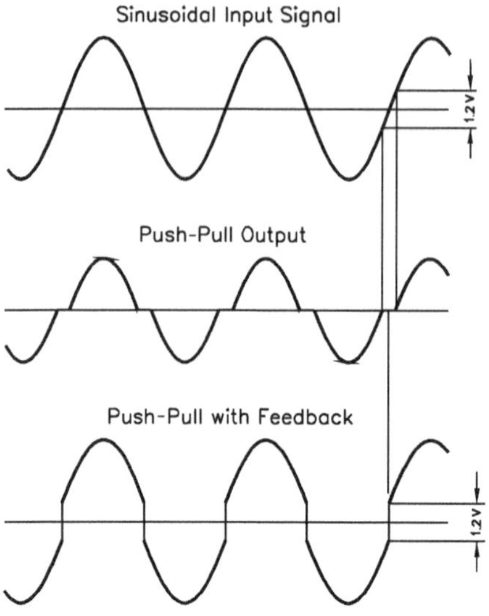

Fig. 11.7. Waveforms in push-pull power amplifier

made as high as 10 megohm, bordering "no load" and herewith "no distortion" conditions for the input signal.

Another argument against transistor-based amplification is the existence of the "dead band" of about 0.6 V where a silicon NPN transistor does not conduct at all. That makes that the basic push-pull circuit blends out the central portion of the signal (Fig. 11.7, center row). But for stronger signals than those plotted in Fig. 11.7, inherent distortion of push-pull circuits becomes far less. For a signal of 10 V amplitude, 0.6 V present no more than 6%, which can be controlled by the degree of feedback through R_1 in Fig. 10.6.

If the tube's "magic sound" is reality or only in the ear of the beholder remains an open question, frequently overruled by the power of money. Tube amplifiers with their multitude of transformers and high-performance vacuum tubes usually demand several times the investment of their transistorized counterparts.

The two faces of the dynamic speaker

The omnipresence of the dynamic speaker in the field of audio transducers does not exclude its presence at the other end of the line – as input transducer, or in short, microphone. Case in point are intercommunication systems, or *intercoms*, where one and the same dynamic speaker takes turns in operating as microphone or as receiver.

What makes intercoms different from telephones is the need to say "end" and press a switch every time you want the other party to come on, while the standard telephone does that switching internally thanks to a dedicated step-up transformer, which windings are laid out in ways to block reflected signals from the receiver's telephone back to the talker. What we thus get are one-way connections from and to either end.

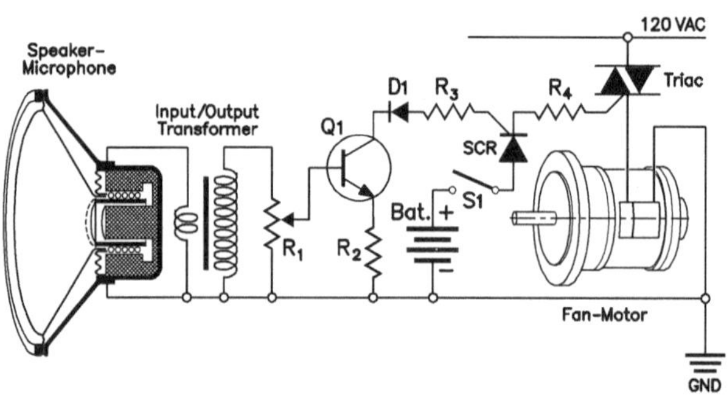

Fig. 11.8. Sound-activated AC motor

The sound-activated motor switch in Fig. 11.8 is only one example out of the multitude of applications of dynamic speakers. It uses a small dynamic speaker as microphone for turning on an AC powered electric motor in response to a clasp of your hands. The actual switching is done by a silicon controlled rectifier (SCR), which supplies the gate current for the triac that turns the motor on and off.

Many aspects of an SCR resemble a switching transistor, except that, once an SCR is turned on by a positive pulse to its gate, it stays on, regardless of the state of the gate. Only an interruption of the main flow of electricity leaves it off until the gate is made positive again. In AC systems, the negative halfwaves provide such interruptions 60 times per second, which makes an SCR act as the AC equivalent of a transistor if you rename the base into *gate*, the emitter into *cathode*, and the collector into *anode*.

However, the diode characteristics of an SCR make that only the positive halfwaves of alternating current make it through, so that an SCR switched alternating current converts into strings of positive pulses which, summed up, amount to only half of the original peak-to-peak voltage. Conversely, the TRIAC (triode for alternating current), in principle an integrated pair of diametrically opposed SCRs, conducts full wave AC, which makes it the semiconductor of choice for power switching applications.

Switching characteristics aside, high current triacs and SCRs are readily available at reasonable costs, while high-power transistors might be fabricated as an arrangement of a number of individual transistors in parallel. This configuration suffers from the inherent differences in gain and sensitivity of transistors. In a combination of, say, 5 transistors with 10 A capacity each, a 100 mA base current could make one transistor conduct 8 A from emitter to collector, while the next conducts 12 A, and the rest something in between. Though the rated capacity of the array would be $5 \times 10 = 50$ A, a momentary overload of, say, 25%, would draw 10 A from the formerly 8 A transistor, and 15 A from the formerly 12 ampere transistor, likely burning it out. The moment that happens, the remaining 4 transistors must conduct in the average $62.5/4 \approx 15.6$ A, enough for triggering an expensive domino effect. A similar surge current would hardly affect a 50 A SCR or TRIAC.

In the circuit shown in Fig. 11.8, the sound-triggered SCR, which controls the TRIAC, will not react to sound signals subsequent to the one that turned it on, which makes the circuit immune to environmental noise. Only the opening of the reset switch S_1 turns the motor off.

The low input impedance of the microphone turned speaker matches that of the primary windings of the up-transformer, which secondary feeds into potentiometer R_1, which settings determine the sensitivity of the circuit by regulating the base voltage of the preamplifier transistor Q_1. The collector current of Q_1 flows through the gate of the SCR and over the then closed reset switch to the positive pole of the battery that supplies the gate current for keeping the triac in the "on" state. The TRIAC remains on until the reset switch is turned into the "open" position.

The resistance of the potentiometer R_1 should match the output impedance of the transformer, typically 800 Ω. Resistors R_3 and R_4 are sized to keep the gate currents of the SCR and, respectively, the TRIAC, within supplier-specified limits.

The TRIAC's capacity must withstand the motor's "turn on" current surge, which can exceed the motor's full-load amperage by 5 times and more.

Piezoelectric transducers

In applications that call for shorter response times than a dynamic microphone can provide, piezoelectric transducers are the answer. Piezoelectricity is the property of certain crystals to respond to applied stress with the generation of an electric field. Conversely, application of a voltage causes dimensional change.

Piezoelectric phenomena belong into the wider field of *electrostrictive effects*, present to certain degrees (10^{-5} to 10^{-7}% of strain) in most all kind of material. But only in so-called relaxor ferroelectrics does that effect reach 0.1%, which, combined with the material's extremely low hysteretic losses, accounts for their use in high frequency applications.

By contrast, the *polarized electrostrictive effect*, commonly known as piezoelectric effect, is more powerful. For instance, a 250 kgf load applied on a 1 cm cube of quartz may produce 13 kV to 14 kV of electric charge.

From Chapter 5, we recall quartz, tourmaline, and Rochelle Salt crystals as the classical stress to voltage transducers in pickup heads of disk players and similar devices. Though Rochelle Salts top all other piezoelectrics with regard to the voltage they generate, their mechanical fragility and hygroscopic properties limit their use.

The lattice of piezoelectric crystals is thought of as bonded pairs of cations and anions, oriented along the crystal's electrical axes. This determines the way crystals have to be cut in order to accumulate electric charge in response to stress.

Ionic dipoles owe their existence to the trend of metal atoms to bind with atoms of nonmetals by sharing some of their outer electrons in an orbital shell surrounding both atomic nuclei. In the SiO_2 (quartz) crystal, for example, each silicon atom "donates" four electrons from its L-shell to a common shell involving both the silicon and the oxygen nuclei. The resulting lack of electrons in the SiO_2 atoms unsettles the previous equilibrium between positive charges in the silicon's atomic nucleus and the number of orbiting electrons, and leaves the silicon atoms as 4+ cations.

On the other hand, the crowding of two extra electrons into a shell around the oxygen nucleus converts the oxygen atoms into 2− *anions*. The resulting ionic bond between 1 silicon ion and 2 oxygen ions as a whole is still electrically neutral, but the different polarization of its components makes it electrically directional.

Figure 11.9 a depicts the directions of the electrical axes E_1, E_2, and E_3 of an SiO_2 crystal with relation to a standard x, y, z coordinate system, which z-axis stands vertical on the drawing plane. Herein, a crystal cut to a right angle parallelepiped with its surfaces in the x, y, z directions (Fig. 11.9 b) has piezoelectric properties.

Pressure in the x direction causes the *longitudinal piezoeffect* (Fig. 11.9 c), with negative charges on the –x surface and positive charges on the +x surface. Their voltage is independent of the dimensions of the crystal. On the other hand, the am-

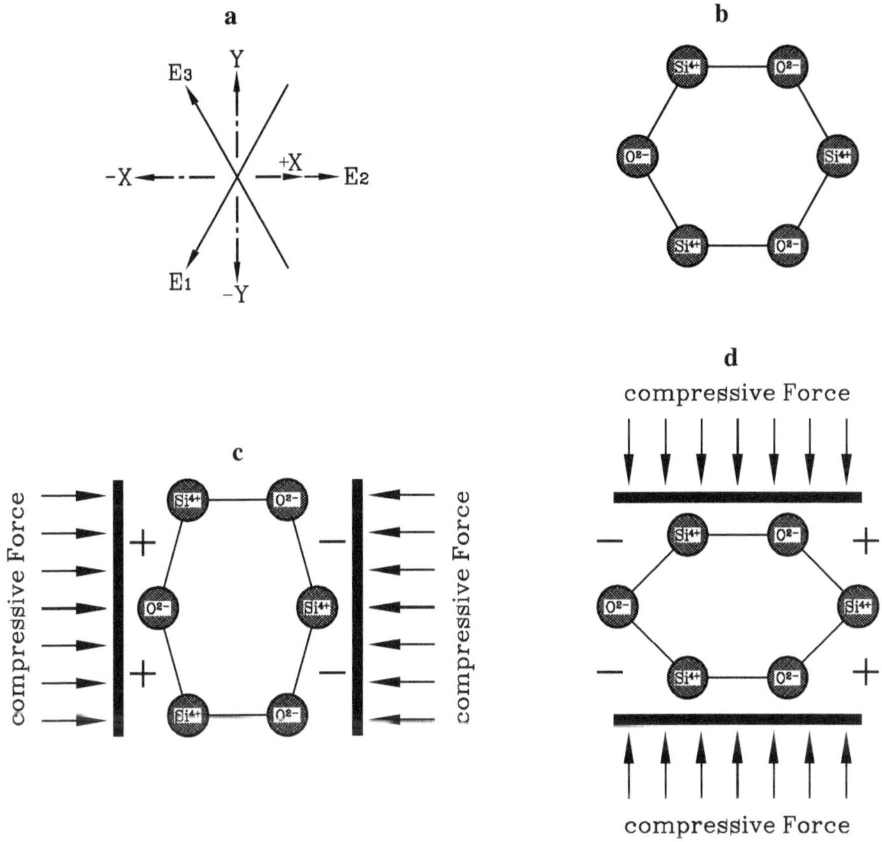

Fig. 11.9. Distortion of quartz crystals

perage (or should we call it microamperage?) drawn from an activated crystal is proportional to the crystal's physical dimensions.

By contrast, pressure in the y-direction (Fig. 11.9d) makes the +x surface of the crystal positive, and the –x surface negative in the so-called *transversal piezo-effect*. In that case, voltage depends on the crystal's physical dimensions.

The open circuit voltage E in response to compressive stress σ on a crystal of length ℓ is given by $E = -g\ell\sigma$, where g stands for the *piezoelectric voltage constant* (not free fall acceleration in this case) in Vm/N. Accordingly, the applied stress σ, the quotient of the applied compressive force and the area it acts upon, has to be expressed in N/m^2 or pascal (Pa), while the length of the specimen must be given in meter (1 m = 100 cm = 1000 mm = 39.370 inches). Further, the conversion from kgf/cm^2 (kilogram-force per square centimeter) or technical atmospheres into absolute units is $1\ kg/cm^2 = 1\ at = 98066.50\ N/m^2$. For lead zirconate titanate, the piezoelectric voltage constant g_{33} is listed as 26×10^{-3} Vm/N. For quartz it is -50×10^{-3} Vm/N. The index 33 indicates identical orientation of the applied force and the electric field.

Ceramic materials that can be made piezoelectrics consist of mini-crystals in random orientation, which must be aligned to develop piezoelectric properties. To a certain point, this resembles our view of ferromagnetic material, imagined as an assembly of magnetic dipoles, which, aligned by an external magnetic field, combine into a permanent magnet.

Just like heating a magnet above the critical temperature of 769 °C disturbs the order of its magnetic dipoles and herewith causes the loss of the material's magnetic properties, piezoelectric characteristics cannot prevail above the so-called Curie temperature, of 133 °C for barium titanate ($BaTiO_3$), 492 °C for lead titanate ($PbTiO_2$), and 350 °C for lead-zirconate ($PbZrO_3$). Gradual cooling of piezoelectric ceramics from above to below their Curie temperature in the presence of an electric field of typically 10,000 volt aligns the dipoles and makes the ensuing piezoelectric characteristics permanent.

Technology of piezoelectrics

Several approaches exist for the conversion of the small changes in size of piezoelectric crystals and ceramics into measurable and usable changes of shape, one of them leaning on the example of bimetallic strips, which we remember as strips of metals of different thermal expansion coefficients, permanently joined by hot rolling. As bimetal strips bend while heated, a sandwich of a thin layer of piezoelectric material bonded to a metal plate, known as *unimorph* (Fig. 11.10, Detail 1), deflects with the application of voltage. In a cantilever configuration, this makes for an electric fan if you add flexible blades at the free swinging ends of the strips and apply a high enough AC voltage between base plate and crystal surface. For effective operation, the resonance frequencies of the unimorphs should match the frequency of the drive voltage or one of its harmonics.

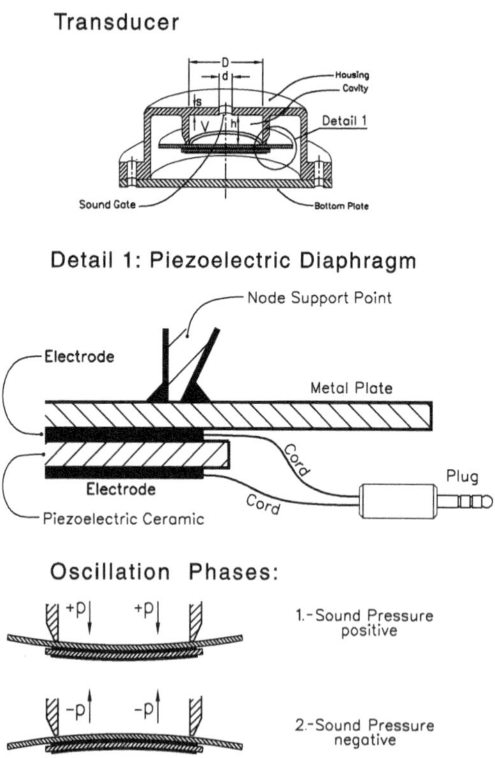

Fig. 11.10. Piezoelectric transducer

Another use of unimorphs is the actuation of relay contacts in response to an applied DC voltage, a job otherwise done by magnets.

Single-pole double-throw (SPDT) switches can be made by applying positive or negative driving voltages, which cause the piezoelectric sandwich to close or open one or an entire set of contacts.

The membrane of the piezoelectric transducer in Fig. 11.10 is a round unimorph. Used as a speaker, the unimorph warps according to the electric signals it receives. As a microphone, it generates a voltage which is picked up by the circular electrodes (Fig. 11.10, bottom).

The resonance frequency f_{res} is given by

$$f_{\text{res}} = \frac{1.648t}{D^2}\sqrt{\frac{E}{\rho(1-\sigma^2)}}\,.$$

Herein stands t for the thickness of the piezoelectric disk, and ρ for its density. E, the modulus of elasticity for piezoelectric ceramics, lies between 600,000 and 700,000 kg/cm^2. σ, the *Poisson ratio*, stands for the quotient of lateral to axial deformation of a test bar under tensile stress and thus quantifies the thinning of the bar's diameter relative to its elongation. For most substances, σ lies between 0.2 and 0.4, and for the usual PZT piezoelectric materials at 0.38. With these material constants (converted in metric units), the square root term in the formula above becomes

$$\sqrt{\frac{6.5\times10^9}{7800\times(1-0.38^2)}} = 986.9\,.$$

For instance, the resonance frequency of a piezoelectric disk of $D=25$ mm diameter and $t=1$ mm thickness results as

$$f_{\text{res}} = \frac{1.648\times0.001}{0.025^2}\times986.9 = 2602.26\ \text{Hz}\,.$$

On the other hand, the cavity of volume V (Fig. 11.10) has the resonance frequency of

$$f_{\text{cav}} = \frac{c}{2\pi}\sqrt{\frac{\text{area of soundgate}}{\text{volume of cavity}\times\text{length of soundgate}}}$$

With $d^2\pi/4$ for the area of the sound gate, s for its length, and $hD^2\pi/4$ for the volume of the cavity, this equation becomes

$$f_{\text{cav}} = \frac{c}{2\pi}\times\frac{d}{D}\times\frac{1}{\sqrt{hs}}\,.$$

For a cavity of $D=25$ mm as above and 8.966 mm of height with a soundgate of $d=5$ mm diameter opening and $s=2$ mm of length, we get with 341 m/s for the speed of sound in air the resonance frequency of the cavity as

$$f_{\text{cav}} = \frac{341}{2\pi}\times\frac{0.005}{0.025}\times\frac{1}{\sqrt{0.008699\times0.002}} = 2602.3\ \text{Hz}\,.$$

The unrounded figure for the height of the cavity has been selected in order to synchronize the resonant frequencies of the piezoelectric disk and the cavity. A program for evaluating those equations can be found at the end of this chapter.

An application of piezoelectric transducers

The *photographic flash trigger* circuit in Fig. 11.11 is meant to freeze short-lived events, such as the passage of a speeding bullet or the implosion of a balloon, in time. It uses a piezoelectric microphone to convert short sound pulses like a shot or the oncoming explosion of a balloon, into pulses of electricity that trigger a photographic flash within a short, adjustable time span.

The signal from the crystal microphone is amplified by the operational amplifier (OpAmp) and reaches the optocoupler's internal light emitting diode (LED) through the capacitor C_2. The ensuing flash activates the likewise internal phototransistor, that provideds gate current to the SCR which, on its part, triggers the flash module. The convenience of an optocoupler[1] to connect the low-voltage signal amplifier and the high-voltage trigger circuit of the flash is the total electrical isolation it provides.

Serial photographs of various phases in the explosion of a punctured balloon, taken with the help of similar circuits, have been published by Andrew Davidhazy from the Rochester Institute of Technology, who also mentions vehicular crash-tests and the view of splintering glass (as from a window pane hit by an errant backyard golf ball).

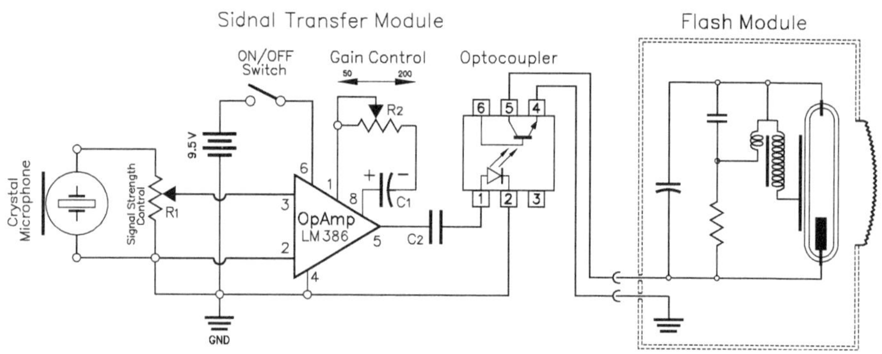

Fig. 11.11. Sound activated photoflash

Ribbon microphone

The ribbon microphone in Fig. 11.12 is an inductive device, consisting of a thin metal strip, mounted under light tension between the pole-pieces of a strong permanent Alnico magnet. As approaching sound waves cause the ribbon to oscillate and cut across the magnetic field lines, the voltage induced between the ribbon's suspension points becomes the audio output of the device.

1 optical signal transmission chip.

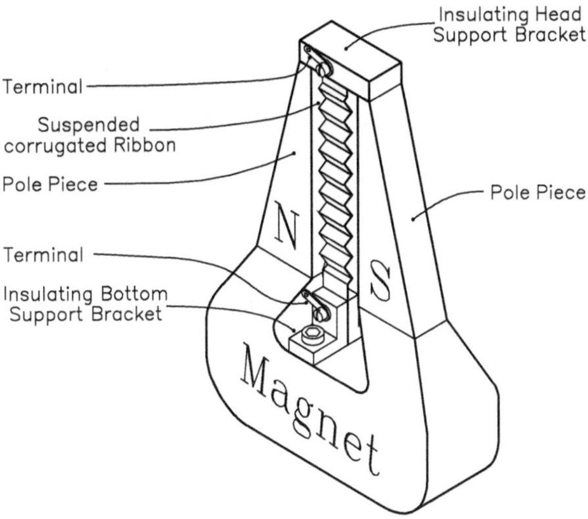

Fig. 11.12. Ribbon microphone (semi-schematic)

This resembles the working of a dynamo, where the wires of the rotor windings cut through the magnetic field lines of the stator. In special applications, such as dynamos powering bicycle lights, permanent field magnets like in ribbon microphones are the rule.

The magnitude of the induced voltage, E, is proportional to the velocity of the ribbon's motions through the magnetic field. With A for the amplitude of the ribbon's oscillations, and x for displacement, we have $x = A\sin(2\pi ft)$ and get for the velocity of displacement $v = \mathrm{d}s/\mathrm{d}t = 2A\pi f\cos(2\pi ft)$.

Due to $\cos\alpha = \sin(\pi/2 - \alpha)$, the shift from $\sin\alpha$ to $\cos\alpha$ is the equivalent of a $\pi/2$ or 90° phase shift between audio in- and output. This explains the often praised particular character of sounds picked up by a ribbon microphone, which some see as a less than perfect reproduction, while others praise the absence of the sluggishness attributed to dynamic microphones. In particular, ribbon microphones are commended for the way they reproduce brass, flutes, clarinets, and song. Their frequency response from 50 to 15,000 Hz comprises most all of the high-fidelity band.

Ribbons with such high degree of responsiveness are made from workhardened pure aluminum of 0.6 μm (0.0006 mm) thickness. Work-hardening makes part of any cold laminating process, and the end-products' hardness depends only on the number of gangs following the latest annealing. Some gain of hardness also happens when the ribbons are sandwiched between tissue and hand peened with a coppersmith's hammer. If you keep that up long enough, the material gets still somewhat thinner, like in the manufacture of gold leaf, but more important, hardens to most any desired degree. Corrugating the ribbon enhances the amplitude of its oscillations, and can be done by squeezing it between a pair of matching spur gears, spaced from each other somewhat above specified center distance so that their teeth don't cut through the sheet. Pitch width varies, but is typically 20 per inch.

The ribbons low ohmic resistance of about 0.2 Ω calls for a step-up transformer to feed into 50, 250, or 600 Ω impedance lines, while the low noise figures of ribbon microphones make their range spanning 120 dB and beyond. But such high responsiveness leads to susceptibility to mechanical stress from accidental impacts and the "blow tests" audio technicians and musicians sometimes use to make sure the mike is turned on and working.

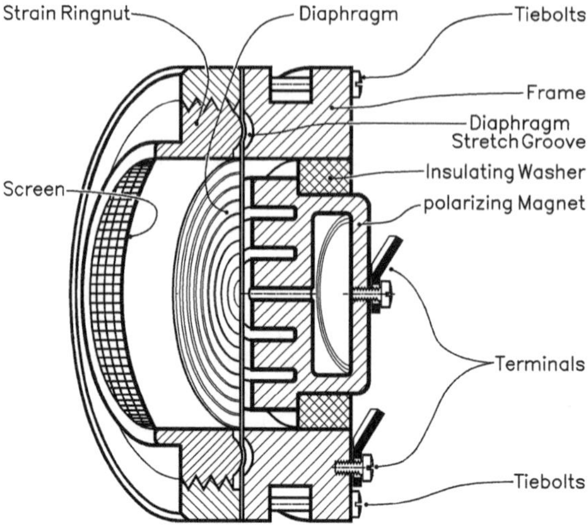

Fig. 11.13. Condenser microphone

Condenser microphones

Condenser microphones as in Fig. 11.13 are often considered the Non Plus Ultra in audio transducer technology. The term "condenser" was common in the early days of electricity for what we now call capacitor. We remember capacitance (in farad, microfarad, or picofarad) as inversely proportional to the spacing between electrodes, the law of physics the capacitive transducer principle is based upon.

The capacitance of a rigid plate with a diaphragm of conductive material in front of it depends on the momentary shape of the diaphragm. A diaphragm of thin enough sheet metal to bend under sound pressure, or better still, a metalized plastic membrane, provides near distortion free sound over the entire range of high-fidelity audio frequencies, from 20 Hz to 20 kHz. Noise figures of capacitive microphones are as low as 60 dB or better. The system works equally well in microphones as in speakers.

For speediest impulse response, the diaphragm can be as small as a quarter of an inch in diameter, which allows to build "pencil microphones" said to reproduce the sound of an orchestra "like being there" – not *despite* but *because* of their minute size.

Like the classic carbon microphone, condenser microphones need an external source of electric power to convert the mike's capacity variations into electric signals. Power supplies can be in-line or located in audio equipment, such as mixing consoles or preamplifiers. Rather than separate power cables from there to the microphone, "phantom powered" audio cables double as conductors for the few milliamperes needed for capacity measurements. However, voltages of up to 48 V are common.

To keep the diaphragm under mechanical stress even in the absence of an input, the condenser microphone uses the flat face of a permanent magnet in lieu of a fixed capacitor plate. The need for prestress becomes understandable by comparison with a wooden board bridging a ditch. The board may easily sustain your bodyweight – and yet – it bends, if only to a minute degree, when a mouse runs across; for a beam under zero stress, even the slightest loads make a difference. Along the same line, the needle of an analog meter oscillates many times around the zero mark and may miss it altogether if only a single hairspring makes it return. Conversely, a pair of prestressed, mutually opposing return hairsprings, places the zero point at the well-defined position where the clockwise torque of one spring equals the counterclockwise torque of the other.

Likewise, the pressure of soundwaves on a prestressed diaphragm shows up as the difference of deflection between the force of the permanent magnet and sound pressure, while a free swinging diaphragm would oscillate in the disturbance-prone zero deflection zone.

The diaphragm is clamped with tiebolts between the two body-rings, and uniformly stretched when the ringnut is tightened down and its bulge pulls the perimeter of the diaphragm into the matching groove in the frame.

An isolating washer centers the magnet in the frame, so that the electrically conductive diaphragm and the mike's body can be grounded, while the magnet gets connected to the "hot wire" from the power supply. A sturdy, vaulted wire screen protects the diaphragm.

Electrostatic speakers

While the electrodynamic speaker uses the force of a magnetic field to convert electrical signals into mechanical oscillations of air, electrostatic speakers employ an electric field to the same end. Here, a very light, electroconductive diaphragm is stretched between a pair of insulating annuli, flanked by electrodes of perforated metal sheet.

Those high-voltage audio signals are being applied in counterphase to the front and rear electrodes of the speaker (Fig. 11.14), while a 2000 to 20,000 VDC bias

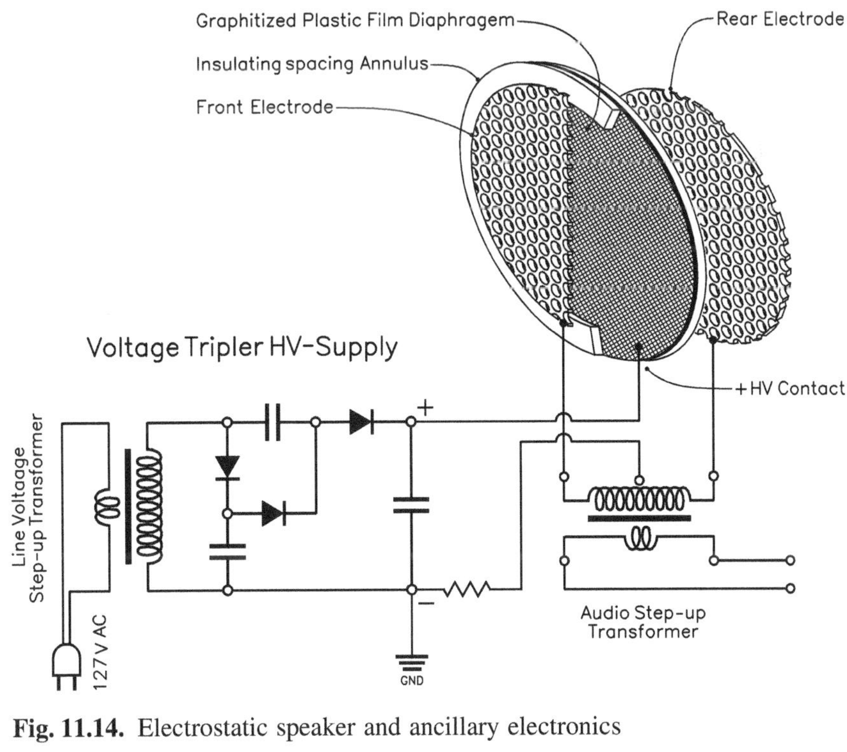

Fig. 11.14. Electrostatic speaker and ancillary electronics

voltage polarizes the electroconductive diaphragm. Stretched out between a pair of insulating annuli, the diaphragm oscillates in step with the audio signals it receives.

High voltage is drawn from a step-up transformer, feeding into a voltage doubler or tripler circuit (Fig. 11.14, left). With audio signals on front and rear electrodes in counterphase, the speaker operates in push-pull mode and is free of the nonlinearity of single-grid speakers.

The critical component of electrostatic speakers is predictably the diaphragm. Consisting of highly elastic plastic sheet with good mechanical strength, such as 5 to 6 μm thick polyester, it is made electrically conductive by rubbing in powdered graphite with cotton balls. The electrodes are perforated metal plates, which openings allow sound-induced air pressure variations to act on the diaphragm. For good sound transmission, the open area of the electrodes should amount to at least 60% the area of the entire plate, though hole size should not exceed 1/4 inch.

The near weightlessness of the diaphragm makes that electrostatic speakers faithfully reproduce all frequencies and amplitudes of sound. This becomes most apparent in the high-frequency band, where inertial forces, proportional to the product of acceleration and deceleration of the swinging diaphragm and its mass per unit area, tend to soar.

A second reason for the near distortion-free performance of electrostatic speakers is that their plastic sheet diaphragms move evenly over most of their surface, while a sheet metal diaphragm deforms into a paraboloid. Resonance frequency of plastic sheet diaphragms can be set by controlling their degree of pretensioning and is usually kept below 100 Hz, so that higher frequencies stay clear of resonance peaks.

An oddball amongst speakers is the LRAD (long-range acoustic device), developed in the aftermath of the terrorist attack on the US *Cole* in 2000 in an attempt to establish a shield of sonar deathrays (or should we call them "deaf-rays?") around anchored military vessels. Essentially a piezoelectric speaker, an LRAD emulates the historical bullhorn on a giant scale. Rather than making a single piezoelectric element act on a diaphragm, it employs a circular field of them, not unlike in appearance to (excuse my burlesque science) mushrooms on a speck of soil as their common base. Used with a microphone, an LRAD works like a super-bullhorn, with the potential of generating up to 150 dB of sound intensity – 60 dB above the roar of a jet engine.

The device got battle-hardened on 5 November 2005, when a throng of pirates, armed with automatic weapons and rocket propelled grenades, tried to board the cruise ship *Spirit* off the Somalian coast. Luckily for the passengers, the assault happened at 5:35 a.m. while they were asleep in their cabins rather than scattered upon the decks, where they could have been shot like sitting ducks. Nevertheless, a rocket-propelled grenade perforated the ship's hull and exploded in a boardside cabin. Anybody in there would have been torn to pieces.

On the other hand, it seems safe to assume that the traditional gun rack we remember from *Mutiny of the Bounty* and similar tales as a must-have piece of furniture in the captain's cabin, was missing on board the *Spirit*; more likely, safety measures must have prevented *any* weapon, even a handgun, from making it on board. So the cruise ship may have had absolutely nothing to match the pirates' armaments, safe for its LRAD bullhorn and the skipper's resolve, likely fueled by remembrances of Michael Durant, the American pilot of a downed helicopter, whose disrobed body had been dragged through the streets of Mogadishu on 4 July 2003, while a jeering crowd lined the sidewalks.

As the cruise ship's communication officer kept the aggressors at bay with full blasts from his 33 inch. LRAD transducer, the skipper, ignoring the attackers automatic weapons, threatened to ram their boats and subsequently headed full speed out into the open seas, leaving the buccaneers behind.

Still, whether LRAD's sonic power rays or the skipper's legally correct "evasive action" saved the day, remains everybody's guess.

Sound tracking

Just as we express the distance between stars and nebulas in light-years, we could introduce the unit of sound-seconds in the measurement of more mundane sets of spaces. With 341 m/s for the average speed of sound in air, the sound-second would measure 341 m or about 1100 feet, or still circa 1/5 of a mile.

Sometimes, we instinctively work with such units, for instance when we use the time lag between seeing a lightening bolt and hearing the thunder for an estimate how far away the storm still is. Likewise, the impact of a log splitter's ax is seen before it's being heard, and the roar from jet engines seems to emerge from a spot way behind the airplane.

With 1460 m/s for the average speed of sound waves in water, the sound-second for underwater listening devices becomes proportionally longer than in air and is the primary unit of underwater piling systems, which rely on the measurement of time intervals rather than of distances directly, though their readouts show such time lapses converted in nautical miles or kilometers.

The Sound Surveillance System (SOSUS) was developed by the U.S. Navy in 1949 in response to snorkel-equipped submarines that had appeared in the waning days of WWII. The snorkel was nothing more sophisticated than a thin-walled steel tube, sealed into an opening in the ship's upper deck. While the boat navigates underwater, the snorkel's upper end clears the waves and supplies atmospheric air to the boat's diesel engines and channels exhaust gases outward.

That simple device extended the range of non-nuclear submarines manifold. While battery powered submarines could navigate underwater a mere 20 miles or so at the speed of 6 miles/h, or 60 miles at 4 miles/h, a snorkel-equipped submarine sailed in 1950 the 8,370 km from Hong Kong to Honolulu in 21 days without having to surface.

SOSUS low-frequency hydrophones countered such unsettling innovations with their ability to spot the sounds of a submarine's diesel engines from hundreds of kilometers. Since then, arrays of detectors have been deployed to estimate a submarine's position by triangulation – a step from simple distance measurement into the principles of cartography.

The strategic importance of underwater ranging peaked during the Cold War, which conversion into a hot (nuclear) war was prevented – so the chronicles report – by the specter of *mutual assured destruction* within the half hour or so it would have taken for a fleet of H-bomb armed intercontinental missiles to cover the distance from the Soviet Union to the United States – time enough for the launch of a similar force from the other side. True believers take the absence of such a catastrophe as proof for the soundness of the idea, while those addicted to formal logic contend that the nonoccurrence of an event isn't proof of the effectiveness of one specific countermeasure. In short, the outbreak of a nuclear war would have proven the hypothesis of mutual assured destruction wrong, but the absence of such a war does not necessarily prove it right; much like your *not* suffering from a common cold while you read this doesn't justify the (appealing) hypothesis that you are immune to this year's strain of viruses. Conversely, finding you with a running nose would constitute proof "beyond reasonable doubt" of your lack of immunity.

But the subtle equilibrium of nuclear forces during the cold war period suffered once again with the event of nuclear powered submarines in the early 1960s. They could round the globe under water, and their range – if laid out in space – would go to the moon and back. In 1962 the USS *Ethan Allen* test fired an A-1 Polaris missile, armed with a W-47 nuclear warhead. The missile's lightweight fiberglass body extended its range to over 2800 miles. Later versions of Ethan class submarines carried up to six nuclear-armed multiwarhead missiles, launchable from under water. Meanwhile, the Soviets got their nuclear-powered missile-carrying submarines in and under water, and a feeble equality was reestablished when *our* destabilizing vectors got equipped with hydrophones to track *their* destabilizing vectors, and vice versa.

Which brings us back to sound ranging. Noise from a ship's propeller can be detected over hundreds of kilometers by training the hydrophone on the hostile boat's *baffles*, a conical zone where turbulence from the ship's screw generates random eddies. While baffle noise guides the tracking submarine, it prevents the hunted from detecting the presence of the pursuer. Only by "clearing the baffles" by sudden sideway turns can it momentarily check whether or not it is being tracked.

Echo sounding

Unlike acoustic listening devices that depend on sound from external sources, echo sounding comprises an autonomous system that generates its proper sound signals and measures their reflections from selected targets.

Echoes from yippee shouts of summit-conquering tourists telling the world of their feat take twice as long to be heard as it takes the sound to cover the distance between the caller and the sound-reflecting rock face. On the other hand, the reflecting wand cannot be too close, because the human auditory system needs at least 0.1 s for readying itself for a new sound excitation after the previous one fades away. Therefore, echo can only be heard at distances of $0.1 \times 341/2 \approx 17$ m and above. Below that threshold, we hear the echo superimposed with the original sound, a phenomenon known as reverberation. Anacroic chambers, used in tests of audio gadgets, are free of such resounding effects due to their spiked wall coverings, that could double as torture chambers in horror movies if it weren't for the soft, porous material they are made from.

Electronic echo known originally as sonar have become ubiquitous as depth sounders in all kinds of vessels, from recreational to the world's biggest oil tankers. Echo sounding instruments include an audio generator, pulse shaper, amplifier, and transducer; the latter mounted in an opening at the bottom of the hull and scanning the ground with a cone of sound waves that, at 3000 meters of depth, covers a circle 100 yards across.

In seawater, the speed of sound depends on temperature, depth, and salinity. Several equations of various degrees of complexity have been developed, the handiest still being the *Mackenzie* formula:

$$c = 1448.96 + 4.591T - 0.05304T^2 + 2.374 \times 10^{-4}T^3 + 1.340(S - 35) + 0.0163D + 1.675 \times 10^{-7}D^2 - 0.01025T(S - 35) - 0.7139 \times 10^{-12}TD^3$$

Herein, T stands for water temperature in degrees Celsius, S for salinity in parts per thousand, and D for depth in meter.

Taking 1460 m/s as the average allows to figure the depth per 1 second delay between emission and reception of a sound pulse as $1460/2 = 730$ meter, or 0.01 s in waters 7.30 m (24 feet) deep. Pulse duration must be kept at a fraction of this time lag as to keep the echoes from interfering with their originating pulses, and yet, each pulse must accommodate a reasonable number of audio oscillations. Cramming, for instance, 12 audio oscillations into 1 millisecond (ms) of pulse width calls for 12000 Hz of frequency. Simple depth sounders transmit at 12 kHz frequency, while lower frequency signals of typically 3500 Hz are used for penetrating the seafloor and show the region's layers of accumulated sediments.

The classical analog readout device for echo sounders consists of a rotating disk with a neon lamp on its perimeter. If T is the pulse spacing and the disk is made to turn by 1 rotation in that same time interval, the pulses cause the lamp to flash at each crossing of the angular zero position. The second flash, triggered by the reflection of the pulse, then happens at an angle proportional to the travel time (down and up) of the sound pulse with relation to T. Halving the speed of the disk's rotation makes that angle proportional to a one-way trip. For instance, for $T = 1$ s, a 30° (1/12 of the full circle) angle between neon flashes means 1/12 s total sound travel time or 1/24 s for the time down to the seafloor. This converts into $1460/24 \approx 61$ meter of depth. Conversely, the slow audio pulses we remember from Stingray movies suggest measurements of great distances, where greater pulse width and lower frequencies can be used.

Apart from marine applications, echo sounding helps to locate buried deposits of raw material, such as metal ores, layers of coal, and principally pockets of oil and natural gas.

Ultra- and hypersound

Much like the spectrum of light spreads beyond the range of visibility, the spectrum of sound frequencies is far broader than the 20 Hz to 20 kHz the human ear is capable of detecting. Dogs' hearing range of up to 40 kHz, is twice that of their human masters, but lags far below the 150 kHz of bats.

Sir Francis Galton (1822–1911), famous for his system of classifying fingerprints, was first to knowingly generate ultrasound with what we now call the dog whistle. It consists of a fine tubing directing a jet of air into a resonating cavity. Though the frequency of the Dalton whistle did not surpass some 22 kHz, this instrument is considered the first ultrasonic generator in history, as it fits into the definition of ultrasound as the frequency band from 20,000 Hz to 10^9 Hz (1 GHz). Beyond lies the domain of *hypersound*, whose theoretically predicted limits border on 10^{13} Hz (10 terahertz). Hypersound is present in fast-moving streams of gases, such as the exhausts of ramjets.

Ultrasound is capable of transmitting far higher power than audible frequencies, because sound pressure is proportional to frequency, while acceleration of the swinging particles grows with the square of frequency. Signals from ultrasound sirens with 200 to 300 watt of power are said to be capable of lighting up a cotton ball in their pass.

The core of most ultrasound generators are 1 mm thick disks of crystalline quartz with 3 MHz eigenfrequency. Their harmonics can be used for the generation of still higher frequencies.

Medical applications of ultrasound lead to treatments that would have been impossible with the use of more conventional means. Most striking is the treatment of gastrointestinal disorders, such as kidney and gallstones, by fragmentation with focused high-energy shockwaves with subsequent elimination of the particulate remainders through the urethra or, respectively, the biliary tree. The idea of crushing the kidney stones by mechanical means and flush out the debris reaches back to Sir Philip Crampton of Dublin (1834), but was firmly introduced only in 1876 by Harvard Professor Henry J. Bigalow, who practiced and popularized this procedure, termed by then litholapaxy.

For the removal of noncancerous grows, such as noncirrhotic liver and pancreatic tumors, a surgical aspirator, named Cavitron, does the job with less blood loss and reduced injury to collateral tissue than the classical scalpel. Essentially, the Cavitron is a combination of an ultrasound vibrator with a suction device, operating at the lower band of ultrasound frequencies.

Conversely, the *harmonic scalpel* uses higher frequencies, typically 55 kHz, to dissect tissue while the heat generated in the process coagulates small blood vessels with minimal damage to adjacent tissue.

But medical procedures make only for a part of the many applications of ultra- and hypersound. Industrial uses include the detection of discontinuities in castings (slag or form-sand inclusions and trapped air bubbles) by observation of their reflections of an ultrasonic beam.

Ultrasonic vibrations allow for emulsifying immiscible liquids, such as oil and water. Aluminum and other hard to weld metals can be joined by pressuring them together with a tool called sonotrode, essentially a metal pin vibrating at 20 kHz to 40 kHz of ultrasonic frequencies. A similar process allows for welding of plastic material.

Ultrasonic drilling of metals and stone is based on the oscillatory motion in the range of 25 kHz and 40–80 μm amplitude of a plunger acting on an abrasive paste or slurry, that in some cases contains microscopic diamonds.

Sonography

Medicinal imaging is the field most affected by ultrasound technology as it dramatically changed the ways of medical diagnosis. It began in 1942 with the publication of Austrian Dr. Karl Theodore Dussik on an ultrasonic investigation of the brain, followed in the 1950s by work of Prof. Ian Donald in Glasgow. Like echo-sounding, sonographic imaging uses the time lag between emission of a signal and reception of its echo to have the ultrasound generator's central processing unit (CPU) estimate the distance of the reflecting object. On the other hand, the intensity of the reflection determines the degree of image blackening.

Medical examinations use 1 to 13 MHz, depending on demands on resolution and penetration. The short wavelength of highest frequency oscillations offer best resolution, but suffer from strong attenuation. 3 to 5 MHz provide highest penetration of deeper tissues.

2-D images are generated by scanning the sound beam over the entire area under investigation, either by periodically tilting the sound transducer into a pattern similar to that of horizontal and vertical deflection of the electron beam in television picture tubes, or by use of a phased array of piezoelectric swingers, arranged similar to the bars of a xylophone. They are sequentially excited by an oscillator-amplifier to come on one after the other.

Since the pictures we use to take with our cameras show the intensity of the reflections of light waves from the object, one would expect sonographic images, made with ultrasound rather than light as signal carrier, to do essentially the same – with the difference that short sound waves penetrate biologic tissues and thus produce pictures deep inside. However, that is not what happens in sonography. While one single variable – the degree of reflection – determines the darkening of a photographic film, sonography adds a second variable: depth of penetration. Still, the intensity of sonic reflections determines the degree of blackening (or whitening) of the image, but depth of penetration, derived from the time delay between incident and reflected waves, is used by the machine's central processing unit to reconstruct a *lateral* picture of the probe, as one would get it from a camera capable of eyeing around the corner. A sonographic image of the eyeball, for instance, shows a lateral cross section of the eyeball rather than of the lens with the retina behind it. Problems like the separation of the retina from the choroid can thus be identified. That's the fundamental difference to other imaging techniques.

X-ray images show bone structures clearly, but soft tissues and body fluids leave only fleeting shadows. Ultrasound, reflected by the boundaries between fluid and soft tissue, and tissue and bone, step in where the imaging power of X-rays ends. Moreover, the added dimension of depth makes ultrasonic imaging unique.

Received by the same probe that emitted the ultrasound in the first place, signal strength of the reflections depends on Z, the opposition to sound flow through the reflecting surface (also known as the specific acoustic impedance of the medium), and relates to the reflection coefficient R by $R = (Z_1 - Z_2)/(Z_1 + Z_2)$, wherein $Z_1 = \rho_1 c_1$ and $Z_2 = \rho_2 c_2$. c stands for the speed of sound in the medium, and

ρ for the medium's density. The *pressure amplitude transmission coefficient* is simply $(1+R)$.

If sound passes from lower to higher impedance, reflected and incident waves are in phase. They suffer a 180° phase shift in the passage from higher to lower impedance.

Densities of muscle and most solid organs, such as kidney, liver, and spleen, lie between 0.99 and 1.08, and sound propagation velocities range from 1500 to 1600 m/s. That places the acoustic impedances of those and similar media between $1500 \times 990 = 1.485\ \mathrm{MN} \times \mathrm{s/m}^3$ and $1600 \times 1080 = 1.728\ \mathrm{MN} \times \mathrm{s/m}^3$ (Rayles). Thus, the highest reflection coefficient we may expect from boundaries between biological substances is $(1.728 - 1.485)/(1.728 + 1.485) \approx 0.0756$. The related transmission coefficient is then 1.0756, showing the high accuracy afforded to get the well contrasted medical ultrasound pictures we are used to.

Most of ultrasonic imaging's popularity comes from viewing the unborn baby during pregnancy, where gender, size, and position of the fetus can be determined. In cardiology, infections of the heart valve (endocarditis) and aortic stenosis can be identified. Gallstones and kidney stones become clearly visible and their location can be determined by measuring it on the image with a ruler like the one you keep on your desk.

In a somewhat different application, Doppler shift in ultrasound is used to determine the speed of circulating body fluids, such as the passing of blood through a heart valve – much like the redshift of spectral lines in the spectra of stars is indicative for the object's radial velocity.

The wavelength of 500 MHz ultrasound in air equals that of the green light (about 0.6 μm or 6000 angstrom), but similarities end there, because sound waves are mechanical oscillations, while light waves are electromagnetic oscillations. In certain applications however, ultra- and hypersound get the better of light as they can penetrate opaque matter.

Hypersound frequencies, starting at 1 gigahertz (10^9 Hz) cannot propagate in air, and suffer strong attenuation in liquids. In solids, secondary emissions of transversal mechanical waves can combine with the longitudinal primary oscillations into the familiar oscillation patterns of the atoms, known as heat.

Hypersound vibrations get transferred to the human body with a handheld sonography probe, called linear array scan head. Externally, it loosely resembles a barber's mechanical haircutter, but houses one or several acoustic transducers. The probe seals to the patient's body with a water-based gel.

Sonoluminescence

Sonoluminescence was discovered in 1934 by H. Frenzel and H. Schultes at the University of Cologne through the appearance of tiny black dots on photographic plates which they developed in an ultrasonically agitated photographic bath. Though the researchers identified flashes of light, generated by the collapse of microscopic bubbles of air in the developing fluid as the culprits, the topic was not followed up until in the late 1980s.

It turned out that single air bubbles trapped acoustically in a water-filled resonance chamber emit short light pulses as they collapse from their original size, of circa 50 μm, down to 0.5 μm in diameter. The continuous spectrum of the emitted light shows the characteristics of black body radiation and reaches far into the ultraviolet, so that the temperature in a bubble has been estimated around 10,000 K.

While most experimenters use a globular flask of quartzglass in their setups, Christopher J. Visker used cylindrical Plexiglas containers with a partially immersed "ultrasonic horn". The latter consists of a piezoelectric disk, cemented to the end of a round aluminum road, which length had been truncated for resonance at the proper frequency of the apparatus (21 kHz). Likewise, the height of liquid in the container equals the wavelength of sound at this frequency. Bubble generation is initiated by pipetting a drop of fluid on the liquid surface.

If properly attuned, the arrangement generates a field of standing waves in the fluid, which displacement nodes, the points of least displacement of liquid elements, trap the bubbles in place, until the approach of the following pressure antinode, the point of highest pressure in the fluid, makes the bubbles collapse. Flashes from imploding bubbles then appear at 40% and 80% of the liquid level along the center axis of the resonance chamber.

Some theories explain the flashes of light simply as the impact energy of the bubble's sudden collapse. Though the resulting temperatures were estimated around 10,000 K, hopes were still high for some time that they could somehow be brought up to the threshold of nuclear fusion, though that hasn't happened so far.

Addendum

GW-Basic Evaluation Program for PZT-Piezoelectric Transducers. File Name: "Polymrph"

```
10   SCREEN 9
20   PI=3.1415926                          π
30   D1=0.025                              'internal diameter of cavity'
40   D2=0.005                              'diameter of sound gate'
50   S=0.002                               'length of sound gate'
60   H=0.008699'                           'internal height of cavity'
70   T=0.001                               'thickness of piezoelectric disk'
80   SY=0.38'                              'σ=Poisson's Ratio'
90   E=6.5E+9                              'modulus of elasticity'
100  RO=7800                               'ρ=density of piezoelectric disk'
110  C=341                                 'velocity of sound'
120  X=SQR(E/(RO*(1-SY^2)))                'math substitution'
130  F1=X*1.648*T/D1^2                     'resonance frequency of piezo-disk'
140  F2=(C/2*PI))*D2/D1)/SQR(H*S)          'resonance frequency of cavity'
150  H=(((D1*D2*C)/10.355*T))^2)*          'height of cavity'
       RO*(1-SY^2)/(S*E)
160  PRINT F1, F2, H                       2602.26 2602.28 0.008699
```

12 Electrical and electronic instruments

In the late 1990s, General Motors ventured into a trial run of 1100 electric cars in compliance with California's legislation, which demanded dominance of electric vehicles by the year 2000. Dubbed EV1, 800 such cars were initially leased for 3 years to third parties, but as soon as the lease expired, got heartlessly crushed along with all the rest of zero-emission cars.

Why did this foray into the kingdom of unpolluted air, intact ozone layer, and freedom from the global warming threat, end in such an undignified demise? Should GM have followed Alva Edison's market strategy of selling the first two-years production of lightbulbs at a loss in order to generate a market for his invention wile fending off potential competitors? After all, why should consumers have coughed up some USD 30,000.00 for an electric car while gasoline cars were available at 1/3 of that price? Let alone the costs for replacement batteries within the first 3 years of use. But even that could hardly have been the reason for the regulators to let go of such a popular law so willingly. So, what happened?

By comparison, the familiar gasoline-guzzling car with an average of, say, 20 miles to the gallon of gas, uses up 5 gallons per 100 miles driven. With 0.68 for the density of commercial gasoline, and 3.785 liter to the gallon, those 5 gallons convert into $5 \times 3.785 \times 0.68 = 12.87$ kg of gasoline. From the heat of combustion of gasoline, 11700 Cal/kg, and the conversion factor of 860 Cal = 1 kilowatt × hour (kWh), the electric equivalent of the 5 gallons (12.87 kg) of gas per 100 miles becomes $12.87 \times 11700/860 = 75$ kWh per 100 miles. However, the output of the internal combustion engine in our cars is limited to the Carnot process value, namely, some 34% of the energy input, while 3% of the remainder get lost in the gearbox and power train; that leaves us with about 32% of 175 kWh "energy on the wheels", namely, $0.32 \times 175 = 56$ kWh.

To compare this performance figure with that of an electrically powered car, we figure backwards from those 56 kWh to find out how much energy must be generated to supply them.

Utilities generate electricity at a somewhat better rate than the engines in our cars, namely, at about 40% and up, but some 7.2% of their power output is lost in transmission through power lines and substations.

On the other hand, electric cars consume extra energy in a number of ways: First of all, present days batteries deliver in the average only 0.80 kWh for every 1 kWh they get charged with, and the car's electric motors can be expected to run at about 90% efficiency. Here again, another 3% go into frictional losses in the cars' transmission train.

Summing it all up, we get $20 + 7.2 + 10 + 3 = 40.2\%$ of losses, which brings the demand of an electric car to $56/(1 - 0.402) = 93.65$ kWh per 100 miles. Generated at 40% efficiency, this becomes $93.65/0.40 \approx 234$ kWh utilities have to come up with. Compared with the 175 kWh per 100 miles of the standard car, this is $234/175 = 1.34$ times higher.

Rather than providing clean air, the use of electric cars in the cities would boost CO_2 pollution by a whopping 34%; however, electricity for charging up those cars would be produced far from the cities where they strive. Call it the counties' contribution to better city air, but it still wouldn't help cleaning the atmosphere or mobilizing voters in the project's support.

On the other hand, taking USD 1.91 per gallon as the expected peacetime price of gasoline, the 5 gallons for a 100 miles drive would cost USD 9.55, while the electric car's 93.65 kWh cost, at the going rate of about 0.055 USD/kWh, USD 5.15. What a bargain that would be! But before you rejoice, consider the following.

The core of the matter is in the question if utilities could generate the electricity for a nationwide fleet of electric cars. If driven 10,000 miles per year, each car would demand some $234 \times 10000/100 = 23400$ kWh per year. And even the equivalent of America's total production of electric power in 2004, of 3,953,407 kWh, could supply power to no more than $3953407/23400 = 169$ million cars. Spread over the country's population of about 300 million, that means $300 \times 10^6/169 \times 10^6 \approx 1.8$ persons per car; which, I guess, comes close to where we stand.

In short, going electric would demand a doubling of America's generating capacity: twice as many coal and oil powered thermoelectric plants, hydroelectric plants, nuclear plants, wind power plants – you name it. An ambitious undertaking indeed. But enough of philosophy.

Back on the solid grounds of electrical engineering, the familiar units of volt and ampere.

Volt and ampere

The units volt and ampere are best understood by comparing the flow of electric energy through wires with the flow of water in pipes, so that the rate of liquid flow per unit of time becomes the equivalent of amperage, and fluid pressure the equivalent of voltage. This model works nicely as long as we don't get into induction, inexistent in mechanics (except maybe in the ethereal spheres of flipping dowsing rods). But first of all, let's propose – no, not a toast – but one more warning label on electrical equipment, reading: "Forget the voltage – mind the milliamps!"

Do it yourselfers who survived an occasional discharge from a car's 15,000 volt ignition system, or the still higher voltage luring on the coating of an oldtime television picture tube, will agree. Conversely, low-voltage sources can be deadly. A tale from the early days of electrification reports on the "absolutely safe" electric installation in a newly built hotel which turned out as shock-proof as the *Titanic* was "sink-proof." Based on the idea that up to 60 volt were hardly felt and thus harmless, the builders converted utility voltage down to that level and rested on their laurels; until the morning when a young woman was found dead in her bed – electrocuted.

Impossible – or not? Sure, death from contact with a 60 V wire is hard to imagine, unless that unlucky woman clasped the bare part of the socket of the light bulb in search for the switch while holding the lamp's grounded body with the other. With 1000 ohm for the average resistance of the human skin between left and right hand, the current would have been $60/1000 = 0.06$ A or 60 mA – enough to cause an involuntary contraction of the muscles in her fingers that shattered the bulb.

The thin sheet metal of the socket could then have cut her skin and touched the tissue below, which conducts electricity several times better than the epidermis, making the electric current through the woman's body build up to some 100 milliamps (0.010 A) and beyond – enough to trigger ventricular fibrillations of the heart muscle.

On the other hand, I have seen European utility workers grasp bare 380 VAC conductors one by one in order to check if they had heated up, and these guys are still alive and kicking.

An amazing contradiction – the more so if we recall that the peak-to-peak voltage in a 380 VAC line is $2 \times \sqrt{2} \times 380 = 1075$ V. But not even that did matter as long as those men stood on the dried-out solid wood floor of the factory building, isolated from ground potential like birds that gather unpunished on the length of electric lines for their yearly migration South. But attempting the same on a rainy day in the open, or while leaning against the grounded body of a machine tool, could be deadly. Likewise, stunt men should not try their high bar performance on electric cables. Publicity would be superb, but the conductors of opposite polarity dangle far too close for comfort.

But enough of the dark side of electricity. The more productive aspect of measuring electric currents and voltage starts with an instrument designed by Jacques-Arsène d'Arsonval (1851–1940).

D'Arsonval galvanometer

The d'Arsonval galvanometer makes part of the group of *moving coil* DC meters which deduce amperage and voltage from the angular deflection of a pivoting coil suspended in a static magnetic field.

The force F (Fig. 12.1) on a single wire in a magnetic field acts at right angles to both the direction of the field and the direction of the electric current (*left-hand rule*). Its magnitude depends on B, the magnetic field strength in tesla, ℓ, the length of wire cutting the magnetic field lines (in meter), and J, the amperage of the current through that wire, according to $F = B\ell J$, which gives the magnitude of the force in newton.

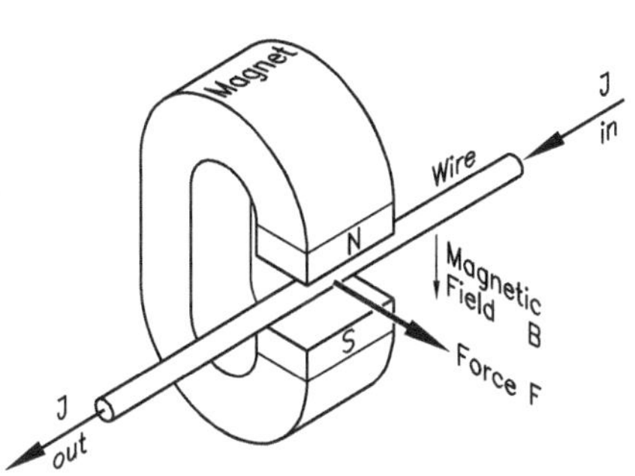

Fig. 12.1. Force on conductor in magnetic field

In any other direction, diverging from that of F by the angle φ, only the component $F \times \cos\varphi$ acts on the wire. That's the reason for the sinusoidal output of alternators, where φ stands for the angular position of the rotating armature.

Likewise, the scale of a moving-coil galvanometer using a regular magnetic field would mirror a sinusoidal function. However, a regular scale can be obtained with a *radial* magnetic field, where $\varphi = 0$ holds true for all possible positions of the spool, because the wires of the coil move orthogonal to the radially oriented field lines throughout. In that case, deflective force F on the spool is independent of φ and proportional to $B\ell J$ only. For a given instrument, magnetic field strength and the length of wires within the field are invariable, so that the force on the spool depends solely on the amperage flowing through the windings.

The d'Arsonval galvanometer in Fig. 12.2 is the classical DC instrument, measuring voltage and amperage alike. In a voltage meter, the spool consists of many turns of thin copper wire, while the ammeter uses thicker wires and fewer coils. Impedance is 1 to 10 megohm for voltmeters, and 100 ohm or less for ammeters.

The radial magnetic field is generated within the enclosed circular opening in the magnet body, which consists of a rectangular Alnico (aluminum-nickel-cobalt)

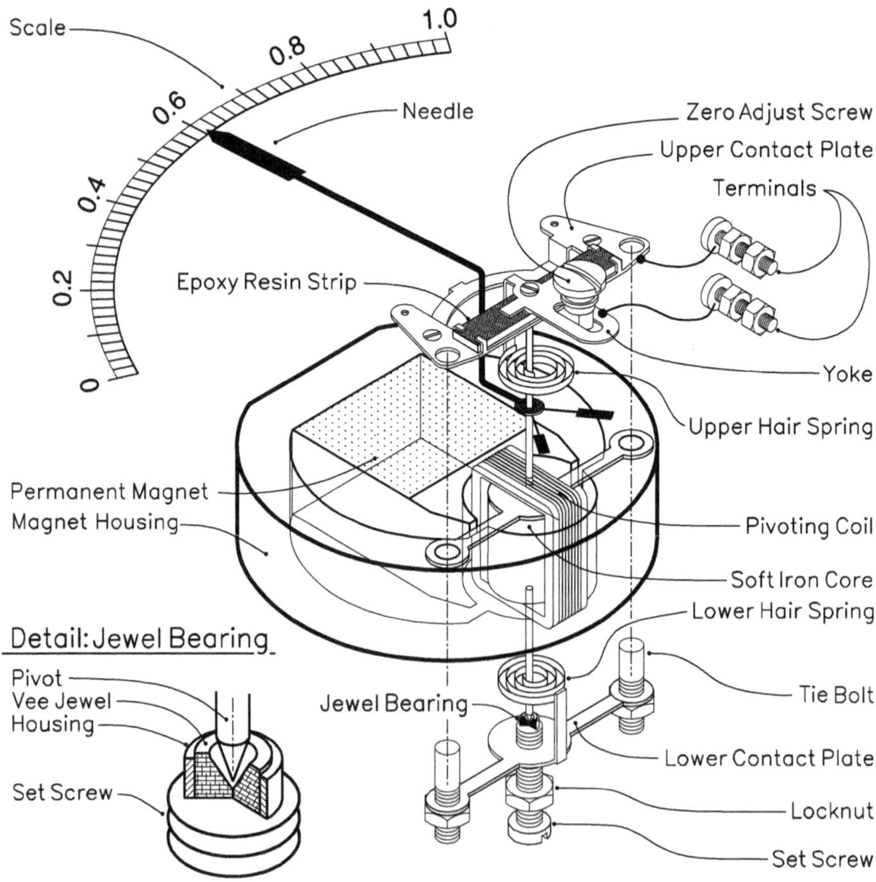

Fig. 12.2. D'Arsonval galvanometer

permanent magnet, housed in a body of ferromagnetic material with a round opening for the spool. This body can be die-cast or made from a package of cut-outs from dynamo sheet.

A soft iron cylinder inside the spool keeps the size of the air gap as narrow as possible and thus maximizes the magnetic flux. Rather than letting it rotate along with the spool, it is bolted to the magnet housing, so that the spool remains the sole rotating element. Its low mass helps the sensitivity of the instrument and prevents sluggish operation.

Along this line, the form for the spool is drawn from thin aluminum sheet, and the wire gauge of the windings is tailored to the necessary for carrying the instrument's nominal amperage. As the presence of the fixed core precludes a straight-through shaft, the spool turns on a pair of opposed semiaxles, attached to the form with isolating bushings. Their pointed ends turn in jewel bearings similar to those in mechanical watches.

Jewel bearings (detail on lower left of Fig. 12.2) are cylinders of synthetic sapphire with a hardness of 9 on the Mohr scale[1] and have a conical cavity with a spherical bottom of about 0.001 inch radius. With the ends of the semiaxles rounded as well, but at a somewhat smaller radius than the tip of the cone, the axles follow epicycloidal orbits in lieu of frictional rotation.

Each hairspring has its inner end attached to one of the semiaxles. Wound in opposite directions, they provide the countertorque in the spool's deflection and bring it back into zero position if no current flows through the spool. Additionally, they are used as the conductors between the free ends of the spool's windings and the input terminals.

The lower hairspring with its inner end clamped or soldered to the lower semiaxle makes that connection through a tongue on the contact plate and the tiebolts (shown in part by their centerlines) to the upper contact plate, which is wired to the negative input terminal.

The upper spring has its outer end attached to a tongue on the yoke, which is wired to the positive terminal and isolated from the upper contact plate by a sheet of fiber-reinforced epoxy resin similar to what we recall from printed circuit boards.

Fine adjustment of the pointer's zero position happens by the turn of a screw reaching through the instrument's transparent front cover. In the process, an eccentric guide pin protruding from the end of that screw adjusts the angular position of the slotted tongue of the yoke.

Some very sensitive laboratory instruments have the spool suspended between fine wires or threads for virtually frictionless rotation. Here again, stable zero positioning can be achieved by twisting one of the strings by a few turns prior to clamping it into position.

The range of d'Arsonval volt- and ammeters can be extended with series or shunt resistors. Users of the ubiquitous multimeter are familiar with multirange scales, such as 0–1.5 V, 0–5 V, 0–15 V, 0–50 V, 0–150 V, 0–500 V, and 0–1500 VDC.

1 the hardness of talc ranges at 1 and that of diamond at 10 on the Mohr scale.

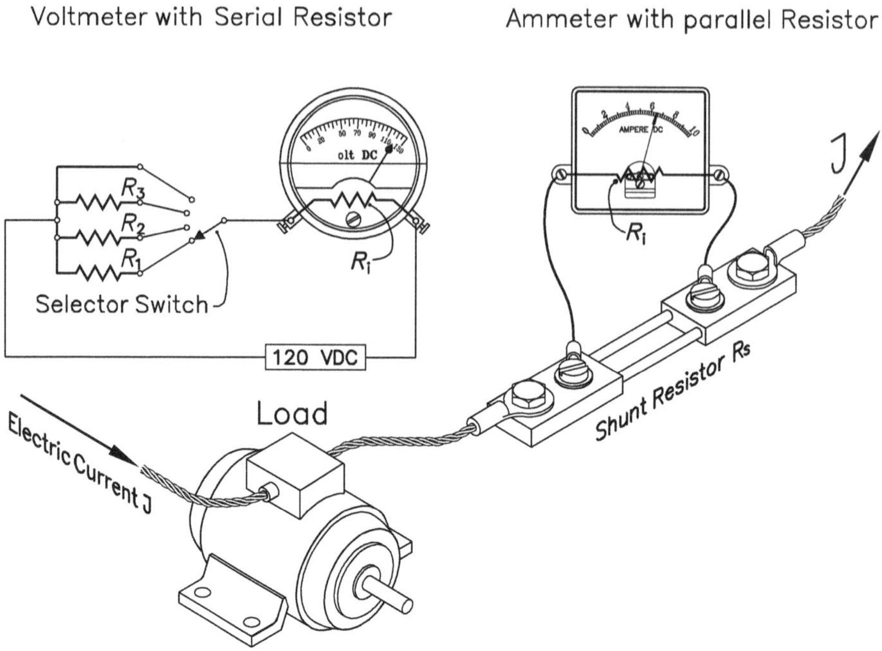

Fig. 12.3. Ammeter scale extension by serial and parallel resistors

Typical amperage scales reach from 0 to 50 μA, 1 mA, 10 mA, 100 mA, 500 mA, and from a separate input socket, from 0 to 10 A.

Voltage range extension by serial resistors is a simple and straightforward process, demonstrated in Fig. 12.3, upper left, for 4 different voltage ranges. With R_S for the ohmmage of the serial resistor and R_i for the instrument's internal resistance, the ratio n of the instrument's original range to the desired range is figured from the fact that the electric current, J, remains the same throughout the system. The coil being an inductive load is of no bearing in DC techniques, so that the internal resistance R_i stands here for the ohmic resistance of the copper wire the coil is made from, and Ohm's law can be used: $E = J(R_i + R_S)$. Further, we know (also from Ohm's law) the voltage drop E_0 over the instrument as $E_0 = JR_i$, which is the readout. To eliminate J, we divide the first equation by the second and get $E/E_0 = (R_i + R_S)/R_i$. With $E/E_0 = n$ for the ratio of the original and the new range, this gives $R_S = R_i(n - 1)$.

For instance, to extend the range of a 6.5 volt instrument to the 130 volt shown in Fig. 12.3 calls for an extension factor of $n = 130/6.5 = 20$. For a meter with 13 kΩ internal resistance, the value of the serial resistor (R_1 in Fig. 12.3) becomes $R_1 = 13(20 - 1) = 247$ kΩ.

Likewise, we can include scales for respectively 65 V and 13 V, using $n = 10$ and $n = 2$, which gets us $R_2 = 117$ kΩ and $R_3 = 13$ kΩ. On a scale of 26 spaces, every mark would represent 5 V on the 130 V scale, 2.5 V on the 65 V scale (that is, 5 V for every one of the longer marks), 0.5 V for the 13 V scale, and 0.25 V for the 6.5 V scale. As shown, scale factors should be selected in ways that accommodate the scale's divisions throughout.

Unlike voltmeters, wired in parallel with the principal circuitry, ammeters are in series with the load, so that an accidentally open ammeter means a general shutdown. Few managers are open to the idea that knowing the amperage of an electric motor is worth the risk of looking at an idle production line while the maintenance department promises an ASAP repairman. A lower-range ammeter with a parallel resistor for range extension is much safer as long as a set of rules such as the following is observed for the connection of those extremely low ohmmage resistors.

Meter and shunt terminals should not be simply screwed one over the other, because a corroded junction could conduct the load current through the ammeter and burn it out explosively.

Therefore, shunt resistors (Fig. 12.3, bottom) have separate terminals for the heavy gauge load-carrying wires and, respectively, the wires connecting the meter. That way, a defective ammeter has no bearing on the circuit's overall functionality, and a bad connection on a load-carrying terminal will not ruin the meter.

With J_i for the amperage through the meter, J_S for the amperage through the shunt resistor, and R_i and R_S respectively for the internal resistances of the meter and the shunt resistor (Fig. 12.3, right), the voltage drop over the two in parallel becomes $\Delta E = J_i R_i = J_S R_S$ from what we derive $J_S/J_i = R_i/R_S$. The scale factor n, defined as the ratio between the new and the old scale, becomes $n = (J_i + J_S)/J_i = 1 + J_S/J_i$ which we rearrange into $(J_S/J_i) = n - 1$. Combined with $J_S/J_i = R_i/R_S$, this yields $R_i/R_S = n - 1$ or $R_S = R_i/(n - 1)$.

For instance, if we want to use a 1 milliamp (1 mA) meter with 120 ohm internal resistance to read up to 1 A, the scale factor would have to be $1/0.001 = 1000$ times, and a shunt resistor of $R_S = 120/(1000 - 1) = 0.12012\ \Omega$ is needed. Any inaccuracy in the manufacturing of the resistor would appear 999-fold in the instrument's readout. That makes shunt resistors industrial precision products, while series resistors for voltmeters are often selected from commercially available precision resistors.

In electric power engineering, the amounts of energy used up by instrumentation is insignificant in comparison with the total of the circuitry. For instance, a 10 V voltmeter of 20 kΩ internal resistance dissipates a mere $E^2/R = 10^2/20000 = 0.005$ watt of energy. In electronics however, an equivalent electric current of $10\ \text{V}/20000\ \Omega = 0.5$ mA could play havoc. For instance, 0.5 mA through the base-emitter junction of a ×100 gain transistor would reappear as 50 mA at the collector, and cause over a 100 Ω load resistor the voltage drop of $JR = 0.050 \times 100 = 5.0$ V. Imagine the reaction of your "cool" power amplifiers and sound boxes if that happened to one of your preamp's transistors! Worse if somebody hooked his voltmeter to the grid of a vacuum tube. The 1 to 10 megohm grid to ground resistor would find itself shunted by the meter's 0.020 megohm internal resistance!

Definitely, the electronic age demanded more than just delicately designed galvanometers. Thus, *the vacuum tube voltmeter*, an electronic instrument by itself, was the answer. It jacked up the meter impedance of typically 20,000 Ω per V to 10 or 11 MΩ by using a differential vacuum tube amplifier at the input. Typically, the jack-of-all-trades, the 6AU7 or 12AU7 twin triode, served this purpose with both anodes on the +B source voltage, and a 10 MΩ grid to ground resistor in one section, and the voltage to be measured on the grid of the other. The meter, wired from cathode to cathode, shows the resulting voltage difference between the two cathodes, which is proportional to the input and pretty immune to external interferences, such as leakage currents and power supply fluctuations.

The vacuum tube voltmeter navigated for many years the waves of the transistorizing tsunami without shipwreck, regardless that the idea of touching such fragile components as transistors and chips with a probe from a device plugged into a 115 V wall outlet gave the creeps to all but the sturdiest characters among electronic technicians. But simple transistorizing the circuit design of the vacuum tube voltmeter proved elusive, because transistors are basically current amplifiers, which use a small base-emitter current to control a far more powerful emitter-collector current. In a 10 MΩ impedance instrument, an input of, say, 1 volt would convert into a miserly 0.1 microampere, which the front-end transistor would have to convert into an amperage high enough to actuate a 200 μA meter. The necessary gain amounts to $200/0.1 = 2000$ times. Stabilizing gains that high against unpredictable variations in semiconductors characteristics due to aging and changes in temperature and external perturbations proved near to impossible.

Relief came with the invention of the *field effect transistor* (FET), a device that, like the vacuum tube, controls the flow of electric current by application of an electric field. As the flow of electrons through vacuum tubes depends on the voltage applied on the grid, the flow of electrons from the "source" to "drain" terminals of a FET depends on the voltage applied to the third electrode, called gate. What happens is that electricity gets funneled through a narrow path of semiconductor material, the channel. Though the channel's physical dimensions are unchangeable, its effective electrical width depends on the gate voltage, while the current demands of the gate are minimal.

Transistorized multimeter

The transistorized multimeter for measuring DC voltage and amperage, AC voltage, and last but not least, resistance, has all that crammed around a galvanometer in a single box in a circuitry scattered with multideck rotary function selector switches, range selector switches, and jacks for ground, volts and amperes DC, volts AC and ohms, and still extra jacks for up to 10 kV DC and AC and a separate 10 ampere DC input socket.

To begin with, the DC voltage stage of a transistorized multimeter (Fig. 12.4) needs an appropriate probe (Fig. 12.4, detail 1) with a 1 MΩ resistor built into the tip, which makes it a high impedance device that needs shielding as not to pick up spurious voltages from the audio band of radio transmissions and a host of other parasitic sources. The shielding is soldered to the ground lug of the plug and shouldn't be grounded anywhere else.

The probe feeds the voltage input to the selector switch of a 10 MΩ voltage divider. With that switch on the 100 mark, we get 10k + 20k + 70k = 100 kΩ for the resistance to ground, which divided by the ladder's total of 10 MΩ gives

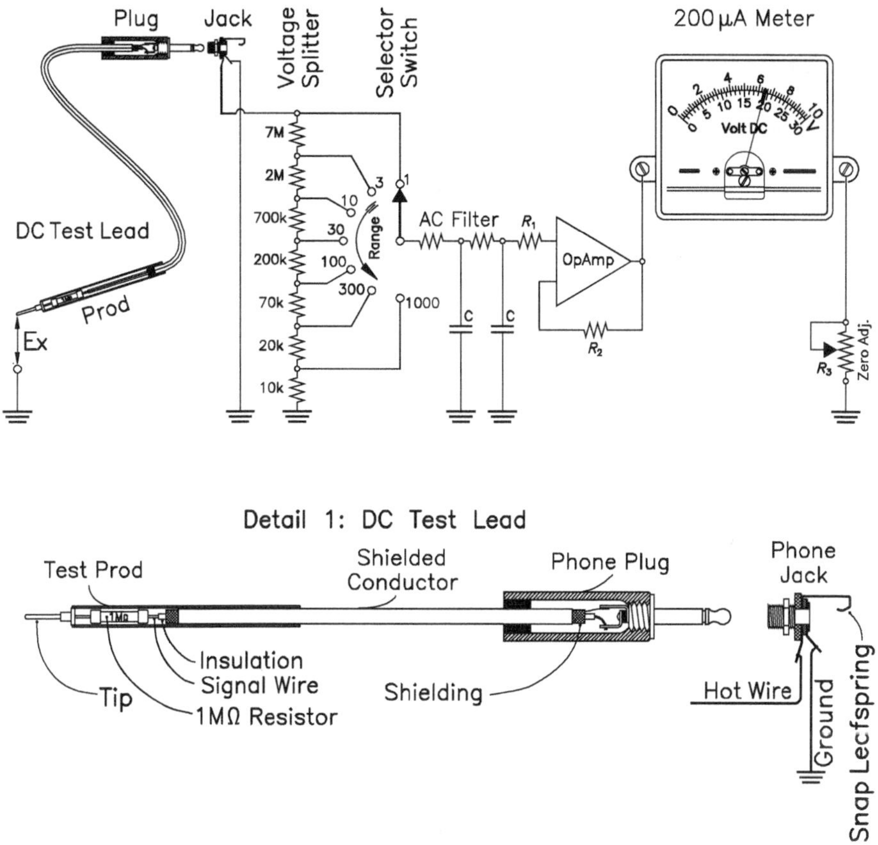

Fig. 12.4. DC voltage measuring circuit in electronic multimeter

$0.100/10 = 1/100$, augmenting the original meter range 100 times. 100 volt input would deflect the needle by 1 volt on the scale.

Next in the signal path is a π-filter to rid the input from residual AC components and spurious leakage currents. The field effect transistor (FET) to convert the signal's impedance from high to low makes part of the operational amplifier, which is why it doesn't show up separately in the schematic. The gain of the OpAmp equals R_2/R_1, the ratio of the feedback resistor R_2 to the input resistor R_1, selected to match meter and input voltage.

Previous to turning the instrument on, the nulling screw on the sight glass of the meter is to be set for zero readout. Once the desired range has been selected on the voltage splitter, power is turned on and potentiometer R_3 is used for zero adjust.

The AC measuring circuit of a multimeter, shown in Fig. 12.5 uses a 2-stage range selector circuit to obtain as many as 11 different voltage ranges in 1 : 3 and 1 : 10 interlaced steps, using no more than 5 resistors in the principal attenuator ladder. A single wire test lead, plugged into a banana jack, leads the voltage into the instrument's front end voltage splitter, consisting of a serial connected pair of resistors of 9990 kΩ and 10 kΩ. With the first range selector switch in one of the

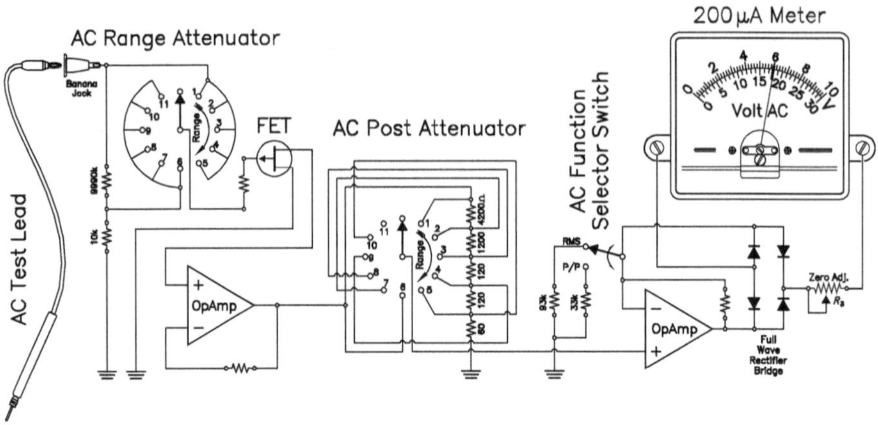

Fig. 12.5. AC measuring circuit

positions from 1 through 5, the signal goes directly to the impedance-matching FET imbedded in the OpAmp. For positions 6 to 11, the input is attenuated by a factor of 1:1000 prior to entering the same path.

Converted to low impedance by the FET and amplified by the operational amplifier (left in the schematic), the signal reaches the second attenuator.

The OpAmp right in the circuit diagram that feeds the signal into the full wave 4-diode rectifier brings signal strength into the range appropriate for the meter. The AC function selector switch provides a choice between readouts in p/p (peak to peak) and RMS (root-mean-square or effective AC) values.

Similar rectifier bridges in power supplies are followed by a pair of filter capacitors, which charge up to peak to peak voltage. Hence the DC output $\sqrt{2}$ times higher than the effective AC input, characteristic for most commercial power supplies. In the absence of these capacitors, the bridge output becomes the average of the rectified AC, which for sinusoidal AC is $2/\pi = 0.637$ times the p/p AC voltage. Compared to the effective AC voltage of $1/\sqrt{2} = 0.707$ times the p/p voltage, this is $0.637/0.707 = 0.90$ or 90%. Accordingly, division marks of average voltage scales are spaced $0.707/0.637 = 1.11$ times wider, and p/p voltage scales 0.707 times narrower than on effective AC scales.

For periodical non-sinusoidal AC waveforms, such relations must be figured case by case from $V_{\text{avg}} = \frac{1}{\pi}\int_0^{\pi} V \times d\varphi$, where V stands for the momentary voltage at phase angle φ. For sinusoidal AC, the integral yields the aforementioned relations, and is simple to solve for pulsed AC. But it must be measured out on waveform diagrams for complex waveforms.

Industries' demands for direct AC measurement by less universal yet robust instruments led to the design of the moving-iron meter.

Iron-vane instrument

The moving-iron meter, called also iron-vane meter, derives its driving force from the mutual repulsion of two soft iron pole plates (vanes) within the field of a current-car-

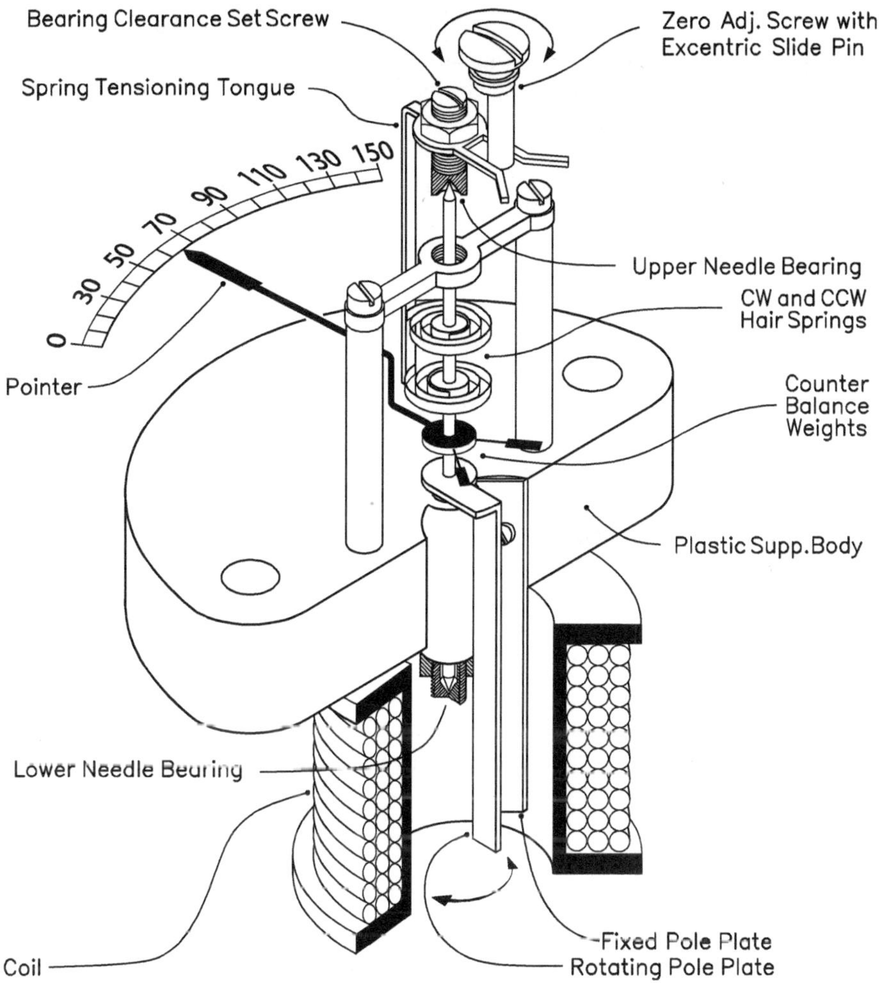

Fig. 12.6. Iron-vane meter

rying coil or solenoid (Fig. 12.6). Homopolar magnetization of the vanes by induction makes them repel each other, and the movable vane to swing around the meter axis.

The magnitude of repulsive forces depends on the field strength B_Z in the solenoid, which is proportional to n, the number of windings, and J, the amperage of the electric current through them, according to $B_z = \mu_0 Jn$. Herein, $\mu_0 = 4\pi \times 10^{-7}$ Tm/A (tesla × meter per ampere) is the permeability constant of space.

Like the d'Arsonval galvanometer, the iron-vane instrument can be fitted as ammeter or voltmeter. For ampere measurements, the spool has few windings of relatively thick magnet wire, providing the appropriately low input impedance. Conversely, voltmeter solenoids need thousands of windings, which added up resistivity results in the desired high input impedance.

Unlike the tilting spool in a d'Arsonval meter, the windings of iron-vane instruments are fixed on the instrument's body, thus dispensing the delicate hairsprings for conducting the measuring current.

In alternating current measurement, the field in the solenoid changes polarity in every cycle, and so does the magnetization of the pole plates. Made of soft steel, they follow the variations of the cycling field with minimum hysteresis losses and continue repealing each other regardless of the magnetic field's polarity.

As usual, the instrument's main shaft is supported between sapphire needle bearings and has its zero point stabilized by a pair of inversely wound hairsprings. Sufficient elbow room for the circular motion of the moving vane is generated by mounting the spool out of center of the meter's rotation axis.

Damping of the needle movement is provided by a vane attached to the main axis (not shown in Fig. 12.6), which rotates in a toroidal sector of rectangular cross section with just enough clearance as to provide a narrow air passage between vane and housing. The width of that leeway determines the degree of damping, which should be enough to keep the pointer from overshooting its final position without making it creep hesitantly toward it.

Like in d'Arsonval galvanometers, zero adjustment is done by manually rotating a screw-headed eccentric pin mounted in the sight glass, until the position of the outer support points of the hairsprings zero the pointer in. Counterweights attached to the rear end of the pointer balance the weight of the front section close enough to make the instruments readouts position independent.

A shortcoming of this universal instrument is the nonlinearity of its scale. Though the field strength inside the solenoid increases proportional to the applied amperage, the degree of magnetization of the pole plates (vanes) follows the typical magnetization function of ferromagnetic material (Fig. 12.7). The curve tends toward the so-called saturation point from where any further increase in the spool

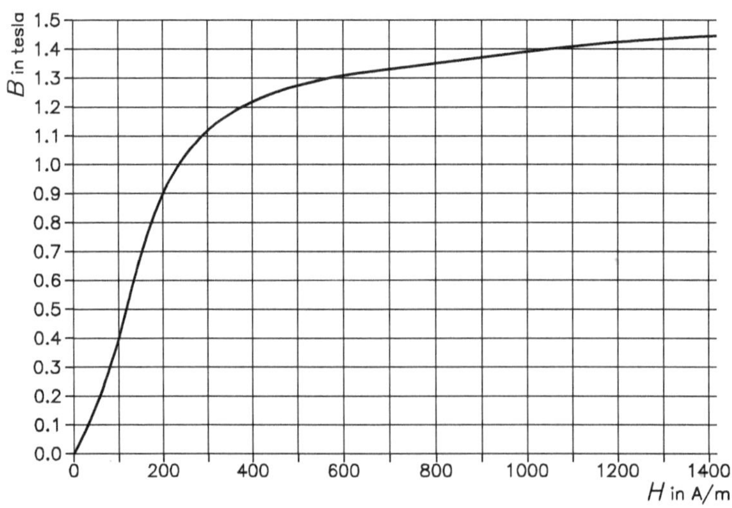

Fig. 12.7. Magnetization curve for H1 dynamo sheet metal

current results in lesser and lesser degrees of gain in the magnetization of the plates, until it merely mirrors the increase of the surrounding field in air. That limits the instruments operating range to the quasi proportional section of the curve between 0 and 1 tesla.

Commercial instruments are available for direct and alternating currents from 15 to 100 Hz frequency and up to 200 amperes and 500 volt. Laboratory versions of iron-vane meters achieve 0.1% of precision. Still, hysteresis and eddy current losses may distort Iron Van Meter readings, a problem absent in the bimetal ammeter.

Bimetal ammeter

A bimetal ammeter measures DC as well as the root-mean-square (J_{RMS}) of AC amperage, regardless of frequency. Herein, the RMS amperage of alternating current equals a DC amperage of identical power output.

As the name suggests, the instrument's driving power is derived from spirally shaped bimetal strips of the kind we remember from temperature sensors, thermostats, and dial thermometers as bonded strips of metals of different thermal expansion, such as iron and brass and several other combinations.

A spiral spring rolled with the metal of lower thermal expansion toward the inside tightens with increasing temperature and unrolls when its temperature sinks. With R for the ohmic resistance of the device, the heat energy it receives from an electric current of amperage J equals J^2R. That points at a scale with toward the right widening spaces between marks, which implies that the higher amperages are read with better accuracy than their humble peers on the left. Such characteristics are beneficial in instruments where the significant values are on the right end of the scale, such as voltmeters monitoring line voltage.

The mechanics of the instrument resembles those of the iron-vane meter in Fig. 12.6 if we remove the solenoid and replace the mutually opposing hairsprings by bimetal strips, of which the one driving the pointer is charged with the operating current, while the other is wound to compensate for possible changes in the temperature of the environment.

The robust mechanism of the bimetal ammeter lends enough force to the pointer movement to push an index toward the highest J_{RMS} since last time's manual reset. Likewise can the pointer's advance be used to actuate a limit switch, a particular that makes the instrument into an on/off controller.

A bimetal ammeter's high power demands of typically 1 to 2.5 VA limits its use to applications where power is abundant. Reading the instrument takes patience, since the pointer may need up to 15 minutes to settle. Therefore, short-term power surges will not show, which may be convenient in some applications but a drawback in others.

All in all, the bimetal ammeter is a rugged industrial instrument for direct measurement of AC and DC amperages that would otherwise afford range-expanding shunt resistors or instrument transformers. It stands up to 1 second surge currents ten times its ratings, and to vibrations that would destroy other instrumentation.

Power meters

Power meters are different from the familiar kWh counters in our backyard insofar as the latters register consumption of electricity over a given period, such as a month, while *power or wattmeters*, such as that in Fig. 12.8, measure consumption per second.

Like in a d'Arsonval voltmeter, the pointer is driven by a spool of a great number of windings, pivoting within a magnetic field. However, the field in a power meter is not that of a permanent magnet, but generated by another spool of fewer windings of heavier gauge wire, accommodating the other spool within.

If the inner spool is connected *across* the load and the outer spool is made to carry the current flowing *through* the load, the resulting deflection of the inner or *voltage spool* depends on the product of voltage and current: $P = EJ$.

While AC wattmeters are built with ferromagnetic cores of dynamo-sheet stampings, AC/DC instruments cannot have cores which remanent magnetism from ear-

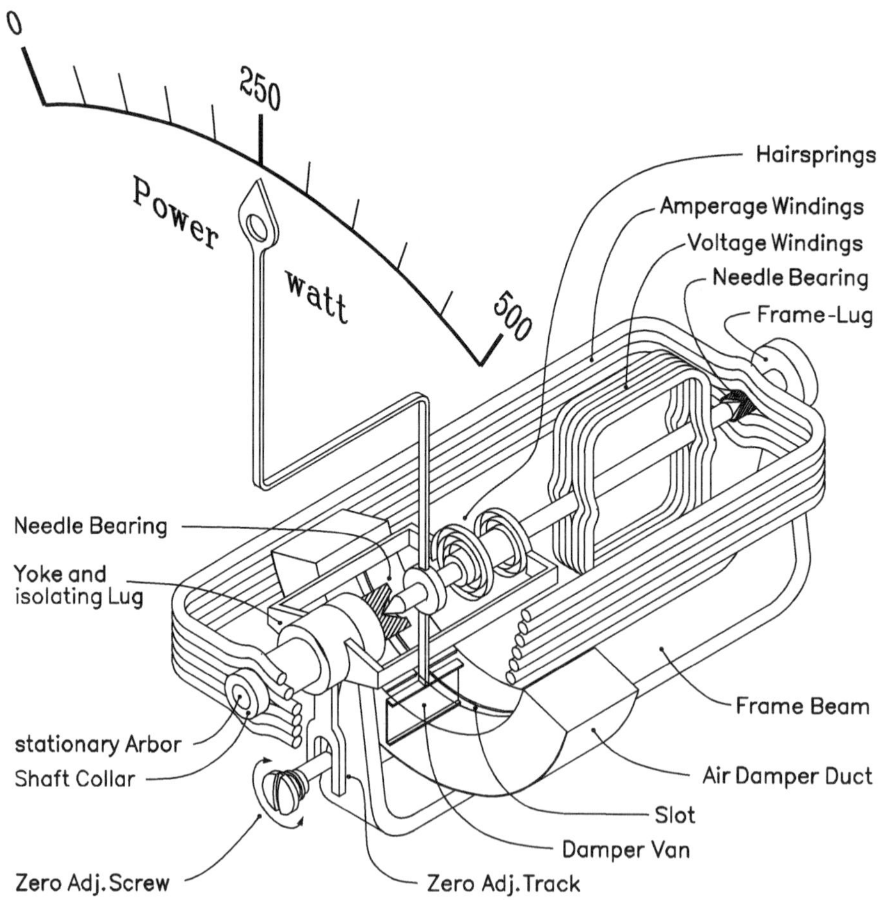

Fig. 12.8. DC/AC wattmeter

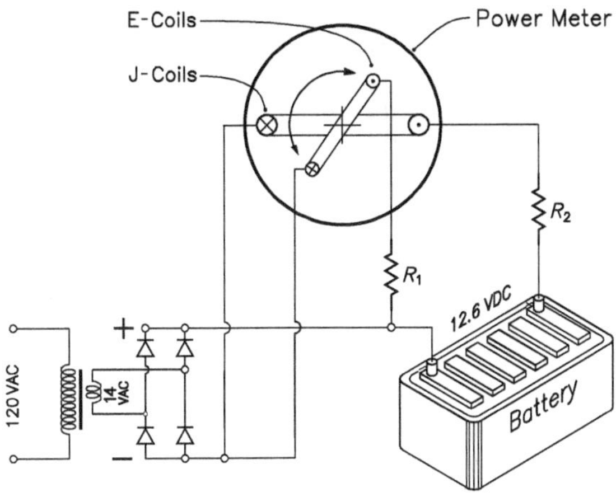

Fig. 12.9. Power meter connections into battery charger's circuit

lier DC applications would show up in the meter readings. In the absence of iron cores, the magnetic field B induced by an electric current J in a spool is proportional to the spool's ampere-turns per meter of spool length, n, according to

$$B = 4\pi \times 10^{-7} \times Jn\,.$$

In the wattmeter in Fig. 12.8, the current-windings constitute the stationary section of the actuating mechanism and house the pivoted voltage-windings within. As usual, mutually opposing hairsprings provide the return torque for the needle and double as electrical conductors for frictionless connection of the voltage coil to the legs of the yoke, supported – yet electrically insulated – by the plastic central support-bushing their ends are molded into.

The yoke's position for zero deflection can be set with the zero adjust screw in the instrument's sightglass, which eccentric tip slides in the slotted stub reaching down from the center bushing. The connections of the voltage coil to the instrument's voltage terminals are the legs of the yoke.

The toroidal air damper in Fig. 12.8 keeps the needle from oscillating while it settles into its final position. Though the vane, attached to an extension of the needle, does not touch the walls of the damper housing, damping is achieved as air friction in the narrow opening between housing and van slows its swing. The vane also doubles as counterweight to balance the needle.

Figure 12.9 highlights the typical circuitry of the DC wattmeter on the example of a battery charging unit. R_1 is the series resistor of the voltage coil, while R_2, in series with the charging circuit, prevents damage from inrush currents as they occur in first-time charges of empty or nearly empty batteries.

Resistance meters

Though the precision of bridge instruments is hard to beat with other systems, direct readout instruments got to dominate the market. A circuit frequently found in multimeters is shown in Fig. 12.10 where R_x, the value of an unknown resistor, is being measured by series connecting it with a resistor of precisely known ohmmage, R_S, the meter, and a power source, read battery.

R_S and R_x constitute a voltage splitter, which outputs the fraction $E_x/E_B = R_x/(R_S + R_x)$ of the battery voltage E_B into the OpAmp leading to the meter. This generates the familiar right compressed scale of resistance meters.

Instead of one single precision resistor, the instrument in Fig. 12.10 employs

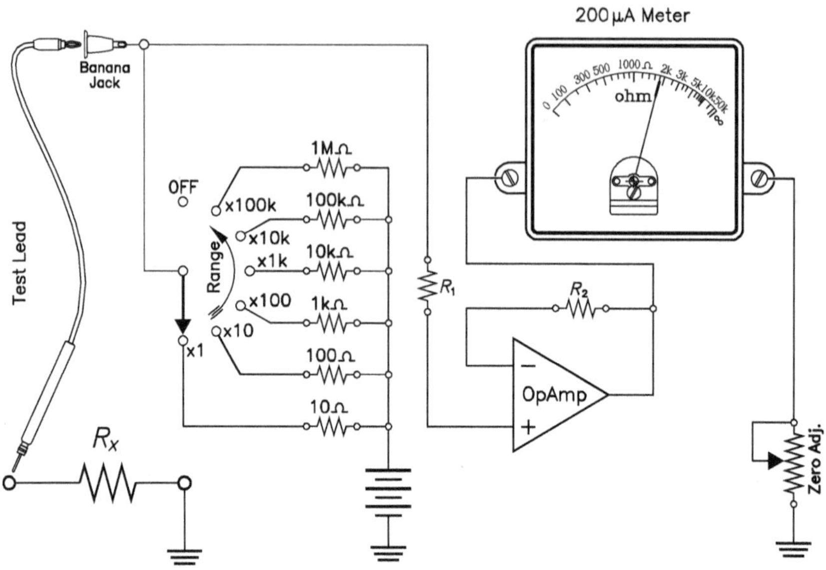

Fig. 12.10. Resistance meter circuit

6 precision resistors in series with the unknown resistor R_x. The range selector switch then connects one out of the six into the circuit, while the zero adjustment potentiometer allows for zeroing in the meter with the input terminals shunted. Minor deviations of battery voltage are equally compensated for, but that shouldn't be seen as a chance to prolong the useful battery life beyond retirement age; the heightened internal resistance of worn out batteries is bound to introduce errors.

The range selector switch should be left in the OFF position whenever the instrument is idle, because some current flows through the meter in all other settings. A discharged battery is a terrible opening for a measuring lecture.

Cross coil instruments

The direct-reading cross coil instruments on dashboards in aircraft and automobiles measure resistance with a pair of crossed spools pivoting within the field of a strong permanent magnet. With the spools at a fixed angle to each other, their torques are mutually opposed, so that the instrument zeros in if the product *amperage times number of turns* is the same for both.

Conversely, with the amperages J_1 in one branch and J_2 in the other, deflection of the armature is proportional to the ratio J_1/J_2. In the scheme in Fig. 12.11 b these currents flow from the battery of voltage E_B over the serial resistors R_R and R_T, which makes $J_1 = E_\mathrm{B}/R_\mathrm{T}$ and $J_2 = E_\mathrm{B}/R_\mathrm{R}$, and deflection proportional to

$$\frac{J_2}{J_1} = \frac{E_\mathrm{B}/R_\mathrm{R}}{E_\mathrm{B}/R_\mathrm{T}} = \frac{R_\mathrm{T}}{R_\mathrm{R}} \,.$$

If R_R is a resistor standard, and R_T the resistance of a temperature-sensitive plati-

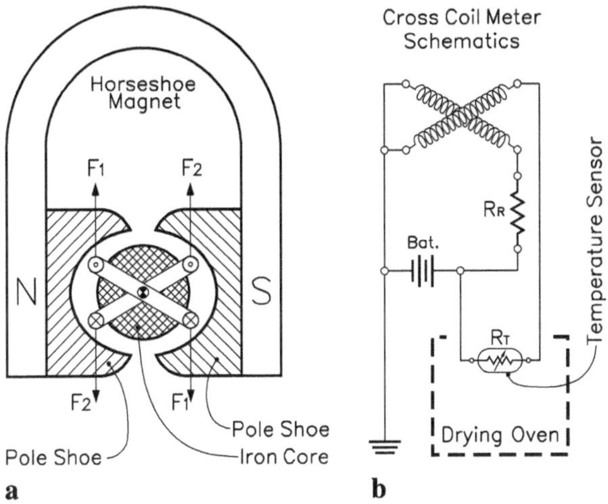

Fig. 12.11. Cross coil comparator

num wire, the temperature of the latter can be read from a scale calibrated in degrees Celsius or Fahrenheit.

Pt resistance bulbs (RTD), consisting of a fine platinum filament encapsulated in a glass envelope, are available with 100 Ω or 1000 Ω at 0 °C (32 °F) for applications such as the drying oven in Fig. 12.11 b. For immunity to changes in the temperature of the surroundings, R_R is made of temperature-stable material, such as constantan.

An interesting solution to linearizing the scale is by profiling the pole shoes of the permanent magnet in ways that the clearance between the inner magnet core and the pole shoes, the interferricum, narrows above and below the centerline (Fig. 12.11 a), so that the higher electromagnetic force on the coil in the narrower sections of the interferricum counteracts the reduction of torque with increasing angle of deflection.

Tenacious oddball

Once Wilhelm Morse's telegraph and Alexander Graham Bell's telephone had demonstrated signal transfer through electric wires, image transmission seemed only one step away. But the telegraph's sending of words as pulsed DC strings and the telephone's transmission of speech as a complex AC-waveform, boils down to nothing but the one-channel transmission of varying voltages. Black-and-white pictures however are an infinite variety of lighter and darker shades over the area envisioned, and their "as is" transmission would take an infinity of separate wires between sender and receiver.

Out of that dilemma grew the idea of cutting the image into an array of minute areas and transmit information on the luminosity of each in sequence – one at a time. The earliest device for scanning images became known as the Nipkow disk, a rotating cardboard or sheet-metal disk with a number of perforations, located in an inwards spiraling pattern. Tangential spacing of the holes equals the width of the image, while the radial pitch of the spiral is made as large as the hole diameter.

Though the principle had been established in Paul Nipkow's 1884 patent application, captioned *Electric Telescope,* the invention slumbered unused while the patent expired, and was picked up no sooner than in 1924 by John Logie Baird, the Scotsman who presented at the London wireless communication fair of 1928 a mechanical TV set devised around a Nipkow disk made to rotate at 900 rpm. The 5 by 4 cm (about 2 by 1 1/2 inch) image consisted of 30 lines, traced by light from

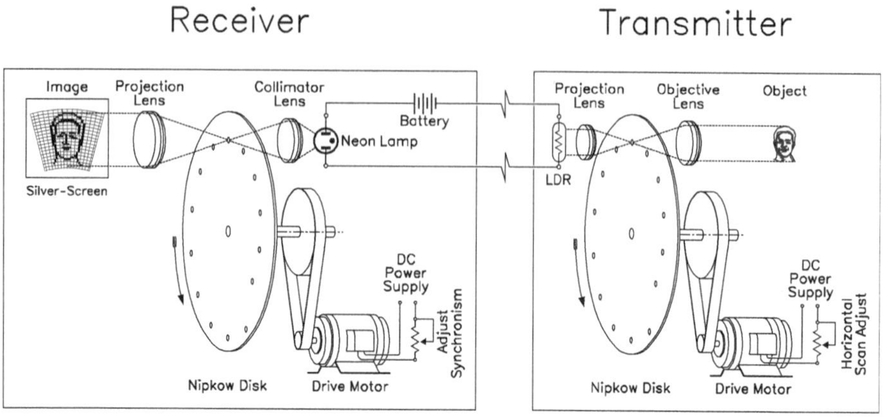

Fig. 12.12. An "electric telescope"

a neon lamp through the pinholes in the disk. Among the demonstration's guests was the proper Nipkow, who, seeing his invention from 40 years ago materialized, issued an emotion-loaded, yet critical report that described what he saw in unflattering terms, such as "a flickering and not easy to discern" picture, hardly visible without a black cloak's shielding from ambient light. Whatever the quality of those early pictures, their principle of transmission is still the same, even in our space age TV sets.

Regardless of its shortfalls, mechanical television proved its potential with a transmission from London to New York on 9 February 1928 and the sales of do-it-yourself kits for construction of a Nipkow disk-based set called Telehor. Successful builders had a choice between the programs from Radio Berlin-Witzleben and the Baird Television Company in London.

Let's get digitized!

Although widely outperformed by the cathode ray tube (CRT) and later by liquid crystal (LCDs) and plasma displays, the rotating disk could be seen as the birthplace of digitizing. Though the term "digital" makes us think "calculator" and "computer", the impact of digital technology on scientific and industrial instrumentation was just as mind-boggling. The accuracy of pulse counts allowed to break through the usual limits of $\pm 2\%$ precision for commercial instruments, and the $\pm 0.5\%$ in high-class laboratory equipment. Pulse counts are absolute, save for a gray zone within the pulse-to-pulse time span, because the period allowed for the count may end immediately after the last step has been triggered, or it can continue until shortly before the edge of the next pulse triggers the counter one more time.

The central processing unit (CPU) of a computer counts millions of pulses without a single mistake. Sort of – let's admit – because errors caused by contingencies, such as voltage surges or intermittedly failing components, could still sneak in. Therefore, much like humans double check their arithmetic, a computer's CPU counts up and down. If the remainder of the two counts is different from zero, an error message is displayed or an automatic recount clicks in. Additionally, some other checks are performed without the user's knowledge; among them the sending of the input's digit sum as an appendix to the proper count, and comparing it with the digit sum of the output. For instance, if the number 2,377,483 is to be transmitted, the computer would sum up the digits as $2+3+7+7+4+8+3=34$, and transmit 34 as an addendum. Here too, any discrepancies between the digit sums of the numbers sent and those received end up as an error message.

Last century's crank-driven calculating machines counted directly in decimal numbers, but electrical technology stopped short of coming up with an equivalent, because electricity has only two *discrete states*: zero voltage and some voltage. The mechanical device for emulating these choices is the *ON/OFF* switch; it supplies zero voltage while open, and full supply voltage when closed. That limits its counting ability to two digits.

In principle, it could be expanded with the use of two switches, which would allow for 4 configurations: open/open, closed/open, open/closed, closed/closed. Labeling switch positions as above, and numbering the possible combinations with 0 through 3, gets the Truth Table (Table 12.1). Combined, the entries in the first two columns yield the first four consecutive digits of the *binary system of numbers*, while the column on the right lists their decimal equivalents: 0 for 00, 1 for 01, 2 for 10, and 3 for 11. Thus, a system of n electric switches could count to $(2^n - 1)$ in binary.

Table 12.1. Truth table

Switch A	Switch B	Assigned value
0	0	0
0	1	1
1	0	2
1	1	3

Digital technology aside, switches can perform a great number of tasks, the most basic being the turn on and off of a load, such as the lightbulb in Fig. 12.13 a by an on/off switch, also called single-pole single-throw (SPST) switch.

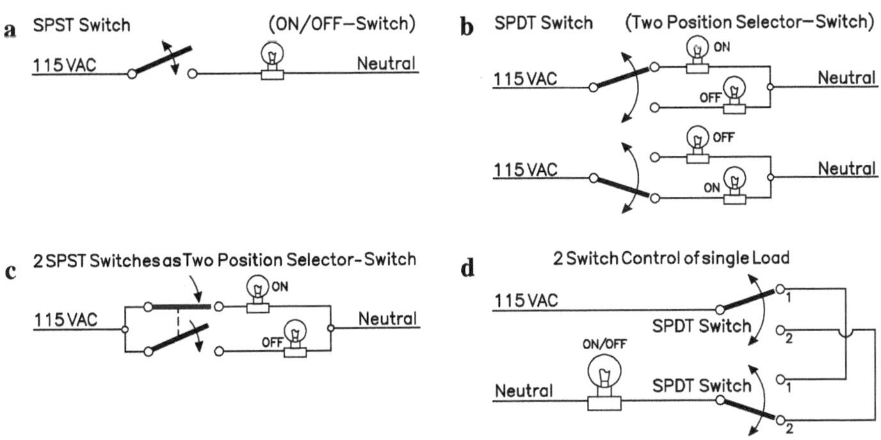

Fig. 12.13. Switch logic

Figure 12.13 b depicts the 2 possible positions of a single-pole double-throw (SPDT) selector switch, used for alternately powering one or the other of two loads. The same task can be performed by the mechanically coupled pair of on/off switches in Fig. 12.13 c, an arrangement frequently emulated in electronics with transistors or vacuum tubes in lieu of on/off switches.

Figure 12.13 d shows a logic circuit for controlling a load from two different locations, popular in domestic installations for turning the lights in a corridor or staircase on and off from both ends. As do-it-yourselfers know from experience, common on/off switches wouldn't do in that case, because the desired effect on the load of *one* switch depends upon the position of the *other.* 2-position switches, wired as in Fig. 12.13 d are the answer. With both switches on 1, or both on 2, the light is on. For unequal switch settings, such as one switch on 1 and the other on 2, or vice versa, it is off.

In control circuits, switches are mostly replaced by *relays*, devices capable of controlling large electrical currents through a low signal current. The capacity of a typical control relay to switch up to 10 amperes with as little as 10 to 20 mA coil current made relays the building blocks of one of the first computers in the agitated history of this machine: The IBM MARK 1 (1944), reportedly sounding like "a roomful of ladies knitting," took 8 by 50 feet of panel space, weighed 5 tons and contained about 750,000 components. The sheer number of parts must have kept maintenance crews as busy as programmers and users.

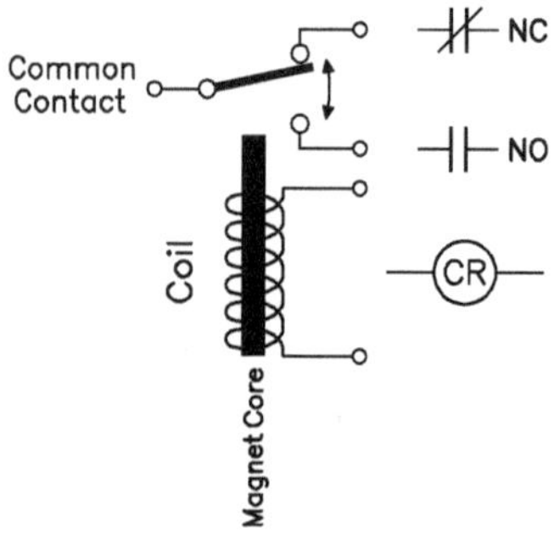

Fig. 12.14. Control relay

The SPDT relay in Fig. 12.14 includes an electromagnet acting on a the hinged armature of ferrous material (not shown), that supports the contacts. With the electromagnet off, the upper contact is on, and the lower one is open. This gets inverted when the magnet is energized.

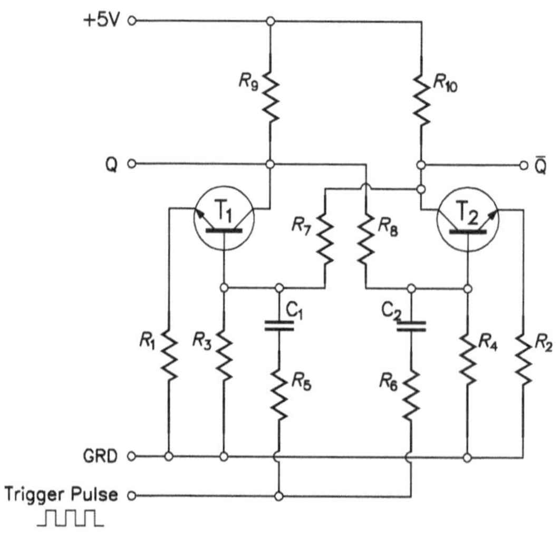

Fig. 12.15. Transistorized bistable circuit

The designations of "normally closed" (NC) and "normally open" (NO), commonly used in ladder circuit diagrams, refer to the deenergized state of the system. All components in the ladder diagram of hard wired logic circuits are to be shown in their deenergized state, even if that leads to representing an unrealistic combination of mutually not compatible switch settings.

The electronic version of the SPDT switch (or 2-position selector switch) is the bistable *flip-flop* in Fig. 12.15, known from the tube era under the name *Eccles–Jordan* (multivibrator) circuit. It is built around a pair of identical switching transistors, interconnected in ways that allow for switching two different loads (R_9 and R_{10}) alternately on and off. Compared to mechanical switches, which yield precisely 0 volt on OFF, and the full supply voltage, say, +5 V, on ON, electronic switching elements suffer from the base-emitter voltage drop and the emitter-collector transmission loss, so that a nominal 0 V output comes out a few tenth of a volt positive, and a nominal 5 volt output is something less than 5 V. In TTL logic devices, 1 is represented by 2.4–5.5 V, and a logic 0 by 0–0.6 V for a +5 V supply voltage.

That's why those two states are usually referred to as 0 and 1, rather than by the voltage of each.

The switching transistors are crosswise interlocked and therefore mutually exclusive; only *one* can be ON (conducting) at a given time, so that the configurations of both OFF or both ON cannot occur. While T_1 is conducting, the voltage drop over its load resistor R_9 is most of the 5 V supply voltage, which brings the voltage at the collector down to near zero and keeps T_2 from turning on.

The next positive pulse from the clock goes through capacitor C_2 to the base of T_2 and turns it on. This brings the voltage at the collector of T_2 down to zero and through R_7 the voltage at the base of T_1, so that T_1 turns off. The device changed state. The next clock pulse triggers the base of T_1 through capacitor C1, and thus initiates another change of state, and so on.

You can manually check a flip-flop by applying a positive voltage to the base of one or the other of the transistors with a piece of flexible conductor in series with a resistor. Nothing happens when you touch the base of the ON-transistor; but touching the base of the other makes it conduct, which shows as near zero voltage at the collector.

The collectors feed into the output terminals Q and $\bar{Q}$. Output at $\bar{Q}$ is the inversion of the Q output: If $Q = 1$, then $\bar{Q} = 0$, and vice versa. Thus it takes *two* clock pulses to bring the flip-flop from the OFF state to the ON state and back. That makes the device a frequency divider by 2. For instance, a 2^{15} Hz (32,768 Hz) clock signal outputs 2^{14} Hz at Q. With two flip-flops in series, output is down to 2^{13} Hz, and with three, to 2^{12} Hz. In a quartz clock, a 15-stage frequency divider outputs the 1 Hz pulses for driving the second hand. From here, a gear train takes over the 1 : 60 reduction for rotating the large hand, and the 1 : 12 reduction for the small hand.

The *ternary flip-flop* in Fig. 12.16 converts monophase DC into three-phase pulsating voltage. The operation of the circuit resembles that of the common flip-flop (Fig. 12.15) insofar as among the 3 switching transistors Q_1, Q_2, and Q_3, 2 must be off in order to provide the base bias for the third to turn on. For instance, if Q_2 and Q_3 are off, their collectors are both at +16 V, feeding over 10 k resistors R_8 and R_{11} into the base bias line that leads over R_3, of likewise 10 k, to the −16 V supply. Since R_8 and R_{11} in parallel amount to 5 k, they constitute with R_3 a 1 : 3 voltage splitter that biases the base of Q_1 with $-16 + (2/3) \times (16 + 16) = +5.3$ V, making Q_1 conduct until the next positive

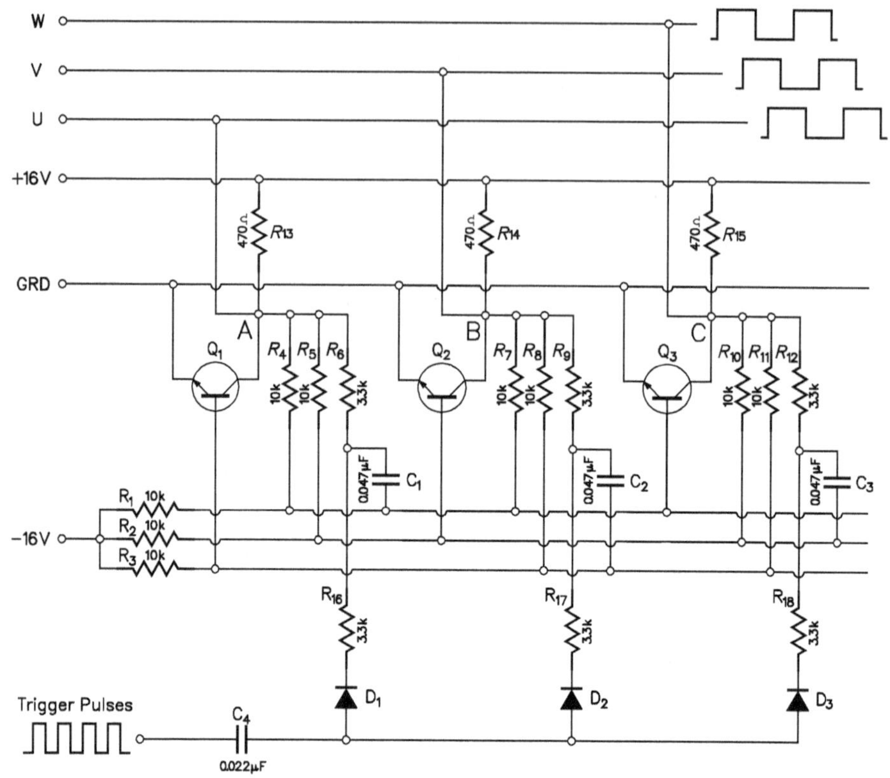

Fig. 12.16. DC to 3-phase AC converting flip-flop

trigger pulse overcomes the negative base bias of Q_2 and turns it on, which makes the base bias of Q_1 into 0 V, and keeps it off.

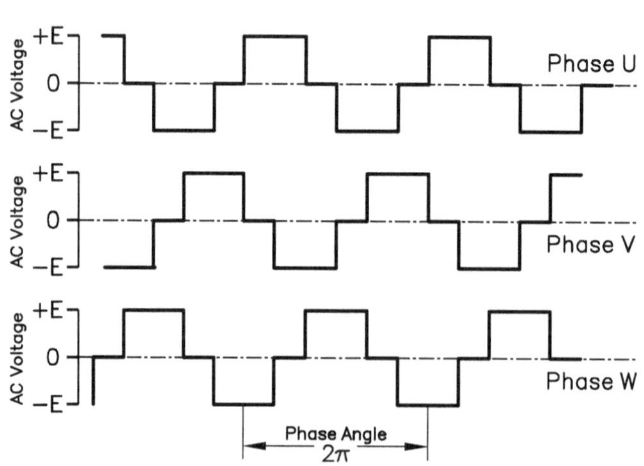

Fig. 12.17. Phase to phase voltage from ternary converter

This way, Q_1, Q_2, and Q_3 come on successively, generating 3 consecutive outputs into the lines marked U, V, W. These outputs are still DC pulses, but they can be converted into "true" AC by using them to trigger conventional flip-flops, which alternately link to the +E and −E supply lines of a center tapped power

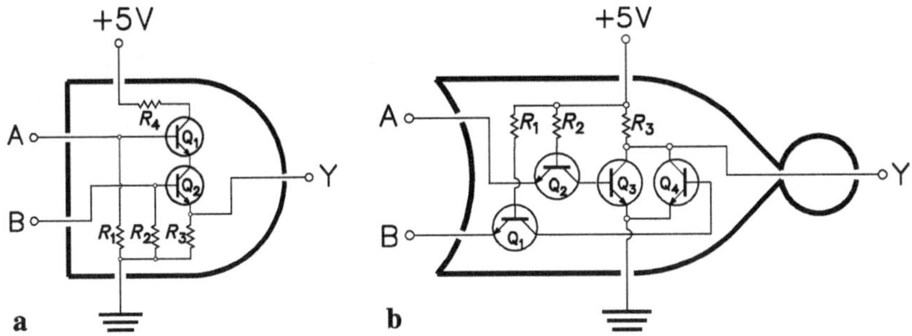

Fig. 12.18. a AND and **b** NOR gates

supply. The phase-to-phase voltage of such a network (Fig. 12.17) does not contain the third harmonic of a sine wave, feared for the hum it generates in square wave-driven AC motors.

Since the device needs 3 trigger pulses to output a single pulse in any of the outputs (U, V, W), the ternary flip-flop could be used to divide a given frequency by 3. Three such devices could already count to $3^3 - 1 = 26$. It would be interesting to ponder where our digital technology would have ended up, had it been built on a 3-based system of numbers instead of the 2-based binary system.

Flip-flops can also be designed with digital switching elements, such as AND and NOR gates (Fig. 12.18). An AND gate yields a HIGH output only if both inputs are HIGH. With either or both inputs on LOW, no output results. Conversely, an OR gate gives a HIGH output if one *or* both inputs are high. The NOR gate, counterpart to the OR gate, outputs a HIGH when both inputs are at zero volt. In all other cases, such as one input ON and the other OFF, or vice versa, or still both ON, the output of the NOR gate is always at zero.

Figure 12.18 shows X-ray images of AND and NOR chips. The AND gate (Fig. 12.18 a) uses a pair of serial connected transistors in emitter-follower configuration. The emitter voltage gets high when A and B are high ($+E$) and with them the bases of both the transistors, making that they conduct. Emitter followers excel by their stability and low output impedance; however, the ratio of the emitter resistor R_3 and the load resistor R_4 in Fig. 12.18 must be high enough to get the output of approximately $ER_3/(R_2 + R_3)$ into the range of the TTL ON state.

NOR gates (Fig. 12.18 b), on the other hand, use a pair of switching transistors in parallel, so that each of them can turn the load ON or OFF single-handed. The device would function with the inputs feeding directly into the bases of Q_3 and Q_4, but the inherently low resistance of semiconductors' base-emitter junction would make this a low input impedance device with high current consumption. Therefore, transistors Q_1 and Q_2 are added as impedance converters.

Gates can be used in flip-flop design much like the transistors in Fig. 12.15. The NOR gates in Fig. 12.19 are cross coupled and thus mutually exclusive. With inputs K and J on +5 V, the flip-flop toggles in response to pulses on CP.

Figure 12.20 shows a binary counter made with 4 JK flip-flops. Pulses applied to the CLK terminal of the first flip-flop in the row make its outputs toggle be-

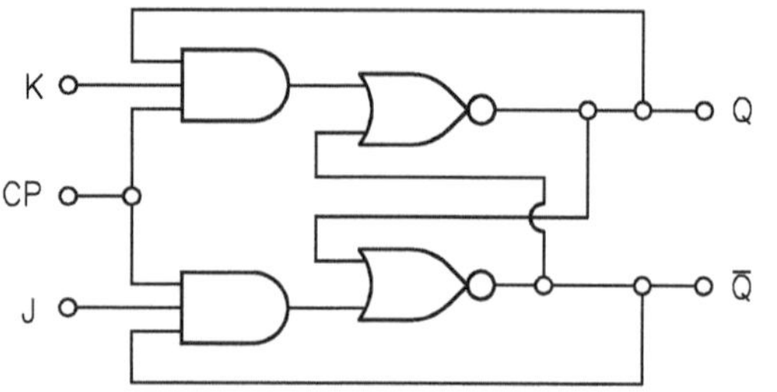

Fig. 12.19. Logic circuit of a JK flip-flop

tween Q = 0 and Q = 1. The small circle preceding the clock (CLK) input symbolizes the characteristic of the flip-flop to toggle at the trailing edge – the clock signal's change from high to low.

Two clock pulses are afforded to bring the output Q of the first stage from 0 to 1 and then back from 1 to 0. That makes the output frequency at Q half of the input frequency (at CLK), which explains the name "divide by two counter." By the same logic, frequency at the output of the second stage is 1/2 of clock frequency, 1/4 and 1/8 at the next two, and 1/16 at the last. When the counting stops, the four LEDs display the total of clock counts in binary code, as in Table 12.2.

The highest number a counter with n stages can handle is $2^n - 1$. That gives $2^4 - 1 = 15$ or a binary 1111 for the counter in Fig. 12.20, which means that all 4 LEDs would be ON at that point. Note that the count starts with zero and not with a one, which implies that the first impulse from the clock resets the first flip-flop while none of the output LEDs comes on. Only the second pulse flips the Stage I so that Q gets high and the fourth LED lights up. The successive configurations of the outputs (LEDs) are listed in Table 12.2.

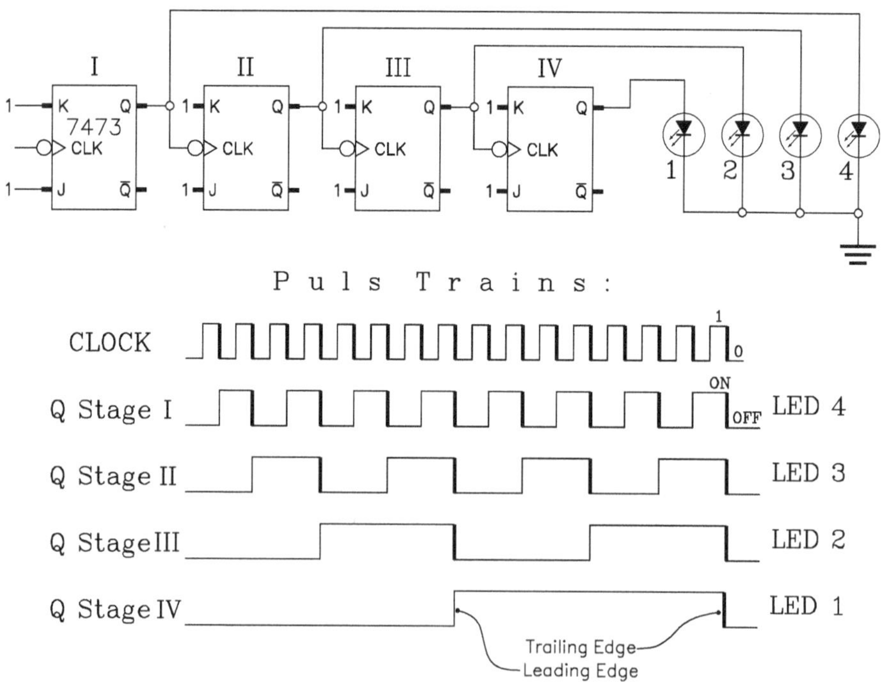

Fig. 12.20. 4-bit ripple counter

Table 12.2. Configurations of outputs for 4-bit ripple counter

Clock pulse nr.	LED 1	LED 2	LED 3	LED 4
1	0	0	0	1
2	0	0	1	0
3	0	0	1	1
4	0	1	0	0
5	0	1	0	1
6	0	1	1	0
7	0	1	1	1
8	1	0	0	0
9	1	0	0	1
10	1	0	1	0
11	1	0	1	1
12	1	1	0	0
13	1	1	0	1
14	1	1	1	0
15	1	1	1	1

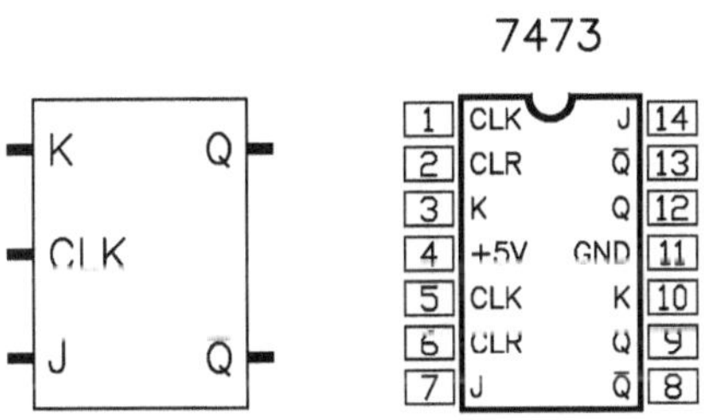

Fig. 12.21. Symbol and chip layout of JK flip-flop

An 8-stage counter goes up to $2^8 - 1 = 255$ counts, hence the 255 symbols in the alphabet of ASCII characters. 20 stages get us above the 1 million barrier, but hard wiring that many stages in computer circuits would be a formidable job. Thanks Heavens for chip technology, which prints all this and much more on a semiconductor substrate at the stroke of a press.

The contact configuration of a 7474 TTL Counter Chip, titled Dual JK Positive-Edge-Triggered Flip-Flop, is shown in Fig. 12.21. As the name suggests, the chip contains two JK flip-flops, so that two such chips would do for the counter in Fig. 12.20. The 7474 runs on about 17 mA per unit and handles frequencies of up to 25 megahertz. Like TTL logic devices in general, input and output impedances are low, which provides immunity to parasitic leakage currents and static discharges. Their low output impedance allows for connecting up to 10 other TTL loads or LEDs directly without an intermediate driver stage.

Counters made from a number of serial connected flip-flops have their shortcomings. The propagation delay in a 7474 flip-flop is 30 nanoseconds (ns) from HIGH to LOW, and 25 ns from LOW to HIGH, a total of 55 ns or 55×10^{-9} seconds. Certainly not long enough to show up in a graph, but accumulating with the number of stages. In a 20-stage counter, time delay from clock pulse to output pulse are up to $20 \times 55 = 1100$ ns or 1.1 μs. The good news is that the counter's capacity increases in powers of 2, while delay times simply sum up. For instance, 40 stages would bring delay time to 2.2 μs, but stretch the counting limits to $2^{40} - 1 \approx 1.1$ trillion counts. Nevertheless, this feed-through of

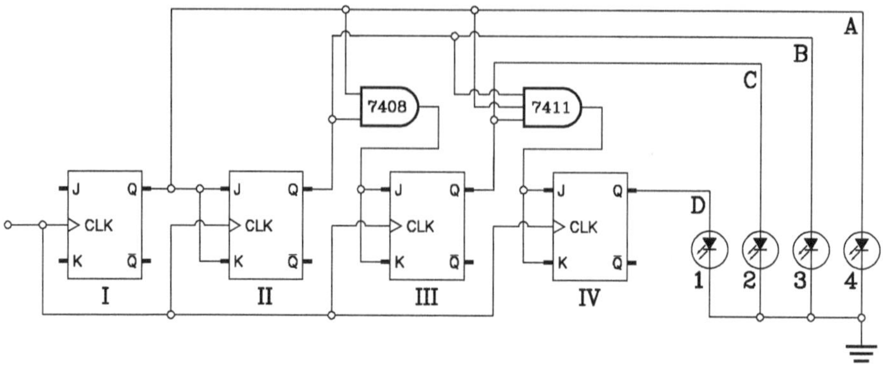

Fig. 12.22. Synchronous binary counter

propagation delays earned that type counter the name *asynchronous* or *ripple counter.*

By contrast, the *synchronous counter* in Fig. 12.22 prevents the ripple effect by clocking all stages simultaneously, and using AND gates to sort out which stage should respond.

Regardless the many appealing characteristics of TTL logic devices, they fell somewhat out of grace with the revolution of handheld, battery-powered devices, such as laptops and game consoles. The C-MOS (complementary metal-oxide semiconductor) family of chips is of the belt-tightening type of electricity consumers. The 74C74 or 4027, equivalents of the 7474 TTL, runs on a scant 8 mA. While TTL devices demand a regulated 5 V power supply, most CMOS chips accept anything between +3 V and +15 V.

In CMOS devices, the transition from high (1) to low (0) happens at half the supply voltage, while in TTL circuits a logic 1 is represented by 2.4–5.5 V, and a logic 0 by 0–0.6 V.

The LED at the end of the tunnel

With the result of our counting at hand, all that remains is the conversion of the binary counter output into legible numbers. As soon as the 1930s, a numerical display named Nixie tube used a miniaturized version of the gas discharge tubes we remember from store- and casino fronts for emulating numerals, but only the emergence of digital electronics in the 1950s made numerical displays paramount. Nixie tubes had several layers of cathodes, each shaped to a number between 0 and 9, mounted above a wire-mesh anode in an atmosphere of highly rarefied neon gas. Appearance-wise, the brilliant red-orange letters of a Nixie tube could still beat today's energy-efficient LED and liquid crystal displays, but a 170 volt tube interacting with our chip-loaded circuit boards looks like a scary proposition.

The "mother of all readouts" became the 7-segment LED chip, shown on the right of Fig. 12.23, which combines 7 elongated light emitting diodes into numerals that resemble their Indo-Arabian forefathers close enough to be read by the

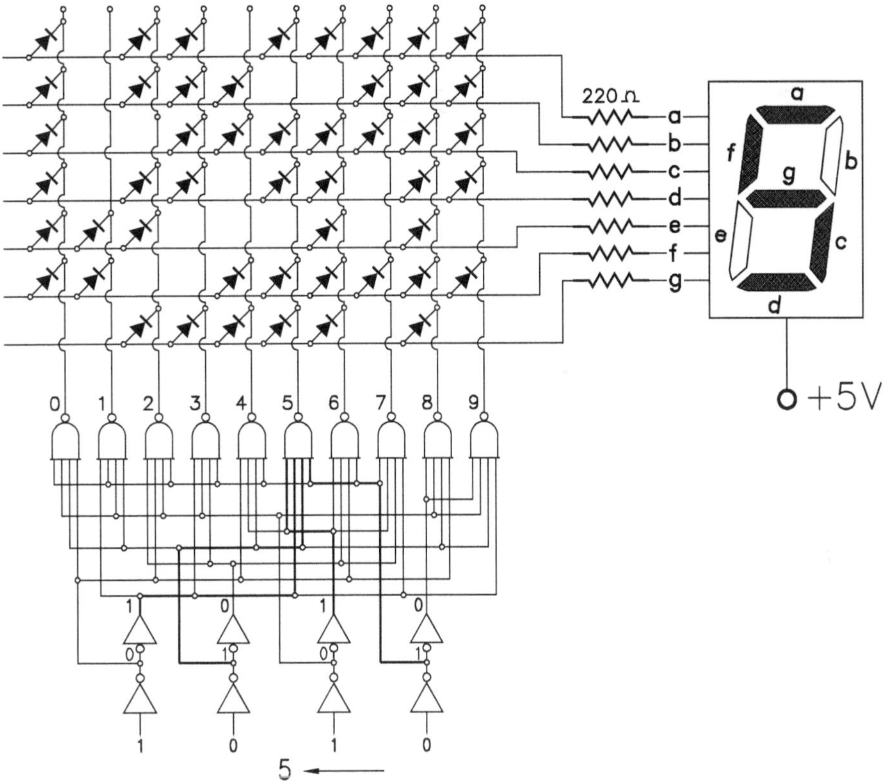

Fig. 12.23. Binary to decimal and decimal to 7-segment LED decoder

owners of electronic calculators as fluently as the letters in the sport pages of their newspapers.

Conversion of the binary output of a counter into a decimal display happens in two steps, laid out in the upper and lower sections of Fig. 12.23. The signal, for example the binary number 1010, passes initially through serial connected pairs of inverters – an arrangement that makes the original of each input, 1010 in our example, as well as its inversion, 0101, available for further processing.

Each of the 10 quadruple NAND output gates needs 4 high inputs to operate. In our example, gate 5 outputs a LOW with all its four pins on HIGH (close to +5 V). Following thick lines in the schematic shows this happening with 1, 0, 1, 0 – the binary equivalent of a decimal 5 – on the chip's inputs. Each of the numbers from 0 (0000) to 9 (1001) has its particular set of conductors converging on the appropriate NAND gate.

The second step happens in the matrix converter in the upper section of Fig. 12.23, fascinating insofar as you can wire it yourself on a protoboard or etch a printed circuit board accordingly. You even may design your own special-purpose matrix, because the system allows the conversion from pretty much everything into whatever else. A matrix similar to Fig. 12.23 can be used to turn on and off selected sets of hydraulic or air-powered valves in the piping of a chemical plant by electrical signals from temperature and pressure sensors somewhere out in the field. The sensor signals go to the vertical inputs of the matrix, while the horizontal legs trigger the relays for powering the valve actuators.

In such humble application as controlling a seven-segment LED, diodes are used to connect the outputs from the 10 NAND gates to the appropriate LED segments. Using wires rather than diodes would generate parasitic cross-connections throughout the matrix and lead to garbled displays.

In our example, five diodes connect the output of NAND gate 5 to the sectors g, f, d, c, and a of the readout. With the anodes of all LED sectors linked collectively to the positive voltage source, current flowing from the +5 V supply through a current-limiting serial resistor and the respective LED sectors toward negative activate the respective sectors of the display chip.

For multidigit readouts, the circuitry shown is repeated as needed, while one of the so far unused gates on the chip is employed for switching between displays.

Digital measurement of analog magnitudes

Digital technology lends itself readily for the measurement of physical quantities, such as temperature, pressure, stress, strain, resistance, voltage, etc., with usually far higher accuracy than that of analog methods.

The primary control elements in all those cases are sensors which are available in two categories: pneumatic sensors converting the measurand into pressure and electrical sensors that convert into voltage or amperage. The former output within the band of 3–15 psi, the latter from 0 to 5 V or, respectively, 4 to 20 mA.

All said, digital measurement boils down to converting sensor outputs into pulse trains and deriving the magnitude of the measurand from pulse count. That makes the precision of digital instruments dependent on the accuracy of the sensor output and the resolution of the impulse counter. For instance, the precision of a pulse count of 100 cannot be better than $\pm 1\%$, because the analog voltage input may stop the count anywhere between the two final clock pulses. To break through the 1‰ barrier, $2^{10} = 1024$ pulses are called for.

Analog-to-digital conversion

There are two approaches to the task of deriving a digital output from an analog input.

1. We can convert the input into a train of pulses for comparison with the output from a clocked digital counter

2. We compare the input, in most cases the voltage output of a sensor, with a digitally generated reference voltage.

The key element in case 2 is the *digital-to-analog* converter (DAC), exemplified in Fig. 12.24 for a 4 digit binary input. The design is straightforward and simpler than that of most of its analog-to-digital counterparts. It employs an R/2R resistor ladder to add "weight" to each of the four binary inputs, D_0, D_1, D_2, and D_3 by halving the analog output from any of the four digital inputs relative to the previous one. This makes D_3 the most significant digit in the ladder, followed by D_2, D_1, and

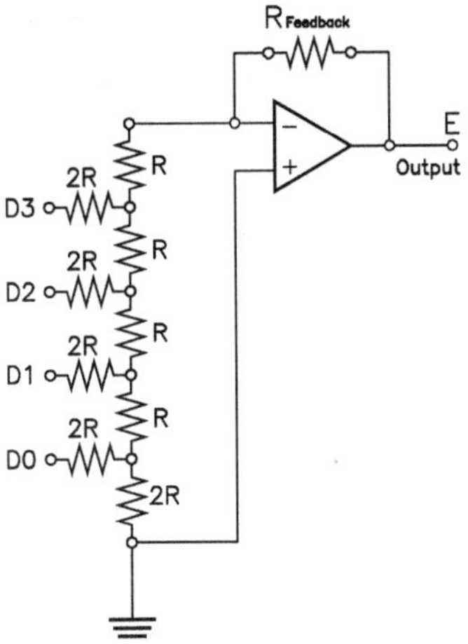

Fig. 12.24. Digital-to-analog converter

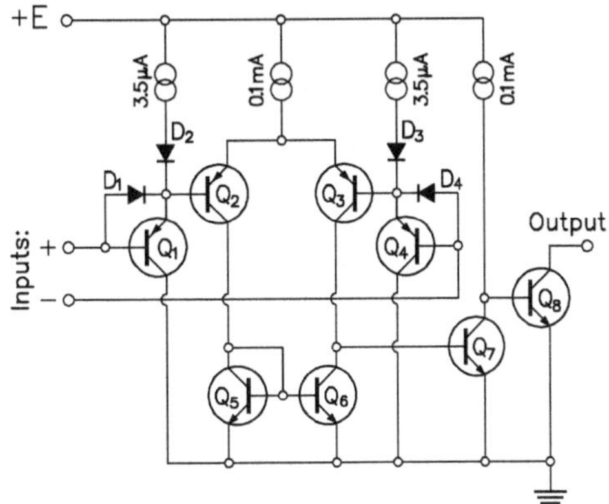

Fig. 12.25. Comparator circuit

D_0. The compounded output voltage is being amplified by an operational amplifier with feedback resistor R_f, which yields negative outputs, unless another OpAmp for inversion into positive outputs is being added. Typical DAC chips work within the 0 to 10 V range and use 12 bits of resolution.

For case 2, *voltage-to-digital* conversion, input voltage is being compared to a voltage taken from a voltage divider chain of resistors, or in a different concept, from the charge of a great number of capacitors in parallel.

The essential component herein is the voltage comparator, a chip that outputs a LOW as long as the input is less than a preset reference voltage, and a HIGH when the input exceeds that voltage. The comparator circuit in Fig. 12.25 employs a differential amplifier, consisting of two pairs of PNP transistors in Darlington configuration (Q_1/Q_2 and Q_3/Q_4), powered by current supplies of 3.5 μA and 0.100 mA. Its output feeds into another differential amplifier, this time of NPN transistors Q_5/Q_6, which controls the final Darlington stage Q_7/Q_8 in an open ended collector configuration.

Input voltages reach the bases of Q_1 and Q_4, respectively, get amplified and bias the bases of Q_2 and Q_3. The arrangement of the PNP transistors Q_2 and Q_3 in series with the NPN transistors Q_5 and Q_6 resembles the class B push-pull output stages of analog audio amplifiers, though the comparators final stage, including the Darlington pair Q_7 and Q_8, operates as a class A amplifier again. Normally, the collector of Q_8 would be connected to +E over a load resistor, but leaving the collector terminal open to feed (sink) into external circuitry makes it possible to interconnect the device with positive voltages other than the +E chip voltage, such as the +5V of TTL logic.

Like in the venerable VTVM, differential amplifiers make the comparator circuit self-contained, while the use of Darlington transistors boosts the circuit's sensitivity to the point where a few millivolts up or down make the difference between a 1 and a 0 output – something paramount for the accuracy of the circuit the comparator is set to control.

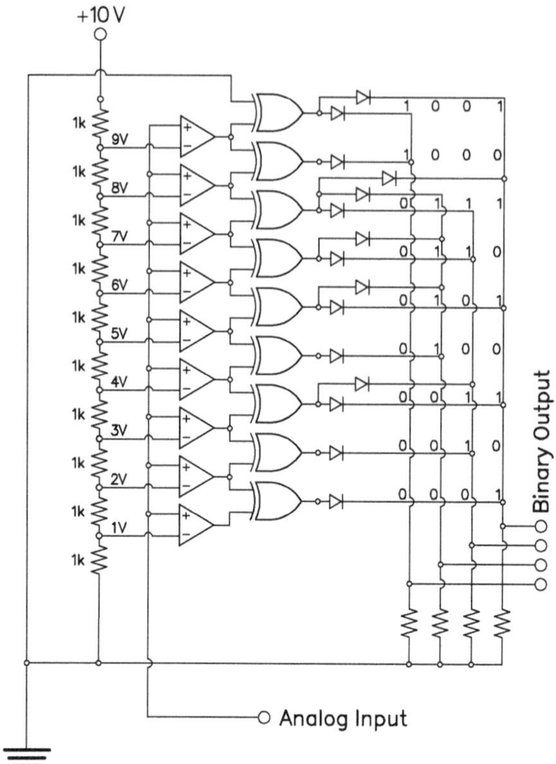

Fig. 12.26. Flash analog-to-digital converter

Comparators are the basic elements of the *flash converter*, the fastest but also the most component-swallowing in the converter family. The circuit in Fig. 12.26 operates by comparing the unknown voltage with the voltages at the nodes of a resistor ladder (*voltage splitter*) between ground and the 10 V regulated +E voltage.

However, the 1 V steps shown in Fig. 12.26 don't provide for anything better than a ±10% accuracy, which would restrict the usefulness of flash converters to low-tech gadgets, using light-emitting diodes in lieu of numerical readouts. Such LED columns still pop up occasionally in toys and hobby projects, such as voltage checkers, where a vertical row of LEDs is being read much like the column of a mercury thermometer. However, demand for voltage dividers of much closer subdivisions than the 1 V exemplified in Fig. 12.26 led to the development of flash converters with far greater numbers of stages.

The components of the converter in Fig. 12.26 are comparators, diodes, and *exclusive OR gates.* In comparison to AND gates, which go HIGH *only* if both inputs are high, OR gates output a HIGH if *either* of the inputs is high. Obviously, the output is LOW with both inputs low, but that still leaves the choice between HIGH or LOW outputs for the case that both inputs are on HIGH.

That kind of ambiguity led to the development of two types of OR gates: the standard OR gate, where both inputs on HIGH make for a HIGH output, and the *exclusive OR gate,* where two HIGH inputs generate a LOW. That boils down to the unique characteristic of exclusive OR gates to output zero (LOW) for *identical* inputs, regardless whether they are HIGH or LOW, and HIGH outputs for *unequal* inputs.

This rule helps to interpret the operation of the converter in Fig. 12.26. All the positive inputs of the 9 comparators at the front end of the converter are connected to the unknown analog voltage, while the negative inputs get the ladder voltages of 1 V, 2 V, 3 V, . . . , 9 V. For the 6.3 volt from a flashlight battery as the unknown analog voltage, all positive inputs up to and including the 6 V terminal of the resistor column are of higher voltage than the ladder voltages of 1 V, 2 V, 3 V, ..., 6 V, so that their respective outputs go HIGH. On the other hand, the 3 comparators above the 7 V terminal of the voltage divider have their positive inputs at lower voltage than their negative inputs, so that their outputs go LOW. Thus, all the exclusive OR gates from the 7 V node up are LOW on *both* inputs, while all the gates below 6 V have both their inputs on HIGH. According to the rule, they all output zero. Alone the gate between 6 V and 7 V gets different inputs (LOW on positive, and HIGH on negative) and outputs a HIGH.

That lone survivor connects over diodes to the second and third binary output line, making them HIGH, while the lines at the far left and the far right are at zero. Altogether, it results in the binary number 0110, the equivalent of a decimal 6, indicating 6 V for the so far unknown analog voltage input. Not surprisingly for a converter of only 10 stages, the decimal fraction of the voltage input does not show.

Useful voltage converters must have far more stages, but the flash converter's multiplicity of components sets a practical limit. So far, 8-bit flash ADC chips with $2^8 - 1 = 255$ comparators are commercially available, and 10-bit flash ADCs with 1023 steps are in the pipeline, promising 0.1% of resolution. But the need to double the number of comparators for every additional bit in the output stands in the way of still further development, much like all the rice of the imperial China didn't suffice to cover the 8×8 squares of the chessboard in ways that every field got twice the number of grains of the preceding one.

On the other hand, all the components of a flash converter operate in unison within the same time span, rather than going through a sequence of clock pulses as other systems do, which makes flash converters by far the speediest of the lot.

Successive-approximation converter

The so-called successive-approximation converter uses a binary series of capacitors to produce a voltage equal to the unknown voltage input. Unlike the resistors in flash converters, which are serial connected, the capacitors in successive-approximation circuits are connected in parallel, so that their charges add up.

The idea rests on the basic law of electricity that links a capacitor's electric charge Q with its capacitance C and the applied voltage E by the formula: $Q = CE$. Inversely, we get the voltage of a capacitor, charged to Q, as $E = Q/C$. For instance, a 0.1 μF capacitor, charged to $E_o = 10$ volt, holds a charge of $Q = 10 \times 0.1 \times 10^{-6} = 10^{-6}$ coulombs. If 50% of that charge is removed, the voltage between the lids of the capacitor drops to $E/E_o = 0.5 \times 10^{-6}/10^{-6} = 0.5$, and E becomes 5 V.

Taking it from there, the software of a successive-approximation converter of, say, 10 V reference voltage, would convert an input of, say, 6.3 V, as follows.

Initially, the analog input, $E_x = 6.30$ V, goes to one of the terminals of the circuit's built-in comparator, while the processing unit channels half of the reference voltage, namely, $10/2 = 5$ V to the other, so that the output goes LOW. That signals the processor to seek the next higher voltage, located halfway through the upper range, namely, at $(10 + 5)/2 = 7.5$ V. With that on the input, the comparator outputs HIGH, which makes the processor to output the next lower figure of $(5+7.5)/2 = 6.25$ V. Being less than E_x, it gets us a LOW. Thus the software tries $(6.25+7.5)/2 = 6.875$ V, resulting in a HIGH. Next check is with $(6.25+6.875)/2 = 6.5625$, still too high. Then comes $(6.25+6.5625)/2 = 6.40625$, somewhat high. One more step yields $(6.25+6.40625)/2 = 6.328$ V, etc. In all this, it shouldn't bother us that the 6.25 V reached in the third step are a much better approximation than the 6.875 V of the fourth. Only we, the omniscient reader, can recognize the closeness of 6.25 V to the sought after 6.3 V. The system, ignorant of the true value of the analog input, cannot identify how good one or the other of its guesses has been. It can only continue narrowing its choices for closest approximation of an identifiable value – 6.328 V in the present case – to the still unknown input.

The method's accuracy is commensurate with the number of steps the program allows for. The 7 steps exemplified above brought the final result within 0.028 V from the input of 6.300 V, allowing for an error of $0.028/6.3 = 0.44\%$. Reaching the high resolution expected from digital instruments calls for far more steps, each consuming the processing time for comparison of E_x with the digitally created voltage, and computation of a new (and better) value. Thus, the capability of the successive-approximation converter to match the unknown analog voltage to any desired degree of accuracy is paled by increasing processing time.

Slope integrating ADC

In contrast to the successive-approximation converter, a slope-integrating analog-to-digital converter (Fig. 12.27) deduces the unknown analog voltage E_x from the time it takes to charge a capacitor, C1, from a constant current source to the level E_x.

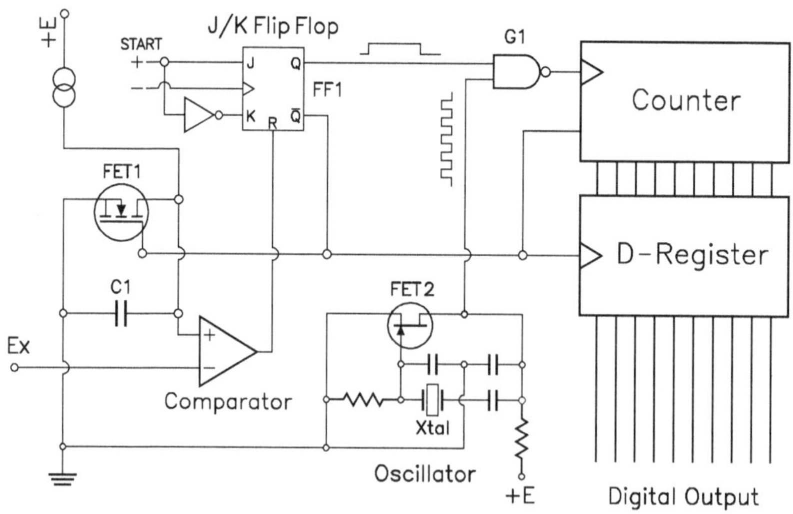

Fig. 12.27. Slope integrating ADC

The measuring cycle begins with the triggering of the flip-flop FF1, bringing its output Q into HIGH and $\bar{Q}$ to LOW. The field effect transistor FET1, which so far shortened C1, stops conducting and lets the capacitor's charging cycle begin.

As long as the output Q of the JK flip-flop FF1 stays high, the NAND gate G1 conducts pulses from the crystal-controlled oscillator to the counter. That goes on until the voltage between the terminals of the capacitor equals the unknown voltage E_x, and the comparator changes its output from HIGH to LOW, switching output Q of the flip-flop FF1 off and $\bar{Q}$ on. Transistor FET1 comes on and discharges the capacitor and keeps it in that state until a new cycle of measurement is initiated with the next triggering of the flip-flop. Most important, the NAND gate G1 stops transferring the oscillator cycles to the counter, which output is then stored in the D-shift register for conversion into a numerical reading, similar in principle to the decoder shown Fig. 12.23.

An offspring of the slope-integrating ADC, is the *dual-slope converter.* Unlike the former, that resets to zero at the end of every cycle, the dual-slope converter has its integrator ramping up and down. As before, the capacitor is charged to the level of the unknown voltage, but rather than shunting it at the end of each cycle, it gets discharged gradually by application of a negative voltage of precisely known magnitude.

With T_1 for the charging time and T_2 for the time it takes to discharge the capacitor by connection to the negative reference voltage $-E_{ref}$, the ratio of the unknown voltage E_x to the known voltage E_{ref} is given by $E_x/E_{ref} = T_2/T_1$. Since T_1 and T_2 are both measured by the frequency of pulses from the same clock, any irregularities in the function of the clock cancel out.

The resulting stability and precision of the dual-slope analog-to-digital converter has made it an attractive choice for circuit designers.

13 Automation – instruments that think

Regardless their phonetic similarity, the terms *automation* and *automatics* stand for different concepts. Automatic machinery is made to produce a given set of parts without help from a human operator, but continue doing so, even if contingencies, such as wear and tear on the cutting bits, make that the machine outputs nothing but rejects.

Automation prevents such shortfalls by comparison of the parameters of the endproduct with those specified. Any discrepancy is made to generate an *error signal* that is used to correct the process. In a different saying, automation adds "intelligence" to otherwise "dumb" machinery or systems.

The same principles guide human behavior as we generate error signals by comparing the factual results of our actions with those desired. For instance, our first drive to a new place of work takes longer than the subsequent ones, because we compare in every trip the desired result – least travel time – with the actual results – time used up – and discover better routes until daily travel time has been minimized.

Likewise, students organize their efforts in response to the difference between the desired results – all A – with the achieved results, reaching from A to – well, let's leave it to them to complete this sentence.

In physics and engineering, the process of controlling an operation through the characteristics of the output is known as *feedback*. An early example from our automobile-centered world is the Borgward Isabella, the late 1950s' dream car of every European teenager in the 14 to 49 years age bracket. The brainchild of Carl F. W. Borgward (1890–1963), then one of Germany's longest surviving automotive manufacturers outside the ivory tower of the really, really biggies, the Isabella included such novelties as a feedback-controlled pneumatic suspension system. So new was the idea of feedback at the time, that a slightly overweighed cop had to learn about it the hard way while filling in a speeding ticket with the hood of the culprit's car doubling as his writing desk. It may not have surprised him to feel the car's front end yielding under the load of his upper body, but that the car then pushed him back, as if shoving an unwanted load off its chest, was not in his books. As so often, the officiality responded with disdain rather than admiration to the offspring of Borgward's creativity, and rather suspected an innocent bystander, the driver in this case, of playing games with the mighty of this world by pressing a camouflaged knob to trigger such unabiding gestures of his car; and it took a flow of truly Carnegian rhetoric to convince him otherwise.

A standard automotive damper lets the suspension springs yield under the impact of a speed bump and other irregularities in the pavement, its oil filling keeps the springs from resiling in ways that could send the passengers flying.

By contrast, Borgward's automation system kept the car's body at the rated clearance above the pavement, regardless of load and bumps, and readily corrected any deviation from this setpoint. Simple as it was, it already contained the entire set of concepts of automation systems.

Herein, *input* is defined as the stimulus from an external source, such as an increase in system pressure under the load of the cop's upper body, what triggers the desired action of the control system.

Output is the magnitude of this action. Borgward's system is of the *closed-loop* variety, since the magnitude of its reaction is determined by the value of the input. Conversely, *open-loop* systems let such adjustment at the criterion of a human operator.

The *controlled variable* defines the state of the system.

And the *manipulated variable* keeps the controlled variable at setpoint level. In the Borgward example, the controlled variable is the vehicle's spacing above the pavement, and the manipulated variable is the dampers internal pressure.

Error message is the difference between the desired and the actual state of the controlled variable, used as *feedback*. Which brings us to a sensitive point of automated control systems: An error must be allowed to occur before the system can start correcting it, so that the concept of an ideally precise feedback controller is like the search for the point where the hyperbola contacts its asymptote.

Thermostats

Thermostats, including the ones that regulate room temperature, are feedback control systems of the ON/OFF type. As the name suggests, they turn the air conditioner on if room temperature is above setpoint, and off if it falls below. Most models perform that function with a bimetal strip, coiled into a spiral, that controls the angular position of a mercury switch.

With the metal of higher thermal expansion on the inner side of the windings of the bimetal spiral, the latter unwinds with rising temperature and tilts the switch to the point where the bead of mercury in the vial eases to the left and interconnects the contacts that power the motor starter of the air conditioner. With the inner end of the spiral soldered to the center axis, room temperature can be set by rotating the latter along with the attached pointer, which indicates degrees Celsius or Fahrenheit on the scale of the instrument.

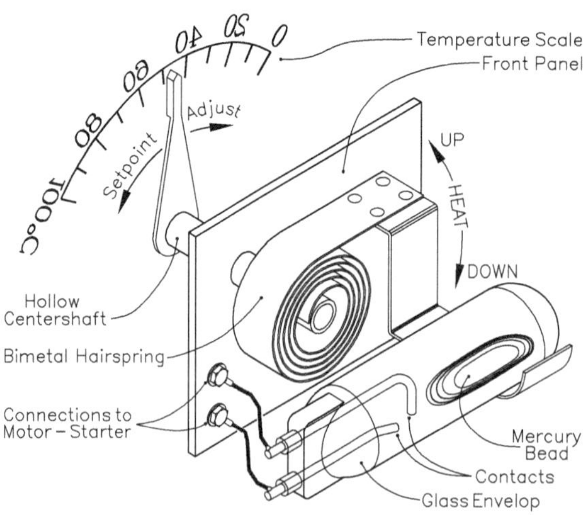

Fig. 13.1. Room thermostat

A certain degree of inertia is called for in the operation of the mercury switch in order to avoid chatter of the motor starter. Luckily, friction between the mercury bead and the glass envelope makes that a slight angular displacement of the bulb does not suffice to help the mercury blob beyond the breakover point. Some extra tilt is needed for getting the mercury in motion and activate the motor starter. Likewise, a similar degree of inclination toward the other side is afforded to interrupt the connection. This so-called *hysteresis effect* can be heightened by giving the glass envelope a slight upward bend around the middle, which the mercury would have to overcome against gravitation to switch from CLOSED to OPEN and vice versa.

In control system lingo, ambient temperature is here the *input*, and *output* is the ON or OFF signal sent to the motor-starter contactor. The *error signal* is the difference between the setpoint and the actual temperature of the bimetal hair spring.

Uses of bimetal thermometers and thermostats are limited to the range of temperatures that leaves intact the crystalline structure of the metals that make up the hairspring. Like in the process of annealing hardened steel parts, changes in the microstructure of steel occur from 200 °C upwards and can be identified by the tarnish on blank species, which with a gradual loss of hardness turn light yellow at 210 °C, tawny at 250 °C, and cornflower-blue at 300 °C. Most commercial thermostats are specified for 0 to 160 °C, where the process of expansion and compression of the bimetal strips remains strictly reversible.

Electronic temperature controllers

For a range of temperatures wider than that for which bimetal thermometers are specified, resistance thermometer bulbs with filaments of nickel for −60 to +180 °C, and of platinum for −220 to +850 °C, encapsulated in glass or ceramic enclosures, are being used as sensors in combination with electronic temperature controllers which measure and control temperature in function of the ohmic resistance of the sensors, given in Table 13.1.

Table 13.1. Ohmmage of standard resistance bulbs

Temperature (°C)	Pt (Ω)	Temperature (°C)	Ni (Ω)	Pt (Ω)	Temperature (°C)	Pt (Ω)
−220	10.41	−60	69.5	76.28	200	175.86
−200	18.53	−40	79.1	84.21	250	194.13
−150	39.65	−20	89.3	92.13	300	212.08
−100	60.20	0	100.0	100.00	350	229.70
−80	68.28	+20	111.3	107.80	400	247.07
		40	123.0	115.54	450	264.19
		60	135.3	123.24	500	280.94
		80	148.2	130.91	600	313.38
		100	161.7	138.50	700	344.59
		120	175.9	146.07	800	374.61
		140	190.9	153.59	850	389.23
		160	206.7	161.06		
		180	223.1	168.48		

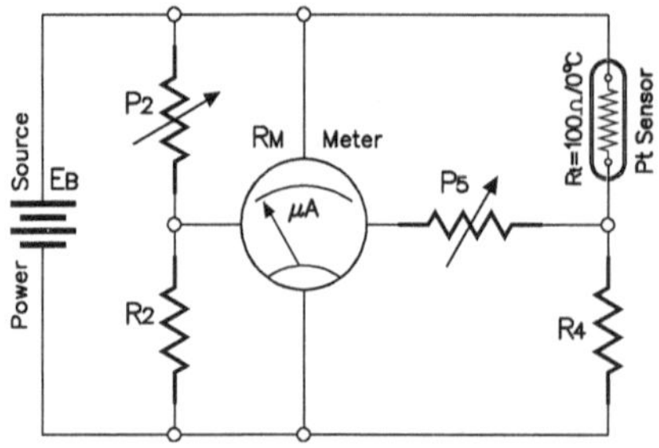

Fig. 13.2. Resistance bridge

Compared to thermostats, temperature controllers offer higher precision and include some additional features, such as *no-signal* shutoff, and *on/off* or *off-only* modes. The latter gives the user a choice to have the controlled system, such as an industrial furnace, either to maintain constant temperature, or to end the heating process as soon as setpoint temperature is reached.

No-signal shutoff prevents overheating in case of sensor failure. Much like light bulbs, resistance bulb filaments wear with age and vibrations, while aging thermo-elements split open at the welded junction of the two unequal metal wires. Unchecked, temperatures would rise in these cases to the meltdown of the heating elements.

Mechanical temperature controllers had that feature permanently built in and would not turn on as long as the pointer lingered around the zero point of the scale. To initiate heating, the controller had to be shunted with an external "start up" switch – preferable of the spring-return push-bottom type, which cannot be forgotten in the ON position.

A Wheatstone bridge-based circuit for measuring sensor resistance is shown in Fig. 13.2. The meter is connected between the junction points of a fixed resistor, R_2, and a potentiometer, P_2, in one leg, and the junction of fixed resistor R_4 and the resistance bulb R_t in the other. The resistance of potentiometer P_2 is set to match that of the sensor at 0 °C. Potentiometer P_5 regulates the sensitivity of the meter and with it the range of measurement. P_5 is to be calibrated for full scale deflection of the meter, while the sensor is kept at the rated upper temperature limit of the bridge.

Rather than physically heating up the probe for this purpose, the sensor is replaced by a dummy resistor which ohmmage can be read from Table 13.1 as 175.86 Ω for a 0–200 °C scale. A 100 Ω dummy resistor is used to adjust P_2 for zero meter deflection unless one prefers to keep the sensor in place and submerge it in a water-ice mixture.

With R_M for the internal resistance of the meter, the meter current J for a given set of bridge resistors (R_1 through R_4) can be derived from

$$J = E_\mathrm{B}\left(\frac{R_t}{R_t + R_4} - \frac{P_2}{P_2 + R_2}\right) \Big/ \left(\frac{P_2 R_2}{P_2 + R_2} + \frac{R_t R_4}{R_t + R_4} + R_\mathrm{M}\right).$$

When the ohmmage of the fixed resistors and potentiometers is introduced, the equation shrinks to a simple fraction that comes handy for figuring the divisions of the temperature scale with your calculator.

Significantly, this formula shows the direct dependence of the meter current J from the power source voltage E_B, which makes a finely regulated power supply a must for the circuit's proper operation.

In the circuit diagram of an electronic temperature controller in Fig. 13.3, a voltage reference of 30 V is generated by serial connection of three REF102A volt-

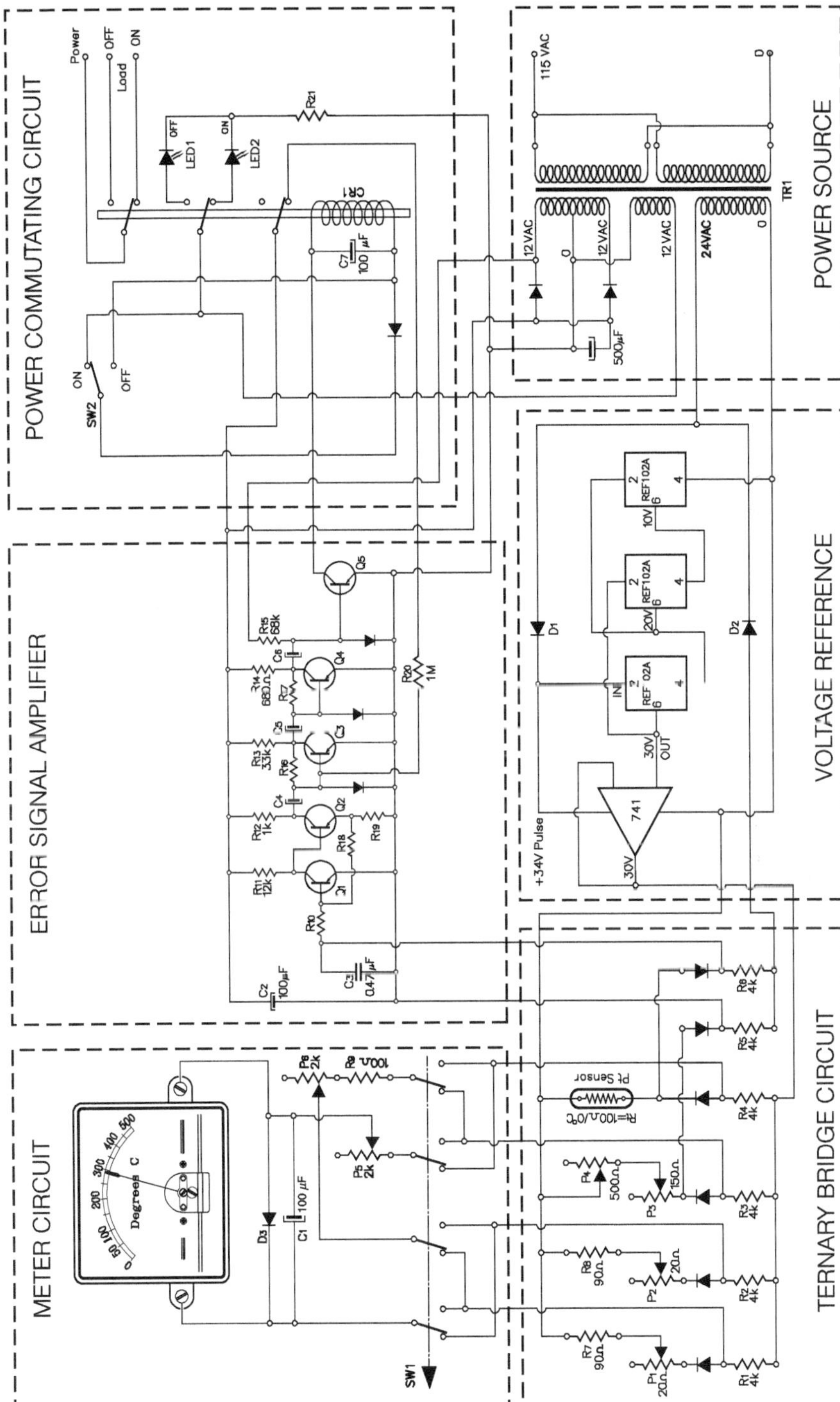

Fig. 13.3. Circuit diagram of electronic temperature controller

age reference chips, controlling an operational amplifier that supplies regulated power to the ternary bridge circuit, containing 3 Wheatstone type bridges: the *set point bridge*, the *measuring bridge*, and the *switching bridge*.

The controller uses the same meter for reading the temperature of the probe and that of the setpoint, wherein the selector switch SW_1 allows for toggling between one and the other. However, powering all those bridge circuits from the same source would cause an uncalled for interaction of setpoint bridge and measuring bridge over the switching bridge and have temperature readings distorted by changes in setpoint temperature and vice versa. To avoid that, temperature and setpoint readings are multiplexed, so that the signal-strength measuring bridges come on with the positive halfwave of alternating current from the power source, while the error signal bridges use the negative AC halfwave. Separation is done by diodes D_1 and D_2.

The setpoint measuring bridge includes precision resistors R_1 and R_3, the latter in combination with the setpoint adjust potentiometer P_4. R_5 and R_6 constitute a measuring bridge between setpoint and temperature sensor, feeding its output into the signal amplifier. That is the section of the ternary bridge that operates in counterphase with the other two in order to prevent mutual interference. One section is off when the other is on, and vice versa.

The 4-pole double throw selector switch SW_1 in the meter circuit allows for toggling between temperature and setpoint readouts. Diode D_3 shunts the meter from voltage spikes exceeding the 0.6 V forward voltage drop of silicon diodes, so that accidental power surges wouldn't burn the meter coil. Capacitor C_1 is to smoothen the pulsating DC power.

The 200 μA d'Arsonval meter used in this circuit is sensitive enough to show temperature directly. By contrast, the controlling function, where minimal changes in the error signal are supposed to switch between ON and OFF, needs a signal amplifier, capable of reacting to extremely small inputs. Yet, it must also handle the high error currents generated when ambient and setpoint temperatures are at their extremes. Therefore, transistors Q_3 and Q_4 are protected from overload by diodes shunting their bases to ground, which limit the emitter-base voltages to 0.6 V. Base bias is provided by the collector-base resistors R_{16} and R_{17}, which also control the negative feedback for stabilizing the operation.

Preamplification is provided by the two front-end transistors Q_1 and Q_2, coupled directly and stabilized by negative feedback through resistor R_{18}.

Power transistor Q_5 supplies the coil of the 3-pole double-throw control relay CR_1, which turns power to the load on and off. Along with the power transistor, this coil is connected to a separate winding on the secondary of the power transformer, so that the intermittent current demands of this relay wouldn't cause voltage fluctuations in other circuit branches.

Accidental zero signal input, caused by breaks in the filament of the resistance bulb or failure of amplifier components, makes that Q_5 turns off due to its positive base bias through R_{15}, and the relay drops out.

Relay "chattering," a rapid sequence of ONs and OFFs when system temperatures linger around the setpoint, is prevented by the bottom pair of relay contacts, which gets locked in ON position through feedback by resistor R_{20}, of 1 MΩ, to the base of Q_3 until the error signals get strong enough to overcome this bias. Other values for R_{20} can be selected to broaden or narrow that lock-in range.

The central relay contacts are used to indicate the state of the system by activating either the blue or the red LED panel-light. And the uppermost set of contacts does what we had been talking all along: It turns power to the load on and off.

The primary of the instrument's power transformer consists of two 115 VAC windings with accessible endings, shown parallel connected for 115 VAC supply. For 230 VAC, the coils must be serial connected.

The secondary of the transformer supplies the instrument amplifier from a center tapped 12 V + 12 V winding that feeds into a full wave rectifier circuit, which along with capacitor C_7 brings the output to 15–16 VDC. Another 12 VAC winding powers the control relay CR_1, and the 24 VAC windings lead into the voltage controller and from there to the ternary bridge circuit.

Operation of ON/OFF mode controllers

In a nutshell, the controller in Fig. 13.3 keeps the load turned on as long as the temperature in the controlled system (such as an oven) remains below setpoint, and turns off when it rises above. That happens regardless of the *magnitude* of the *error signal*, the difference between setpoint temperature and system temperature.

As shown for a muffle oven in Fig. 13.5, the temperature sensor is positioned close to the thermal center of the controlled system, switching the heating elements on and off regardless of the temperature of the heater elements themselves. This causes sensor temperatures to trail heater temperatures, which pass the setpoint long before the controller turns off. The excess heat energy stored in the heater elements and their supports causes the temperature inside the muffle to continue its raise even while power is off, until the temperatures of the heating elements and the muffle meet somewhere in between. This point of equilibrium lies above the setpoint temperature – the system overheats.

Conversely, oven temperature keeps sinking even with power back on until it levels with the slowly rising temperature of the heater elements. Only then starts the actual reheating of the system in its integrity.

This phenomenon, caused by the *thermal inertia* of the controlled system, depends in the case in point on the mass and thermal constants of the materials the oven is made from and on the surface area of the resistance wires.

The ohmic heat dissipation Q (in watt) of a given resistor R is $Q = J^2R$, and the resistance of a wire of cross-sectional area A and length ℓ is $R = \rho\,\ell/A$. Herewith, the resistance and with it, the heat energy dissipated by a given length of wire, is proportional to the fraction ℓ/A.

For instance, a resistance wire of 100 m length and 10 mm^2 cross-sectional area generates the same heat energy as 10 m of 1 mm^2 wire, because $100/10 = 10/1 = 10$, but the latter would weigh and cost $100 \times 10/10 \times 1 = 100$ times less. Pricewise that sounds too good to be true, and in the sober light of design engineering, it is. The degree of heat transfer from a hot wire into the environment is proportional to the surface area of the wire, in our case to $\sqrt{10} \times 100 = 316$ vs.

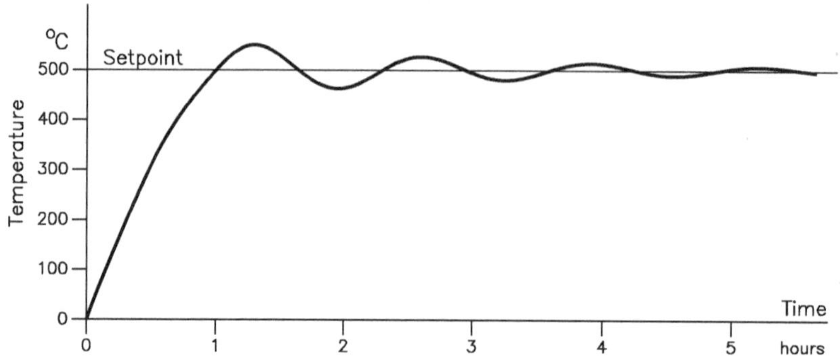

Fig. 13.4. Temperature fluctuations in ON/OFF control

$\sqrt{1} \times 10 = 10$, a ratio of $316/10 = 31.6$. In comparison with the long, thick wire, the short thin wire would thus release 31.6 times less heat and retain the rest until it melts and evaporates.

A rule of thumb calls for keeping the power below 4000 watt per square foot of wire surface area as to keep the differences between heater and muffle temperatures reasonably low. However, in cases where the desired muffle temperature gets near the rated limits of the heater wires, lower wattage-to-surface ratios must be used. Allowable wire temperatures are 1000 °C for most commercial Cr-Ni resistance wires, and up to 1350 °C for highest quality Fe-Cr-Al alloys. The closer we get to those limits, the closer must temperatures be regulated. For instance, a 50 °C overshoot in a 1300 °C rated oven could trigger a meltdown.

The results of the alternating "too little–too much" operation of ON/OFF controllers is illustrated in Fig. 13.4: The greatest overshoots happen at the beginning of the operation, while not only the inside of the oven but also the cold walls absorb heat. As the walls get hotter, temperatures tend to stabilize.

In applications where the controller is used only to shut off the oven at a set temperature, such as the baking of ceramics and steatite articles, the user unwittingly compensates for the overshoot by setting the turn-off temperature somewhat below the desired process temperature, leaving it to the thermal inertia of the system to do the rest. The amount of this kind of correction varies with the model of the oven in use and is being set empirically.

For continuous use, process temperature tolerances must be equal or greater than the expected fluctuations. It helps to preheat the oven to somewhat below the desired temperature and keep it there until the walls got their share of heat, so that the low amplitude fluctuations, shown toward the right of Fig. 13.4, are already prevalent during the first ON/OFF cycle.

Another approach to minimizing fluctuations is in installing *two* sensor elements (Fig. 13.5), one for instance at the center of the muffle and the other close to the heaters, and assign to each a certain degree of participation in the process of regulation, called *authority*.

Figure 13.5, bottom, shows the installation of two 100 Ω resistance bulbs in an industrial muffle furnace and the respective circuit diagram – a bridge with sensor S_1 in one of its upper legs, and a 100 Ω resistor in the other. The other two legs of

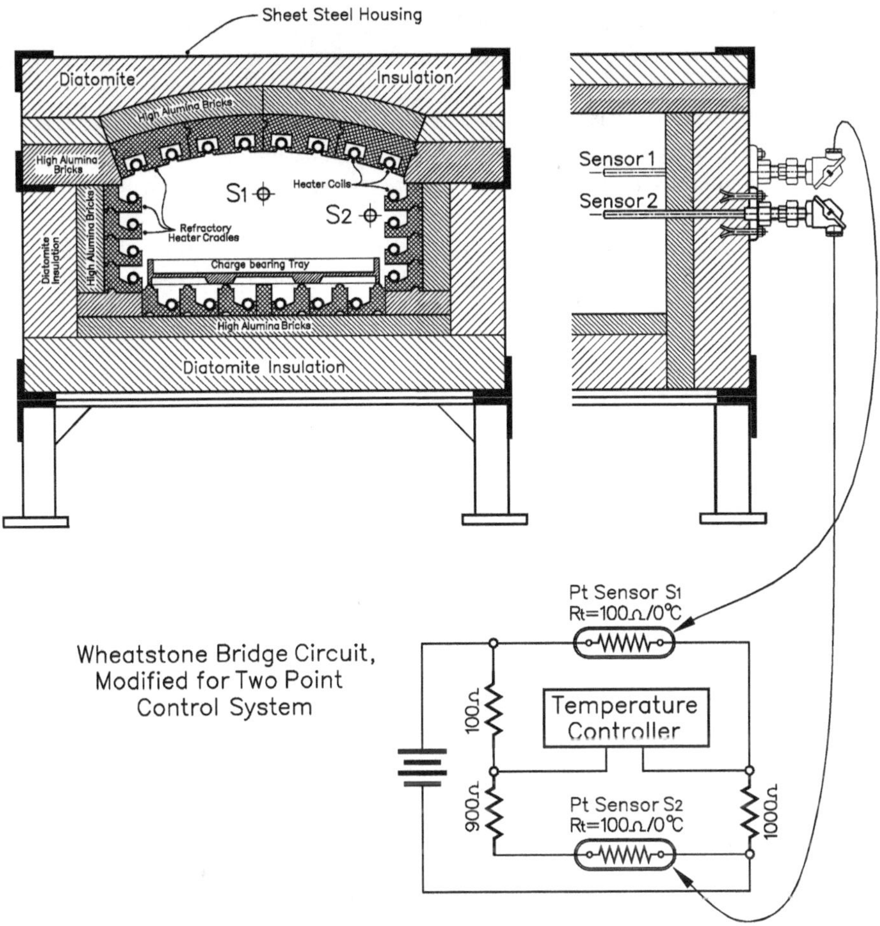

Fig. 13.5. Industrial oven with 2-point ON/OFF temperature control

the bridge include sensor S_2, the one close to the heaters, in series with a 900 Ω resistor, and a 1000 Ω resistor in the opposite leg. Consequently, any change in the temperature of S_2 will influence the bridge output $100/1000 = 0.1$ or 10 times less than the same change in S_1, which leaves the authority of the former at only 10% of the latter.

Thermal insulation of the muffle in Fig. 13.5 is provided by layers of diatomite bricks (marked by wide-spaced hatching), excellent thermal insulators due to the air trapped in their internal voids. Among thermal insulators, non-circulating air is second only to vacuum (we remember from Dewar vacuum flasks).

However, diatomite should be kept below 900 °C and therefore needs to be insulated from the muffle by layers of high alumina bricks (narrow hatch).

The heating spirals' cradle modules are of refractory.

The charge rests on a tray of high chromium steel, supported on raised sections of the bottom heater elements. Since the charge, that may include tools, molds, stamps, and dies, claims most of the muffles' free space, sensor S_1 has been shifted to above center.

Sensor S_2, with its 10% authority, locates right in front of one of the heating spirals. At that short distance, radiant heat transfer dominates over conduction and convection and tends to bring the temperature of the sensor close to the temperature of the wires.

A sensor's authority can be set to values different from 1 : 10 by replacing the 900 Ω resistor with one of lower or higher ohmmage, and changing the 1000 Ω resistor in the other arm of the bridge accordingly.

Luminosity control

The circuit in Fig. 13.2 can be used to check illumination if we replace the resistance bulb R_t by a photoresistor, and set the potentiometer P_2 for the resistance the phototransistor would have if exposed to the desired degree of illumination. A nulling meter in that bridge would show zero at the desired degree of illumination and indicate low light or high light by left or right deflection of its pointer.

Though we hardly ever have problems with the intensity of illumination as long as it comes powered from the closely controlled utility networks, it may be called for in other cases, such as surgical operation centers in locations where power comes from mobile motor or generator compounds or is drawn from the networks of minor power plant whose voltage may fluctuate up and down when a major consumer kicks in or out. Simple ON/OFF control would do nothing better than turn the light-source periodically on and off. However, employing two rather than one single ON/OFF controller in combination with a variable voltage transformer allows to confine brilliance in a narrow band.

Of the pair of ON/OFF controllers in this setup, the application diagram in Fig. 13.6 shows only the output relays, marked CR_1 and CR_2, for three different scenarios: cycle 1 for the case of excess brightness, cycle 3 for lack of brightness, and cycle 2 when the desired brightness has been reached.

1st cycle. *Illumination below setpoint* of both, the High and the Low controller. CR_1 as well as CR_2 are on, the servomotor turns the slide contact of the Variac toward higher voltage.

2nd cycle. *Illumination correct*, that is, between the setpoints of the High and the Low controller. CR_1 is off, CR_2 is on. With both motor terminals connected to the negative pole of the DC supply, the servomotor stalls.

3rd cycle. *Illumination too bright*. CR_1 as well as CR_2 are turned off, which inverts the motor connections from $+/-$ to $-/+$. The transformer's cursor lowers output voltage.

For best stability, the motor should have a small flywheel installed on its shaft, which inertia keeps it running after turn-off long enough to bring the cursor close to its midscale position.

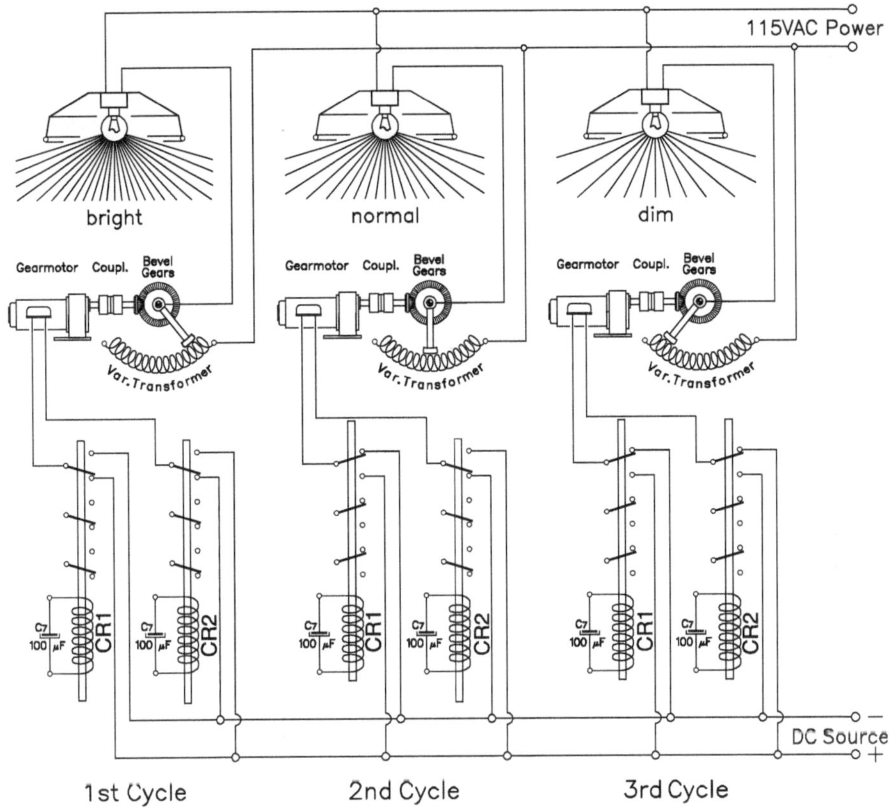

Fig. 13.6. Variac illumination control

CO_2 control in exhaust gases

The complete combustion of hydrocarbons, such as heating oil, kerosene, gasoline, and diesel oil with oxygen yields carbon dioxide gas (CO_2) and vaporized water (H_2O). Internal combustion engines and turbo fans use air as their supply of oxygen, so that some 4 parts of nitrogen per 1 part of oxygen are funneled along without actually participating in the combustion process, except for the generation of some impurities of NO_2 and other nitrogen compounds in the exhaust gases.

With C_mH_n as the chemical symbol of hydrocarbons in general, the process of combustion reads

$$C_mH_n + \left(m + \frac{n}{4}\right)O_2 + (4m+n)N_2 = mCO_2 + (n/2)H_2O + (4m+n)N_2 \,.$$

For a certain hydrocarbon, say, propane (C_3H_8), the values $m = 3$ and $n = 8$ allow for figuring the composition of the exhaust gases from this equation as $C_3H_8 + 5O_2 + 20N_2 = 3CO_2 + 4H_2O + 20N_2$. Using the atomic masses of 12 for C, 1 for H, 16 for O and 14 for N, the exhaust gases contain by weight

$3 \times (12 + 2 \times 16)CO_2 + 4 \times (2 \times 1 + 16)H_2O + 20 \times 2 \times 14N_2 = 764$ mass units, or in percent by weight $3 \times 44/764 = 17.28\%\ CO_2$, $4 \times 18/764 = 9.42\%\ H_2O$, and $560/764 = 73.30\%\ N_2$.

CO_2 contents below 17.28% would thus sign incomplete combustion due to lack of air. The missing fraction of CO_2 then comes as the highly toxic carbon monoxide (CO), which generation affords only half as much oxygen as that of carbon dioxide.

For gasoline and diesel oil, which are mixtures of several hydrocarbons, the formula is computed separately for each component and the concentrations of the combustion products are then added up.

The CO_2 content of a gas mixture strongly influences its capacity of heat conduction.

For example, the heat conduction coefficient k in W/cm °C for CO_2 is 0.015; CO, 0.023; nitrogen, 0.024; oxygen, 0.024; air, 0.024.

Since k lingers around 0.024 W/cm °C for all exhaust gases, except for CO_2 with its 0.015, the CO_2 content of a mixture is determinant for its degree of heat conduction.

The CO_2 meter and controller in Fig. 13.7 operates on this particular. This convenient instrument is built around a standard Venturi tube that can be inserted into an industrial gas line wherever convenient. Gas samples are being drawn through fine channels from the narrowest point of the Venturi restriction and, respectively, the expansion zone, into the flanged-on sample chamber. Continuous renovation of the sampling is granted by the pressure gradient between those two locations.

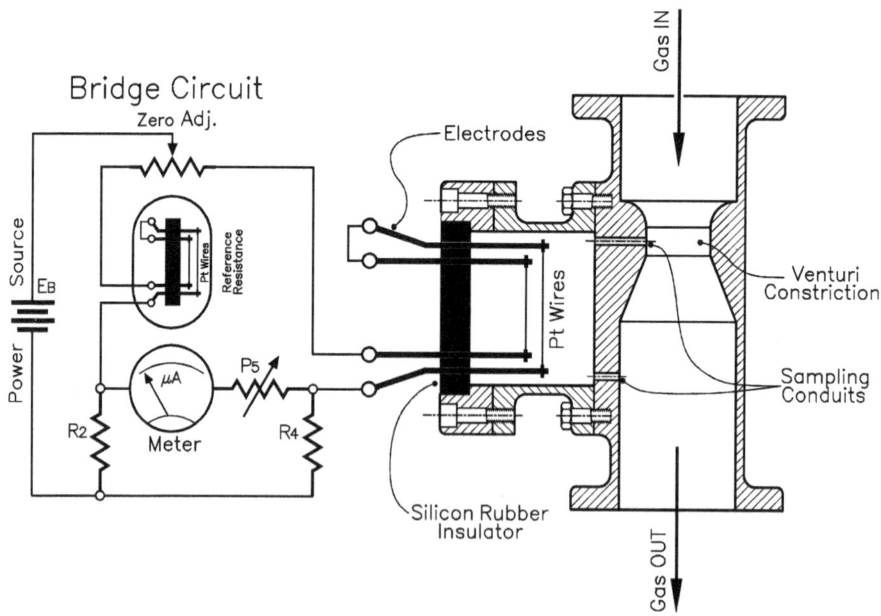

Fig. 13.7. In line exhaust gas analyzer

The degree of heat conduction is derived by comparison of the temperatures of a pair of thin platinum wires in the sample chamber with an identical pair housed in an external reference probe. Both make part of a bridge circuit which overall impedance is kept low enough as to make the Pt wires heat up.

With heat conduction of carbon dioxide gas lower than that of the other components of the mixture, wires in the sample chamber will get hotter than those in the reference chamber as a function of the gas sample's CO_2 content.

The resistors R_2 and R_4 of the bridge circuit must be of sufficiently low ohmmage as to conduct the amperage afforded to keep the platinum wires in the sample and reference chambers at high enough temperatures for reliable measurements of their heat transfer into the chambers environments. Higher temperatures promise higher measuring accuracy, but must be kept below the point where the Pt wires lose tensile strength and get flabby.

The zero-adjust potentiometer (top in the schematic) should be wire-wound to withstand the amperage of the bridge currents.

Sulfur dioxide measurement

Depending on its place of origin, crude oil may contain up to 6% of sulfur, which in the process of combustion binds with oxygen to sulfur dioxide, SO_2. Exhausted into the atmosphere, sulfur dioxide is absorbed by water droplets in clouds and generates sulfurous acid and other sulfurous compounds, a solution we got to know as acid rain. Acid rain, defined by a pH value of less than 5.6, causes damage to tree growth and adds to the acidity of soils.

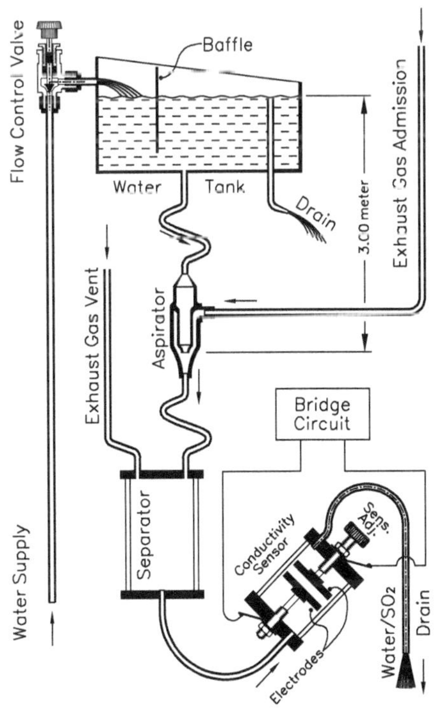

Fig. 13.8. SO_2 checking setup

The simple device in Fig. 13.8 allows for checking the SO_2 content of exhaust gases in general, including the discharge from sulfur combustion chambers in the paper industry, by dissolving the gases' sulfur dioxide content in water and measuring the electric resistance between a pair of face-to-face mounted electrodes awash in the resulting aqueous solution.

Samples are being drawn from the exhaust piping with an *aspirator* that resembles the ejector vacuum pump safe for the water supply, that consists in the case in point of a level-controlled reservoir of constant overhead. The aspirator doubles as mixer of the gases under test with the jet of pulverized water.

The resulting solution enters the *separator*, which allows the nonabsorbed gas to escape while a saturated solution settles at the bottom from where it proceeds into the sensor container with the mutually facing electrodes.

Constant overhead in the aspirator nozzle is essential and has been standardized as 3.00 meters of water column. This boils down to rigorous level control in the water tank, obtained by overfeeding the tank just enough for the drain pipe letting off the excess continuously. A baffle between water inlet and outlet keeps waves caused by the inflow from disturbing the liquid surface around the drainpipe.

The electric part of the measuring circuit resembles the bridge in Fig. 13.7, but must be AC powered to avoid electrolytic separation of the solution in the sample chamber. The sensitivity adjustment (Sens. Adj.) screw regulates the spacing of the electrodes and is set for full meter deflection for a saturated SO_2 solution in the sensor. Other ranges can be set at user's criterion.

Humidity control

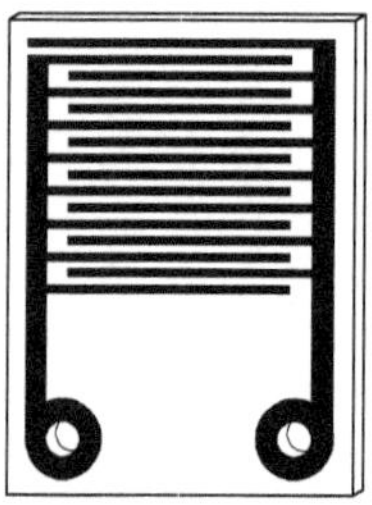

Fig. 13.9. Resistive relative-humidity sensor

Mechanical humidity meters use the moisture-dependent expansion of a strand of degreased human hair (preferably taken from "blondes" – the saying goes) as the sensing element. The instrument's mechanism for converting the minute length changes of the wisp of hair into a readable deflection of the pointer includes bimetal elements for temperature compensation.

Amongst electronic *relative-humidity sensors*, which derive signal currents either from changes in the degree of heat transfer, capacity, or impedance, the latter (Fig. 13.9) are the most simple and most popular types. Their function-

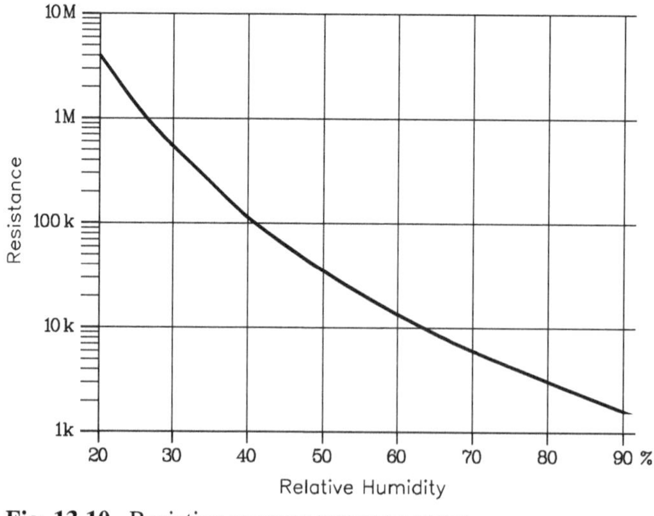

Fig. 13.10. Resistive sensors response curve

ing rests on the hygroscopic characteristics of lithium chloride and the related changes of the compound's impedance with absorbed humidity (Fig. 13.10). Pure lithium chloride conducts electricity from 11% water content upward. However, conductivity starts at 1% if polyvinyl alcohol is used as solvent.

Hygroscopic elements, such as that in Fig. 13.9, usually consist of a plastic substrate with two interlacing gold leaved grids stamped upon. Here again, power supply is symmetrical AC rather than DC, in order to forestall polarization of the sensor. For consistent results over a reasonably wide range of temperatures, the electronics of humidity controllers include a circuit tailored to compensate for temperature-dependent changes in sensor resistance.

Relative humidity of the air we breathe – expressed by the fraction (in % by weight) of water vapor per volume unit of air (kg of water in every m^3 of air) – is critical for our well-being, in particular in air-conditioned environment. Of highest importance is the humidity content of air in certain industrial installations, in particular in the spinning and weaving industries, where air conditioning is frequently left to an ingenious system of interaction between air flow, water particle injection, and natural heating or dissipation of heat by evaporation.

Elsewhere, the operation of dryers and drying furnaces is being controlled in relation to the relative humidity of the exiting air. And so is the humidity of inert gas envelopes that prevent oxidation of metal parts in heat treatment, crucible casting, and electric welding.

Moisture control

Moisture control is a must in textile production, which includes the three basic processes of *spinning*, *weaving* (or *knitting*), and *finishing*.

Raw cotton bales are broken open and fed into the breaker and the scutcher for cleaning. Then follows the *carding* and *drawing* process, separating the fibers into a blanket of loose parallel strands, straightened by drawing and combing.

Previous to the proper weaving operation, the warp is pulled through the sizing (binder) and dried in the slasher, which strengthens the yarn and keeps it straight. Only then follows the familiar weaving operation, where a great number of parallel threads, the wrap, is passed lengthwise through the loom to be interlaced with crosswise running threads, the weft, fed from the fly shuttle[1] on its way transversally through the width of the wrap and back.

As the yarn exits the slasher, its moisture must be continuously controlled in order to warrant consistent weaving operation, because excess moisture could cause mildew to form, while exceedingly dry yarn turns brittle.

Moisture content of the yarn is derived from the electrical resistance of the layer of yarn between slasher roll and detector roll (Fig. 13.11) and is controlled by varying the speed of the slasher.

1 a tool passed back and forth through the shed and stores weft yarn while weaving.

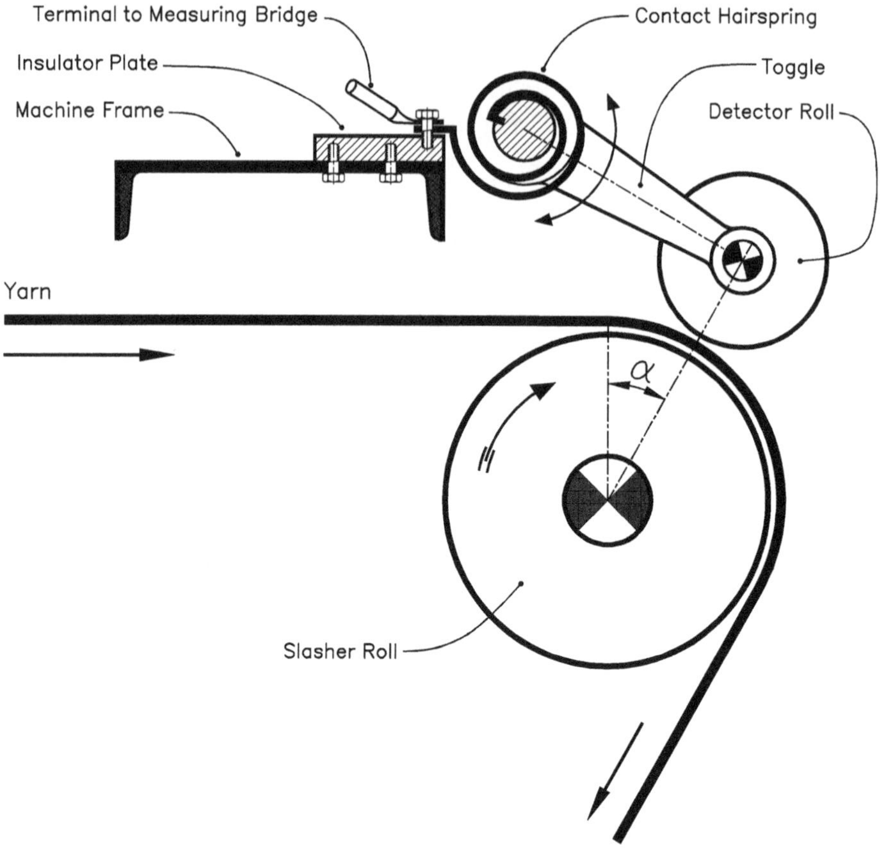

Fig. 13.11. Slasher control pickup

If the machine frame is connected with "ground" of the resistance measuring bridge, the mount of the detector roll must be isolated from the rest of the machine and made to connect into the circuit over the hairspring that makes the detector roll idle on the warp.

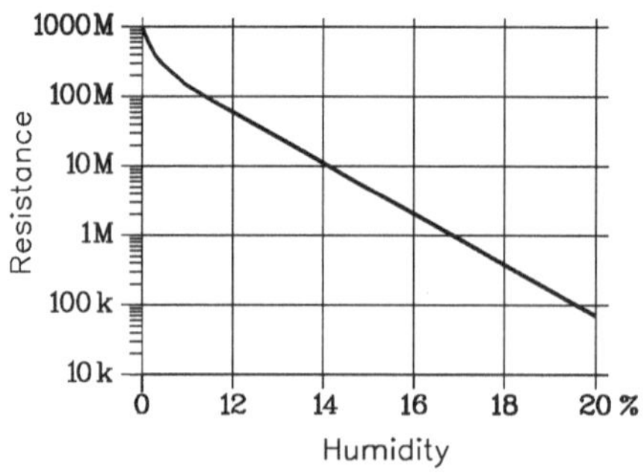

Fig. 13.12. Specific resistance of rice in function of moisture content

In the paper industry, the kraft running through paper finishing machinery needs continuous control, because excessive moisture makes the paper mottle, darken, or even crush in the calender rolls, where pressure combined with a measured amount of slipping is paramount for the finish and quality of the outgoing product. Too dry paper – on the other hand – comes out underweight and poorly finished.

Though the moisture content of cereals, such as grain, rice, beans, flower, etc. is regulated by government agencies, the standard moisture sensor consists simply of a pair of electrodes, inserted in the granular product.

The resistance between the electrodes is usually a logarithmic function of the product's moisture content, as exemplified in Fig. 13.12 for rice. Here too, specific resistance is defined as the resistance in ohm (Ω) between two electrodes of 1 square centimeter area each, spaced by 1 cm (0.394 inches). Conductivity is the inverse of specific resistance, and fittingly measured in ohm, or siemens per meter.

Conductivity of fluids and liquids

Conductivity of fluids and liquids are measured in a bridge circuit (Fig. 13.13) that compares the flux of electric current between a pair of submerged electrodes with the flux between an identical set of electrodes in a sealed probe, containing a liquid of reference, such as pure water or other choices from Table 13.2. This standard too is submerged in the liquid under test not too far from the location of the sensor, so that both remain at identical temperatures. That's an essential feature, because liquid conductivity depends strongly on temperature.

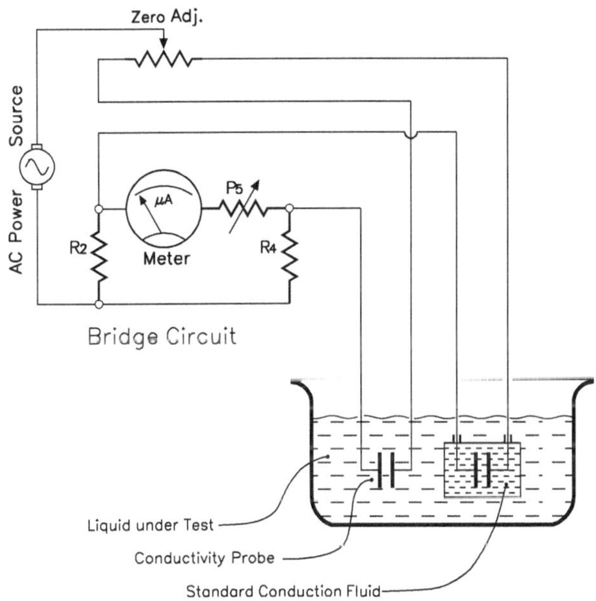

Fig. 13.13. Conductivity of liquids measurement setup

The meter in the circuit diagram of Fig. 13.13 can be calibrated to show conductivity of the liquid on a non linear scale, or the bridge output can be used as the error message in an automated control system.

Here again, the circuit must be powered by alternating current in order to avoid electrolytic dissociation of the liquid under test. Table 13.2 exemplifies conductivity values for some reference solutions.

Table 13.2. Conductivity of liquids

Liquid	Conductivity $1/(\Omega \times \text{cm})$
Chemically pure water, 0 °C	1.58×10^{-8}
30% sulfuric acid, 15 °C	0.7028
Saturated table salt solution, 15 °C	0.2014
N potassium salt solution, 15 °C	0.09254
20% hydrochloric acid, 18 °C	0.7615
31% nitric acid, 18 °C	0.7819
20% phosphoric acid, 18 °C	0.1129
20% sodium hydroxide, 15 °C	0.3270
20% potassium chloride, 18 °C	0.2677

Some control theory

While ON/OFF controllers keep the controlled variable (temperature, pressure, etc.) at or close to the setpoint but regulate merely by turning the power supply periodically on and off, a proportional controller keeps the system activated all the time and corrects any deviations from the original settings by initiating a measured counteraction. Nevertheless, it remains a "dumb" device (no offence intended), insofar as it has no fixed setpoint and thus has no way of checking if this action was too much, too little, or just right. The more sophisticated methods of *proportional* plus *derivative* control, or better still *proportional* plus *integral* plus *derivative* control make up for such shortfalls.

The functioning of proportional control can be expressed by the simple equation

$$W = P(T_{\mathrm{P}} - T)\,,$$

where T_{P} is the desired process temperature, and T the actual process temperature. W stands for the energy applied by the control process to the controlled variable, and P is the proportionality constant, called in that case the *proportional gain of the controller.* The equation applies likewise to other process variables, including pressure, humidity, luminous intensity, etc., if we replace T by the appropriate symbols, such as p, h, and I_{L}.

The overshoot problem can be mitigated with a controller that takes the *speed of change* of the process variable into account by introducing a derivative term into the equation:

$$W = P\left[(T_{\mathrm{P}} - T) + D\frac{\mathrm{d}(T_{\mathrm{P}} - T)}{\mathrm{d}t}\right],$$

where D is a second proportionality constant, called the *damping constant*. The term $\mathrm{d}(T_{\mathrm{P}} - T)/\mathrm{d}t$ stands for the speed of process temperature change. A rapid decrease of temperature would thus add more energy into the process than a slow loss. Proper selection of D allows for making the process asymptotically approach the desired value, and keep it there free of oscillation.

Still better results can be achieved by adding a third control variable, based on the amount of energy invested so far into the process:

$$W = P\left[(T_{\mathrm{P}} - T) + D\frac{\mathrm{d}(T_{\mathrm{P}} - T)}{\mathrm{d}t} + I\int(T_{\mathrm{P}} - T)\,\mathrm{d}t\right]$$

The third proportionality constant, I, is called the *integral gain parameter* or *reset level*. The systems that cater this equation are collectively named *proportional-integral-derivative* (PID) control systems.

Let's relax. This is a user friendly theory we apply every time we hit the ON button of the cruise control in our family car, where the math-work is done in silence by a central digital processing unit, resembling a dedicated computer. The primary setpoint is given by v_{P}, the car's speed at the moment you turn the cruise control on, and is kept there by proportional control principles as long as the load on the engine does not change. Correction requested by the central processing unit in the form of an analog voltage signal actuate a vacuum motor, a little device under the hood that gets its power from the partial vacuum at the air intake of the engine and adjusts the position of both the throttle and the gas pedal until the error signal is nulled.

The integral term $\int(v_P - v)\,dt$ of the equation helps in this process as it detects long periods of slightly too high or too low speed. At constant speed, the integral is simply the error signal – setpoint speed minus actual speed – times the time interval of its presence. At variable deviations, the value of the integral can be found as the area subtended between the car's true speed curve and the straight line showing the set speed. In other words, integral control adds weight to the error signal. Long duration errors weigh heavier on the controller output than shorter ones.

This way, all is well unless it ends up at the onset of a hill, where demands on engine power rise dramatically. Though proportional control alone would eventually take care of that, we, the almighty consumer who paid the price of the car, want it to continue at the rated speed and "neither snow, rain, nor gloom of night" should change that. Here is where the derivative term $d(v_P - v)/dt$ of the equation kicks in. This term is actually the car's acceleration or deceleration, as it divides a velocity difference by the respective time difference or, in a different interpretation, the slope of the time–velocity curve. This fraction remains at zero as long as the car rolls along at constant speed, but springs to life at the loss of speed on an upward slope and via the vacuum motor opens up the throttle until the set speed is resumed and the term is nulled again, but this time at a wider throttle opening.

Proportional control

Ideally, proportional control systems correct the settings of the controlled element, such as a valve, proportional to the magnitude of the error signal, and thus keep the controlled variable at a steady, unwavering level.

One of the earliest and battle-tested proportional control systems is the speed governor on James Watt's steam engine. Controlling and stabilizing the rotational speed of steam engines is essential; first to prevent the flywheel from being torn apart by centrifugal forces, and second to keep the machine running at its highest efficiency.

The control function of the speed governor in Fig. 13.14 is based on the centrifugal force of a pair of iron balls mounted on hinged levers which make them rotate in proportion with the speed of the machine's flywheel. The magnitude of centrifugal force F is $F = mR\omega^2$, where m stands for the spinning mass, R for the distance of its center of gravity from the axis of rotation, and ω for angular velocity, equal to $\pi/30$ times rpm. The gain of centrifugal force with the square of rpm explains the high sensitivity of a system where centrifugal force actuates the steam valve that throttles the machine's steam supply.

As centrifugal forces thrust the masses outward, the lever mechanism converts this move into a downward displacement of the speed controller's central spindle, which – via the rotary joint – positions the steam valve. Mounted under

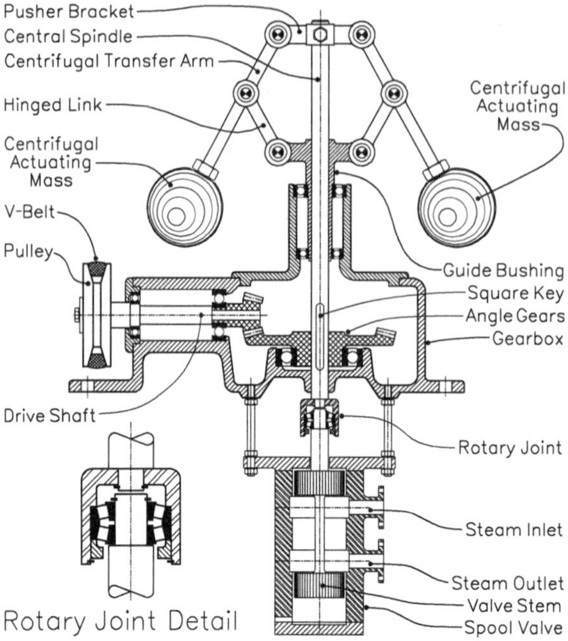

Fig. 13.14. James Watt's type speed governor

the bottom plate of the device, this valve admits just the proper amount of steam into the cylinder to keep the machine running at rated rotation. In Fig. 13.14, the valve is shown fully open, but it would close gradually as the weights move upward and force the spindle down.

The anchor points of the lever mechanism that transfers the displacement of the centrifugal transfer arms, which support the weights, to the pusher bracket on the upper end of the central spindle, are located on the guide bushing. Kept in fixed position relative to the machine frame by a pair of ball bearings inside the flanged bushing on top of the gear box, the spindle positions the valve stem over a rotary joint. The latter consists of a pair of back-to-back mounted conical Timken bearings, housed in the bell-shaped bushing on the lower end of the spindle, while the inner rings sit on a recess on the stem of the spool valve.

This arrangement differs from James Watt's original concept, which employs a throttle valve in line with the steam piping, activated over an extra lever mechanism by the governor's spindle displacements. Likewise, the V-belt drive between spindle and flywheel must have been a flat leather belt on the original.

Pneumatic control systems

As recent as at the turn to the 20th century, compressed air was still considered a viable contestant to electricity as a means of energy transfer. Though it didn't prevail in long-distance systems, compressed air as energy carrier survived in most industrial plants as a factory-specific system supported by air piping for distribution of air from a central compressor with storage tank to a variety of pressure tools, air-motors, air cylinders, linear actuators, air powered valves, etc.

Air power gains importance in hazardous environment, where electric sparks in switching operations could trigger explosions. This includes factories of explosives and fuel gases, and in particular mines, where methane gas leaking in from the surrounding earth blends with atmospheric air into just the right proportion for highest flammability, where explosions become most likely. Although "explosion proof" rated electric components are widely available, compressed air is the radical and foolproof alternative.

In control engineering, pneumatic systems are attractive in particular when the controlled or the controlling variables themselves are gas or liquid pressure, so that pneumatic control can be done with fewer conversion elements than electrical control. For instance, level control in open tanks is straightforward and can be done by a common pressure switch, commanded by hydraulic pressure at the exit pipe of the tank. Low pressure triggers the switch to turn on the pump motor, while high pressure makes it turn off. For liquids other than water, pressure equals the *hydraulic head times density*.

For sealed tanks, however, gas or vapor pressure in the space above the liquid – such as steam pressure in boilers, vapor pressure in liquid gas cylinders, or externally applied pressure in autoclaves – adds to gravitational head pressure. The pneumatic control system in Fig. 13.15 is designed to maintain constant liquid level in a pressurized vessel by controlling the settings of the pneumatic product admission valve shown in the lower right of Fig. 13.15.

Gravitational head pressure in the vessel (left in Fig. 13.15) is derived from the pressure differential between upper and lower spuds, shown on the right side of the tank. The two pressure signals are subtracted in the *differential pressure converter*, which includes a cylindrical housing divided by a diaphragm into an upper and a lower chamber. Total tank pressure from the lower spud is channeled into the bottom chamber of the converter, head pressure into the upper chamber. While

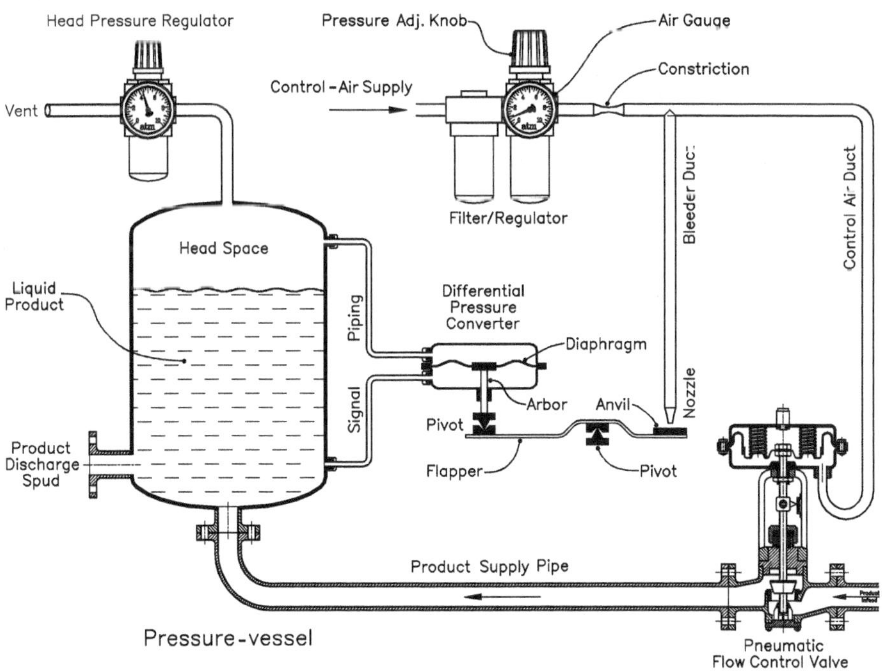

Fig. 13.15. Pneumatic level control in pressure vessel

the former pushes the diaphragm up, the latter pushes it down. The resulting pressure differential shows as the magnitude of displacement of the arbor – or call it piston, if you wish – which reaches from the center of the diaphragm down through a stuffing box to the force balance seesaw lever.

This unit is the heart of the system and also illustrates the working principles of most pneumatic pressure controllers. It resembles a seesaw insofar as it includes the *flapper*, or pivoted beam, which left end is weighed down by the tip of the arbor of the differential converter, while its right end carries the *anvil*, a flat tungsten carbide bit in the path of a jet of compressed air from a constant flow supply.

The latter includes a dedicated compressor with a precision pressure regulator. Conversion from constant pressure p to constant flow happens in a calibrated restriction in the piping. Air speed v at the discharge side of the restriction is $v = \sqrt{2p/\rho}$, where ρ stands for the density of the air.

The airstream is now divided between the actuator chamber of the product valve (Fig. 13.16, lower right) and the nozzle blowing at the anvil of the flopper. The farther the flopper yields to the dynamic pressure of the impacting airstream, the greater the portion of the available air escapes through the nozzle, and the less pressure is left to actuate the product valve. That's what happens when raising liquid level in the tank increases pressure in the lower chamber of the differential pressure converter and lifts the pivot holding down the left leg of the flapper, so that the other leg tilts downward and opens up the gap between anvil and nozzle.

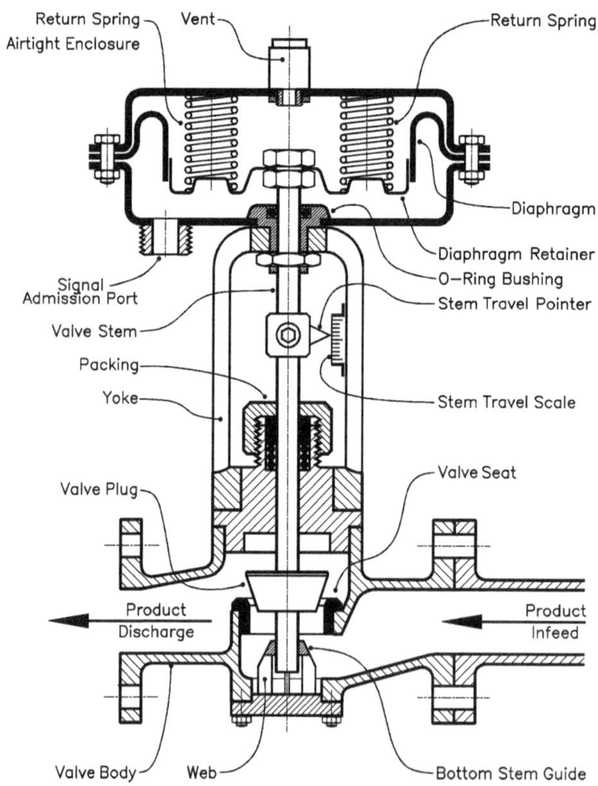

Fig. 13.16. Pneumatic flow control valve

The *product supply valve*, the component the fuss is all about (Fig. 13.15, lower right), determines the amount of liquid entering the storage tank, which should equal the amount withdrawn.

The product supply valve, actually the final element of the control system, called *manipulated compo-*

nent in control-specific lingo, is shown in greater detail in Fig. 13.16. Its design includes the valve as the passive component, and the valve-actuator as the active component. The relation between the valve's degree of opening and the volume of liquid passing through depends on the longitudinal section of the valve plug, which can be *linear, proportional, or logarithmic*. The ideal plug profile varies from application to application and must often be chosen by experience.

For best stability, the valve stem is guided on top and bottom. Hereby, the lower end guide-bushing is supported by four webs which slender shape leaves ample space for the passage of product.

The settings of the valve actuator are based on the force balance between the rolling-membrane with the attached spring support tray, and the force of the return springs. The spring support tray, a metal sheet stamping with 2 or more embossments to keep the compression springs safely in place, is center connected to the upper end of the valve stem. A vent on the upper side of the housing keeps pressure in the space above the diaphragm equal to ambient pressure, regardless of the diaphragm's rising or lowering.

As described, the system excels by its lack of sophistication and robustness of its elements, made to stand up to rough handling and the often demanding industrial environment in the processing industry, including acidic or caustic atmospheric conditions, vibration, extreme temperatures, and occasional overloads. Components such as pressure regulators, differential pressure converters, and pneumatic flow control valves are readily available, while the section containing mechanically movable parts, including the flapper and its auxiliaries, make part of proportional pressure controllers. The latters are available with circular or strip-charts, and often offer digital readouts obtained from a built-in pressure-to-voltage converter.

Other pneumatic control components are sequencers to set opening and closing times of a number of air-valves per operating cycle, which carry one cam for every valve on a common driveshaft. Cycle time is being set by adjusting the speed of rotation of the camshaft, and each particular valve's opening and closing times are set by adjusting the angular position of the respective cams.

Pneumatic timers, which ON/OFF periods are derived from the time it takes for a measured amount of compressed air to escape through a small opening in a miniature air cylinder.

Pneumatic spool valves (Chap. 8), with pneumatic actuators.

Further, position triggered valves, flow sensing flapper valves, vacuum sensing valves, and air operated electric switches.

Among the final elements of a control system rank pneumatic actuators, air cylinders (Chap. 8) supplemented by actuators that stop dead the moment the position of a magnetic ring on the piston reaches that of an externally mounted magnetic field sensor, used in rotary work tables and pneumatic sheet metal strip feeders for stamping dies.

Further, this group includes air-powered rotary actuators (Chap. 8), clamp cylinders for holding work-pieces securely in place, and twist clamp cylinders, which rotate out of the way by 90° after release.

Finally, control systems employing a blend of pneumatic, electric, and electronic elements, are often ideal in a field where digital electronics are becoming predominant. In automotive engineering, digital control steers valve timing in motors, throttle opening, fuel–air mixture, etc., while the venerable vacuum motor still populates the world under the hood, and pneumatic dampers, though digitally controlled, are in charge of the riders' comfort.

Flashback to the roots of automation

While the engines in our cars and the steam engine operate through the reciprocal motion of one or several pistons, rotary primary movers, such as the steam turbine or jet engines, convert the kinetic energy of the driving media – normally the products from combustion of hydrocarbons with oxygen from air – directly into rotational energy. In both cases, a jet of superheated steam or, respectively, exhaust gases is made to change direction by up to 180° along the curvature of the turbine blades, so that the reactive force of the jets thrusts the blades backwards. An idea that goes back to Heron of Alexandria's *Aeolipile*, which principle is illustrated in Fig. 13.17. Although those days technology was far from building anything resembling a steam turbine as we know it, Heron's device got as close as it could. Instead of the blades, Heron used a pair of angled copper tubing, press-fitted on a hollow globe of copper sheet, the rotor. One of the weakest links in all that must have been the stuffing boxes that then couldn't have been machined with the necessary degree of precision – if at all –, while the gaskets were no doubt of nothing more flexible than leather.

Nevertheless, Heron's turbine went on rotating, first in reality and later in the chronicles into our days, regardless of the merely 90° deflection of the jets and the

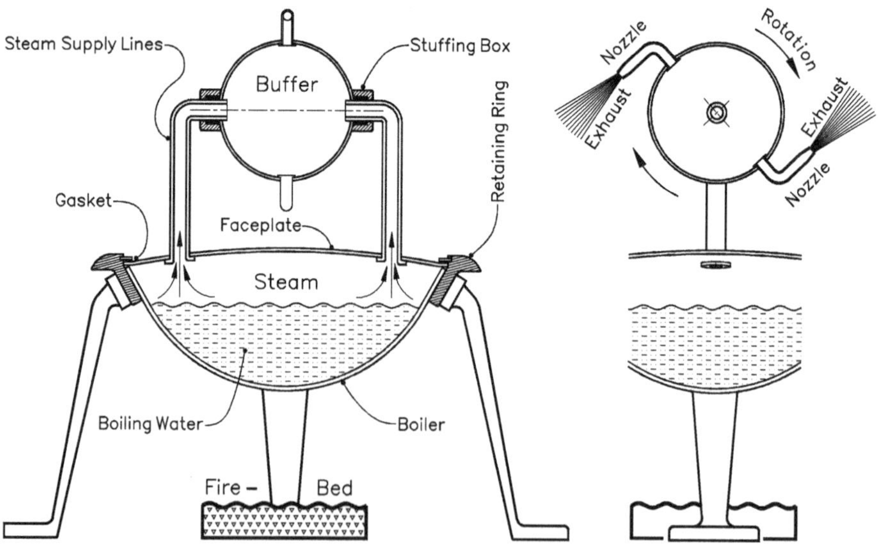

Fig. 13.17. Heron's of Alexandria harbinger of the steam turbine

inherently low steam pressure, let alone the cloud of leakage steam that must have shrouded the inventor and his brainchild.

Though the idea of rebuilding the Aeolipile from a toy into a true prime mover may never have crossed the inventor's mind, his is one of the first devices in history that converted thermal energy into kinetic energy. And if you want, you could see a proportional control system hidden in it: Steam pressure and with it, rotational velocity control by the amount of steam leaking out at the stuffing boxes.

Bibliography

Arfken GKP (1984) University Physics. Academic Press, New York

Beyer WH (1987) CRC Handbook Mathematical Sciences. CRC Press, Boca Raton

Bolton W (1996) Instrumentation and Control Systems Newnes/Elzevier Ltd., Linacre House, Oxford – UK

de Silva GMS (2001) Basic Metrology for ISO 9000 Certification. Butterworth-Heinemann, Linacrehouse, Jordan Hill, Oxford OX2 8DP, A Division of Reed Educational and Professional Publishing Ltd.

Deutsches Institut für Normung e.V. (1996) DIN Geschäftsbericht 1995/96

Dubbel H, et al. (1994) Handbook of Mechanical Engineering. Springer-Verlag, Berlin

Gasvik KJ (2002) Optical Metrology. John Wiley & Sons Ltd., The Atrium, Southern Gate, Chichester, England

Gröber-Erk (1933) Die Grundgesetze der Wärmeübertragung. Verlag von Julius Springer, Berlin

Haberland G (1941) Elektrotechnische Lehrbücher. Max Jänecke Verlagsbuchhandlung, Leipzig

Haeder W, Gärtner E (1980) Die gesetzlichen Einheiten in der Technik. Deutsches Institut für Normung e.V., Berlin

ISTE (2007) French College of Metrology, Metrology in Industry: The Key for Quality

Kurtz M (1991) Handbook of Applied Mathematics for Engineers and Scientists. McGraw-Hill, New York

Lide DR (1999) CRC Handbook of Chemistry and Physics. CRC Press, Boca Raton

Loxton R, Pope P (1990) Instrumentation – a Reader. Chapman & Hall, London

Mache H (1921) Einführung in die Theorie der Wärme. de Gruyter, Berlin

McCoubrey AO (1991) Guide for the Use of the International System of Units – The Modernized Metric System. National Institute of Standards and Technology, Spec. Publ. 811

Richter H (1953) Aufgaben aus der Technischen Thermodynamik. Springer-Verlag, Berlin

Thomas O (1933) Astronomie, Tatsachen und Probleme. Verlag Das Bergland-Buch, Wien

Tuma JJ (1976) Handbook of Physical Calculations. McGraw-Hill, New York

Walker PMB (1976) Cambridge Dictionary of Science and Technology. Cambridge University Press, Cambridge

Webster JG (1999) The Measurement Instrumentation and Sensors Handbook. CRC/IEEE Press, New York

For further reading

Industrial Metrology by Graham T. Smith (Springer – Jul. 28, 2002)

Springer Handbook of Mechanical Engineering
by Grote, Karl-Heinrich; Antonsson, Erik K. (Eds.)
2009, XXVIII, 1580 p. 1822 illus., 1551 in color

Metrology in Industry: The Key for Quality
by French College of Metrology, Feb. 4, 2008

Metrology and Properties of Engineering Surfaces
by Mainsah, E.; Greenwood, J. A.; Chetwynd, D. G. (Eds.)
Kluwer Academic Publishers, Dordrecht, 2001

Machining Dynamics
Frequency Response to Improved Productivity
Schmitz, Tony L., Smith, K. Scott
Springer, Berlin, Heidelberg, New York, 2009, 310 p. 125 illus.

Basic Metrology for ISO 9000 Certification
by G. M. S. de Silva
2002, Butterworth-Heinemann, Linacre House, Jordan Hill, Oxford OX2 8DP

Frontiers of Characterization and Metrology for Nanoelectronics
2007 International Conference on Frontiers of Characterization and Metrology for Nanoelectronics by Seiler, D. G.; Diebold, A. C.; McDonald, R.; Garner, C. M.; Herr, D.; Khosla, R. P.; Secula, E. M. (Eds.), 2007

Practical metrology by K. J. Hume (English Language Book Society, 1965)

Numerical Data and Functional Relationships in Science and Technology Landolt-Börnstein Online, Springer Link

How Apollo Flew to the Moon by W. David Woods (Springer Praxis Book/Space Exploration)

Index of names

Subject index

F

G

M

U

V

SPRINGER NATURE

GPSR Compliance

The European Union's (EU) General Product Safety Regulation (GPSR) is a set of rules that requires consumer products to be safe and our obligations to ensure this.

If you have any concerns about our products, you can contact us on ProductSafety@springernature.com

In case Publisher is established outside the EU, the EU authorized representative is:

Springer Nature Customer Service Center GmbH
Europaplatz 3
69115 Heidelberg, Germany

FSC
www.fsc.org
MIX
Papier | Fördert gute Waldnutzung
FSC® C083411

Kolibri 360 GmbH
Ferdinand-Jühlke-Straße 7
99095 Erfurt, Deutschland
produktsicherheit@kolibri360.de